PRIVATE TELECOMMUNICATION NETWORKS

The Artech House Telecommunication Library

Vinton G. Cerf, *Series Editor*

PRIVATE TELECOMMUNICATION NETWORKS

Bruce R. Elbert
Hughes Communications
Galaxy, Inc.

Elbert, Bruce R.
    Private telecommunication networks.

    Bibliograhy: p.
    Includes index.
    1. Telecommunications sytems.   I. Title.
    TK5101.E412   1988   384.3'3   88-7366
    ISBN 0-89006-316-8

Copyright © 1989
ARTECH HOUSE, INC.
685 Canton Street
Norwood, MA 02062

International Standard Book Number: 0-89006-316-8
Library of Congress Catalog Card Number: 88-7366

   10   9   8   7   6   5   4   3   2   1

To my wife, Cathy

Contents

Preface

Telecommunication is that division of the information technologies which deals with the carriage of information from its point of creation to one or many other locations where it can be used. A telecommunication network is a structure or web of communication links that connects such locations together. This approach minimizes the number of links, while allowing any point to be reached. As an overall statement, we consider networks which interconnect distant locations in a metropolitan area network or a wide area network. Public networks have always been operated by government-sanctioned organizations (either private companies or government enterprises) to satisfy the needs of the general public. Businesses, institutions of learning, and government agencies are becoming aware that private telecommunication networks are of vital importance to the successful accomplishment of organizational goals and strategies. This book is written at an introductory level to help such organizations as well as individuals better understand and apply telecommunication technologies and services. Demand for such a treatment is best exemplified by the fact that the breakup of the former Bell System now frustrates those purchasers of telecommunication services who were accustomed to "one-stop shopping."

If you are an individual trying to gain an understanding of the complex and evolving picture of private telecommunication, this book is expected to satisfy your needs. In many companies, telecommunication is split between voice (telephone) and data (data processing or information processing); in some cases, video services (audio-visual) are covered by yet another organization. The networks that provide these services may then be fragmented and overly costly. One of the primary goals of this book is to show how to integrate these various capabilities and thereby better serve the organization's needs, including that of cost containment. Many individuals who work in these fields are employed by noncommunication companies, but have technical backgrounds gained either in school or on the job. What has been lacking is a consolidated treatise which unites all of the relevant components, but maintains a focus on the private telecommunication environment.

The professional field which integrates all applications is that of telecommunication management. It is now an established career with many entrants coming through university programs at major institutions around the U.S. This book can be used as a text or reference in a course dealing with the architecture and strategy of applying telecommunication in the business context.

From a technological perspective, the emphasis within the book is on digital communication approaches. The voice, data, and video worlds are clearly going digital. Therefore, telecommunication people need to understand how digital switching and transmission technologies can be properly exploited. In particular, we cover in considerable detail the techniques for integrating voice, data, and video signals on a common digital facility. The complementary roles of fiber optics and satellite communication are elaborated in a number of instances throughout the book. This is because each approach has advantages which can be exploited under appropriate circumstances.

Individuals who work for telecommunication equipment vendors and service providers (such as common carriers) should be able to make good use of the material in this book. Marketing, in particular, requires an understanding of the user's or customer's perspective so that the vendor can best serve their needs. A hard sell in today's competitive environment will just not work; rather, it is the vendor who better understands and serves the customer who will succeed. The user perspective in the book should also help strategic planners working for vendors who wish to find new opportunities for products and services. This is a particularly fertile field because the environment is so dynamic, with no single company dominating the marketplace as AT&T once did.

The chapters of the book develop topics as needed, providing an organized flow. This will generally make it easier for the reader to proceed sequentially. Technical topics are introduced in certain of the chapters, but mathematical formulas and derivations are avoided. Instead, theory and technical data are presented in the form of graphs and tables. The book is therefore aimed at a person with an interest in telecommunication and some comprehension of transmission and switching, but does not require a background in science or engineering. Considerable effort, however, has been made to ensure that the technical information, where presented, is accurate and timely. We have attempted to take a view toward the future so that the material will not quickly grow stale. For example, the introduction of integrated services digital networks (ISDN) is discussed along with the potential of this promising capability.

Chapter 1 is a general introduction to the field of private telecommunication networks. We develop the environment of modern business communication, where strategic units can exploit both the technology and service capabilities now available on the market. The nature and role of government deregulation and the divestiture by AT&T of the local telephone companies are reviewed in detail. We give some concrete examples of the kinds of networks that strategic units are currently implementing and using. As a starting point for private networking, the public telephone and data networks must be carefully considered as a source of capability.

Consequently, the nature and availability of public network services is the first major area covered in detail in Chapter 2. We review the service offerings of local telephone companies, long-distance carriers, public data networks, and other classifications of specialized common carriers (satellite and terrestrial). These aspects are often overlooked in other texts, but telecommunication professionals are finding that public telephone and data networks are becoming more attractive to businesses, even as private networks proliferate. For example, the long-distance carriers are continuing to reduce rates, and they have introduced flexible services like software defined networks, which make switched long-distance service very attractive in a wide variety of applications. Also, it is useful to compare how an organization's needs can be met with offerings of common carriers as well as through private facilities. This provides an educated basis for performing trade-offs before making a decision on the architecture of a private network.

The ways in which telecommunication networks are used in business applications are covered in Chapter 3. This takes the reader through user capabilities in voice, data, and video communication, identifying operating equipment and services that are applied on the customer premises. On-premises telephone systems like key systems and PBXs are detailed. A section on telemarketing reviews the facilities and approaches which allow a centralized sales staff to serve a diverse customer base. The data communication field is elaborated, providing a foundation in the leading environments, including OSI, IBM, and DEC. We also review ancillary services like paging, cellular radiotelephone, and remote printing. Taken together, these applications amount to a "bag of tricks" which the telecommunication professional can apply in a variety of business situations.

Chapter 4 contains the core technical material of the book, detailing the elements and structure of modern digital telecommunication networks. The emphasis is on integrating voice, data, and video services for efficient use of a digital backbone, such as that provided with fiber optic cable systems. We supplement this material with discussions of digital satellite communication networking techniques using international earth stations and domestic VSATs. The material is tutorial in nature so that a technical background is not needed to gain a practical understanding. Readers who have a solid background in digital communication systems will find this to be a well organized review of technologies applicable in a commercial environment, as opposed to those which are merely interesting in a technical and scientific sense. In keeping with the tutorial theme, we define and explain the "buzz words" commonly used in the industry.

Chapters 5, 6, and 7 deal with the management aspects of implementing and operating a private telecommunication network, which is the type of material usually omitted from books of this genre. These topics are vital, however, because they can dictate the success of a networking project. The network management problem is defined and elaborated in Chapter 5. These management techniques and systems provide on-line monitoring and control of network resources. If properly implemented, a network management system, which runs in parallel with the actual traffic network, provides control of network facilities that are either owned by the organization or taken as services from common carriers. The conceptualization and design of the network is covered in Chapter 6, which also deals with economics. The economic aspect is necessary because we either create a network for cost-reduction reasons, or we must make trade-offs between networking approaches for which economics plays an important role. This chapter will be particularly helpful to readers who wish to understand how to estimate user requirements and then convert them into a network architecture. Chapter 7 covers implementation, which deals with buying telecommunication hardware and services. This chapter will help a telecommunication manager decide how to organize his or her staff for the procurement process, how to request and evaluate proposals from vendors, and then how to manage the actual implementation of a network capability.

In Chapter 8, we consider how an organization pursues a business strategy through the use of private telecommunication. This chapter employs the concepts of competitive strategy and competitive advantage, outlined by Professor Michael Porter of the Harvard Business School. Readers involved in business strategy will find this information useful for justifying a private telecommunication network, where the need exceeds simply reducing costs. Many organizations have discovered that a properly configured and implemented network helps achieve business goals (e.g., increased market share, greater profits, lower risk of competition). Supplementing these general principles, we introduce the specific recommendations of Peter Keen of the International Center for Information Technologies. Dr. Keen relates information technologies, particularly telecommunication, to the competitive position of an organization in its respective industry. We include some case studies of strategic networks to make the discussion more vivid.

Moving from the area of competitive strategies, we complete our discussion with a review of success factors in telecommunication networks in Chapter 9. The telecommunication management needs a guiding star as it defines, implements, and operates a telecommunication environment for an organization. One of the critical factors often overlooked is that of education. The field of telecommunication management is now an accepted profession, and educational institutions offer programs which will develop qualified participants. Formal education is supplemented with current information made available through trade publications, books such as this one, and short courses or seminars. We find that, because this field is so dynamic, the telecommunication professional must constantly undergo retraining. To cap the final chapter, we look into the future in an attempt to grasp the possible directions for the evolution of the information industry.

This book is a synthesis of hundreds of sources of information, including technical and business books (see the Bibliography), trade shows, industry publications, seminars, and discussions with users and vendors. The original idea was developed for a seminar on the new telecommunication technologies that the author conducted at the Japan-America Institute of Management Science. That seminar covered perhaps only a quarter of the topics in this book, but nevertheless required a careful investigation of the structure of private telecommunication networks. The result of this study was the realization that a private network must include capabilities from the more advanced public networks, both voice and data. In addition, the author has enjoyed employing the principles of competitive strategy and competitive advantage as outlined in Chapter 8. Readers will also notice that satellite communication receives considerable coverage throughout the book, being an area in which this author has several years of experience. Not long ago, private networks were synonymous with satellite communication because of innovations by COMSAT, Hughes Aircraft, Satellite Business Systems, Arco, Sears, Citibank, and others. With the advancements in terrestrial telecommunication technologies, private networks are less satellite oriented, but satellite communication still plays an important role, as demonstrated by companies like Holiday Inns of America, J. C. Penny, and WalMart.

Acknowledgments are due to many people and organizations without whose contribution this book would not have been possible. Quite obviously, my experience with Hughes Communications, Western Union, and COMSAT provided valuable exposure to telecommunication technologies, private networks, and user needs. Interactions with friends and associates at these companies have always been useful and valuable. There are too many individuals to name, but I would like to express appreciation to Jan Kaechele for wise counsel. While they may not have realized it, telecommunication managers with whom I have spoken have provided vital insight into the problems and opportunities of this turbulent field. Some of the individuals include Larry Lake and Gary Badberg of American Express, and Hartmut Berger and Lance Ede of Electronic Data Systems. People who work for equipment suppliers, including Audry MacLane of Network Equipment Technologies, Gordon Heinrich of Fujitsu America, Linda Fischer of GTE Strategic Information Systems, and Stan Galkin of Codex provided vital guidance during various interactions. Professor Norman Abramson of the University of Hawaii, who offered me the opportunity to participate in the aforementioned seminar, has continued to be of great help to me in developing material for publication. I credit my comprehension of business strategy to the professors at Pepperdine University under whom I studied. These individuals include Frank Largent, former director of the Presidential/Key Executive MBA program, Len Korot, Tom Dudley, Rusty Rostvold, Dick Kaehler, Dan Williams, and Kurt Motamedi. Rich Murray of Nolan, Norton and Company reviewed Chapter 8, and he provided useful comments and suggestions.

At an earlier point in my industrial career, I was influenced by telecommunication professionals at Bell Laboratories and with the Post and Telecommunication Directorate of the Government of Indonesia. In the latter case, I would like to express my particular appreciation to Mr. Wikanto and Mr. Sutanggar Tengker. Underlying all of this was my four years as an Army communication officer in the Signal Corps, an exposure where I learned that the most important thing to do is to get the communications in. (I have never forgotten this point!) Important lessons in microwave systems engineering were provided prior to my transfer from the U.S. to Vietnam by J. Deygout, then a colonel in the communications branch of the French Army. Ironically, both Deygout and I applied his techniques in that same far-away land, but during different engagements.

To my mother, the savvy business woman, and my late father, the quintessential technician, I am deeply grateful for providing the environment to grow and to learn. Special appreciation goes to my wife, Cathy, to whom I dedicate this book. As she has once again helped and motivated me to write a book, I am sure that this will not be the last time.

Chapter 1
Introduction

These are exciting times in telecommunication as rapid advances in digital technology are bringing new capabilities into the marketplace. It is not enough, however, to be merely intrigued by "gee whiz" scientific advancements and technological innovations. The key to success is in their meaningful application in the business context. Users and managers of telecommunication networks, however, must be conversant in and comfortable with all of the dimensions of modern information systems. These systems generate, manage, and transmit information in audible, visual, and numerical forms. As the capabilities of these systems grow, users have greater potential for growth in their own businesses.

Managers of telecommunication for companies and government agencies in the U.S. know that the days are gone when they could pick up the telephone and call on a single entity to take care of all their communication needs. The industry which supplies these services and equipment is in a state of evolution — some say a *revolution* — as new technologies and new competitive realities cause tremendous change in its structure. Consequently, commercial and government organizations which employ telecommunication networks must also change the way they perform their respective strategic roles.

Undoubtedly, the best telecommunication system in existence in the world prior to 1984 was invented, built, developed, and operated by the Bell System of the American Telephone and Telegraph Company (AT&T). The technology of the telephone was exploited to its fullest as local service was enhanced through the availability of long-distance service. Being the largest privately owned company in the world, AT&T seemed capable of almost anything it endeavored to do. Not the least of its accomplishments was the position of its stock as a highly rated investment. The recognizable Bell logo appeared everywhere, and the concept of universal telephone service was a reality. The monolithic Bell System continued until 1984 when its breakup was ordered by a federal court. An age of turmoil then began.

Private companies always had specific needs which could not be filled by standard public telephone service, and the old AT&T did not always respond to these needs. Companies could lease equipment from AT&T to handle local office requirements in the form of key telephones and *private branch exchanges* (PBXs). Automatic exchange facilities, allowing a company to dial between extensions within a building, were introduced as a service called CENTREX. When a company needed to connect multiple locations in the same region and in other regions, however, the situation became less congenial. This, of course, was not unexpected since AT&T's economies of scale did not easily go hand in hand with customization of services. Competition was restrained by government regulation, but during the 1970s, this regulatory hold and protection of AT&T began to be dismantled. The first competition in telecommunication was in the area of international telex and telephone service, where companies like Western Union International, RCA Globecom, and ITT Worldcom competed aggressively with one another. Domestic competition, however, did not start until the arrival of a new long-distance company called Microwave Communications, Inc. (MCI). Likewise, the only supplier of telephone instruments and switches in the U.S. had been AT&T, but a Canadian manufacturer, Northern Telecom, introduced more advanced digital switching equipment which companies bought for use as internal PBXs. There was thus significant competition to AT&T before the Bell System was broken up in 1984.

This turbulent environment forms the backdrop for this book, which is intended to explain the "how" and "why" of private telecommunication networks. One of the key aspects is the technology of modern telecommunication. Therefore, to understand private telecommunication networking, we must be familiar with the range of possibilities from digital communication techniques, hardware, and software. Technical papers from engineering journals, however, do not provide adequate explanation for those who intend to apply the technology as opposed to understanding what makes the black boxes "tick" internally. This book was written to overcome this "information gap." Besides providing the needed operational tutorials, we provide considerable detail on how to blend the various technologies to meet the changing needs of corporations and government agencies. To begin, we identify the context within which private telecommunication networks have meaning and utility.

1.1 OVERVIEW OF CORPORATE TELECOMMUNICATION

Conceptually, a private telecommunication network is a web of communication pathways between various types of electronic transmitting and receiving devices. The network concept is illustrated in Figure 1.1 in the form of an alterable and flexible cloud. The types of entry devices connected to the network are the familiar pieces of telephone, computer, and video apparatus found in offices, and even homes, around the country. The private network cloud is attached to other clouds representing *local area networks* (LANs) within buildings and campus environments, and public voice and data networks which transfer information to points around the country and even the world. The public network connection or *gateway* is required in almost all cases because people within the private network must communicate with remote parties, such as associates, friends, customers, suppliers, or the government. Connecting our private telecommunication network involves both the electrical interface (e.g., signal levels, bandwidths, timing, frequency) and the procedural aspects (i.e., setting up calls, addressing information for delivery to the proper user, arranging for billing of services). The ideal telecommunication network, however, is one that is easily used and adapted, much like the cloud illustrated in Figure 1.1. The reality is that connecting devices to the private network and connecting this network to others is often a complex problem. Add to this the constant changes undergone in the development of new user devices, such as personal computers, speech recognition, video conferencing, facsimile, and electronic mail. One of the key objectives of this book is to explain what exactly is inside of the private telecommunication network cloud.

The kind of organization which needs particular private telecommunication network capabilities is referred to herein as a *strategic unit*. A strategic unit of a major corporation engages in a specific line of business, such as automobile manufacture or the delivery of health care. In the case of government, examples of strategic units would include the armed forces, the customs department, and a public university system. A strategic unit thus has a purpose to be fulfilled in which telecommunication can play an important role. Without an appropriate telecommunication network, the unit may not be able to implement its strategy and engage in its chosen line of business or activity.

Some strategic units are large and some are small. In addition, a corporation or government agency can have several strategic units within its span of control. This gives rise to the possibility of combining the telecommunication needs of a number of strategic units to achieve an economy of scale. To do so represents a strategic decision in itself, because differences in business strategy between units will cause the particular needs to be opposed. There could be a trade-off of efficiency for the particular purposes when this type of aggregation is done to consolidate network requirements of various diverse users. Another concern is that when the telecommunication networks are integrated, the businesses no longer can be easily separable. The economies of scale could be substantial, however, if we consider the greater volume of communication "traffic" that would be offered to the network.

We often address the telecommunication manager, who is the key person in the strategic unit with the responsibility to plan, implement, and manage the telecommunication "plant." In many organizations, buying telephone service was

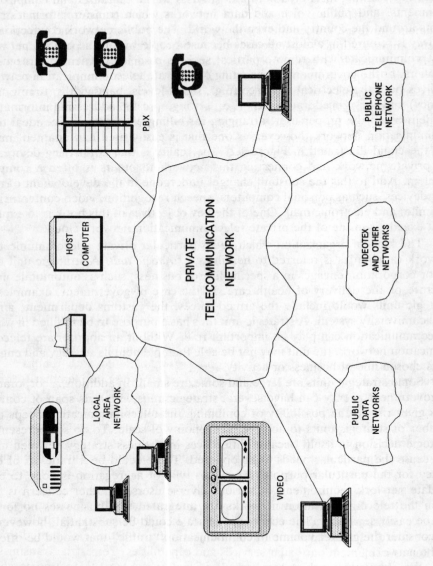

Figure 1.1 The private telecommunication network environment.

merely part of the procurement function wherein items such as office supplies, printing, and other necessary support services were purchased and provided. Changes in the needs of strategic units and the telecommunication business environment, however, caused the telephone buyer's job to expand, and the position of telecommunication manager emerged. Another important aspect is *data processing* (DP) and *data communication.* Larger companies own mainframe computers, which typically require communication facilities to connect remote entry points and terminals into the central sites. The DP manager often took responsibility for obtaining the necessary telephone lines and interface equipment. Consequently, we now have the duality of the important roles of telephone and data communication management. The need for overall telecommunication management is heightened by the increasing cost of procuring services from outside sources. These costs represent over 60% of the telecommunication budget of a typical strategic unit. An economical approach is to combine telephone and data communication requirements into a single integrated network. Integration of voice and data, however, is difficult for organizations that have not yet combined these two functional areas.

1.2 THE NEW DIGITAL INFRASTRUCTURE

Much of the strength of modern private telecommunication networks comes from the use of computerized switching and high capacity digital transmission. Recognizing that any information signal (voice, video, written, or numerical) can be represented by a stream of digital data, the telecommunication systems of the Western world are being digitized. A comparison of bandwidth requirements, expressed in kilohertz (kHz) and kilobits per second (kb/s), for familiar forms of communication is presented in Table 1.1. The low end is dominated by teletypewriter (now becoming largely extinct in modern organizations) and low speed data for user devices such as video display terminals and personal computers. Moving up the scale, we encounter digitized versions of analog voice and video signals. The highest speeds, and consequently the widest bandwidths, support television transmission for normal broadcasting and ultimately for *high definition television* (HDTV) images.

This new digital infrastructure supports and surrounds the telecommunication network environment, as illustrated in Figure 1.2. The capstone to the structure is the private telecommunication network. A variety of providers are tied to one another to implement particular services and capabilities. Terrestrial transmission systems are available, which can be used to bypass the local telephone company (telco) monopoly. A similar role can be played by satellite communication for long-haul links, which, along with bypass, is illustrated as part of the infrastructure. The structure and associated elements together permit builders of private telecom-

Table 1.1 Bandwidth Requirements for Various Forms of Communication

Form	kHz	kb/s
Morse code		0.05
Teletypewriter (telex)		0.1
Low-speed data or fax	3.0	4.8
Voice (telephony)	3.3	32
Audio (sound)	15	386
Computer channel		512
Television (compressed)		768
Television (broadband)	4200	45,000
Television (high-definition)	6000	135,000

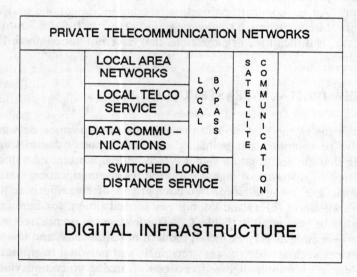

Figure 1.2 The new digital infrastructure and private telecommunication networks.

munication networks to expand their horizons beyond that which is available to the general public.

The digital form of communication had modest beginnings as a way to put more telephone trunks on a pair of wires. For more than twenty years, voice frequency information has been digitized over the wires; the circuit established at the central offices and through the local subscriber loops, however, was still analog in nature. This hybrid approach, shown in Figure 1.3(a), is still used in public networks due to the extensive investment in analog subscriber loops. In time, telephone instruments which convert speech directly into digital form for connection by digital switches to digital transmission facilities will be common place; see Figure 1.3(b). For some applications, data and graphic information can be added and routed essentially through the same system, possibly from the telephone instruments themselves. Note that digital trunk transmission (between exchanges) is used in both schemes; in the all digital approach, however, conversion from analog to digital format is performed within the telephone instrument. These developments are reviewed in the following paragraphs.

1.2.1 Digitization of Long-Distance Networks

The primary motivation for telephone companies to convert to digital is to compress voice traffic so that local and long-distance transmission facilities can be used to their fullest capability. Since it costs about the same amount of money to install and operate a 600-channel analog coaxial cable system as does a digital fiber optic cable system with 100 times the capacity, digital fiber optic facilities can provide cheaper service. Trends in the cost of transmission are provided in Figure 1.4 as a function of the investment cost per voice channel. This is a measure often used in the telecommunication industry for comparing transmission technologies. As the channel capacity is increased, the cost per channel declines, primarily because a fixed facility is being used to carry more telephone traffic. Three long-haul systems, each providing a link 2000 miles in length, are compared. We see that microwave radio and satellite communication provide good economic performance for capacities in the range of 100 to 10,000 channels. A fiber optic cable, however, is capable of further reducing the investment cost per voice channel, simply because of its capacity for digitized traffic. At 1000 channels, fiber is nearly four times as costly as microwave; the same fiber system when fully loaded, however, is one-quarter the channel cost. In general, transmission systems with greater and greater capacity generally provide the means to reduce the cost of carrying a telephone conversation. We must, however, determine whether and how these fiber facilities can be filled to produce these favorable cost characteristics.

On both ends of the system are electronic devices called *multiplexers* (reviewed in Chapter 4), which process the traffic, assembling and disassembling it for transmission and reception, respectively. The digital form of multiplexers is inherently more efficient and flexible, and the cost per link of the electronics is on a par with the analog version. During the current transition from analog to digital, information conversion occurs within the multiplexer. When the telephone instru-

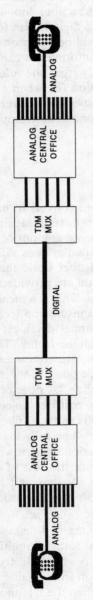

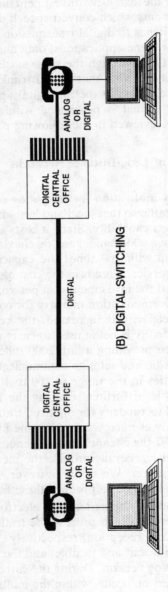

Figure 1.3 A telephone connection from instrument to instrument.

(A) ANALOG SWITCHING

(B) DIGITAL SWITCHING

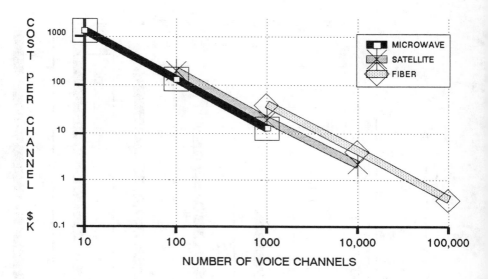

Figure 1.4 Trends in investment cost per channel for long haul transmission media (microwave, satellite, and fiber optic cable).

ment takes over this job, the multiplexers of the network will become nearly invisible within the structure of the transmission and switching systems. Another important feature of digital multiplexing is the ease with which voice and high speed data may share the same communication facility.

The favorable economics of fiber optic cable and digital multiplexing are forcing the retirement of older analog transmission systems, particularly coaxial cable. During 1987, the first coast-to-coast fiber optic paths were established by US Sprint. AT&T and MCI, the other major long-distance carriers, have subsequently completed similar connections. These links now connect every major city in North America; a similar trend is evident in Western Europe. The rapidity with which fiber optic facilities have been installed is shown in Figure 1.5, and US Sprint's nationwide network is shown in Figure 1.6. The local telcos are also rapidly installing fiber within the major metropolitan boundaries.

This commitment of long-distance carriers and local telcos to fiber optic cable is primarily to provide conventional telephone service, thereby employing the digital exchanges (switches) that they individually own. The structure of digital networks is such, however, that major units of capacity can be separated and leased to other organizations for their bulk private telecommunication needs. There is a hierarchy of building blocks of digital capacity where the standard unit of 1.544 Mb/s is the T1 channel (also referred to as the DS1). Table 1.2 presents the principal levels of the digital hierarchy employed in North America. Across the national

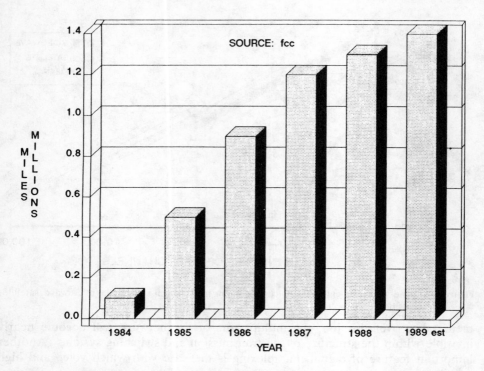

Figure 1.5 Long distance fiber optic growth (total fiber-pair miles).

transmission network, the smallest standard unit is the DS0 channel, which provides 64 kb/s of capacity. The standard T1 channel carries a bundle of 24 DS0s to transmit a like number of digitized voice channels, using standard 64-kb/s pulse code modulation (PCM). The number of voice channels can be increased to 48 or 96 by using digital bandwidth compression. All of the long-distance companies and many local exchange (telco) carriers offer T1 service. As discussed in Chapter 4, the T1 channel is a versatile pathway for integrating data and video with voice to provide a wide variety of communication services. Magnifying this value is the current new generation of T1 multiplexers and digital switches, any of which can be installed for use in private telecommunication networks. In Europe, the standard is 2.048 Mb/s, which is built up from 32 individual channels at 64 kb/s each.

The growth of intercity fiber optic capacity and its attendant low cost per unit of bandwidth have made an even higher level of the digital hierarchy attractive for the largest private networks. A DS3 has 45 Mb/s of digital capacity, enough to carry 28 T1 channels or one full-motion color television channel. Most private network operators would not need the capability of the DS3, but others would indeed sign up for some of these channels if it were appropriate.

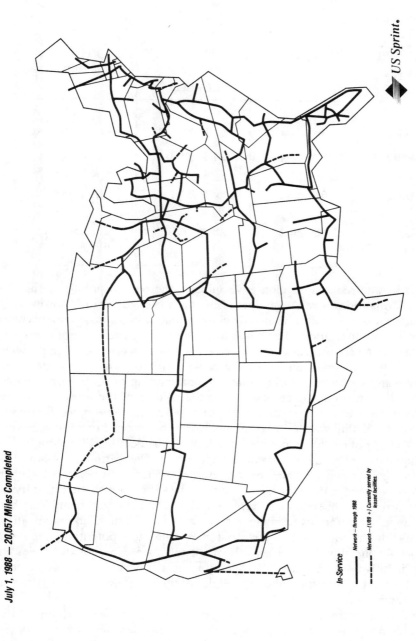

Figure 1.6 US Sprint's fiber optic network. (Illustration courtesy of US Sprint.)

Table 1.2 The Digital Multiplexing Hierarchy currently in use in North America

DEFINED LEVEL	RATE	RELATED TO OTHER LEVELS	TYPICAL USE
DS-0	64 kbps	(Root)	Voice, 56 kbps data
DS-1, T-1	1.544 Mbps	24 X DS-0	Voice, data, and video teleconferencing
DS-2	6.0 Mbps	6 X DS-1	High capacity trunking
DS-3	45 Mbps	28 X DS-1	High capacity trunking, video

In this new telecommunication environment, the cost of point-to-point capacity in bulk quantities is relatively low. This is a two-edged sword, however. If an organization relies on fiber transmission between major locations, it can succumb to total loss of communication when a link in the network is broken. For example, fiber links have been cut on occasion when construction crews (working on projects unrelated to telecommunication) severed a cable with a back hoe or some other piece of mechanical equipment. The unintentional cutting of a fiber optic cable and its dramatic effect on service have led to the coining of the term "back hoe fade." In a microwave radio system, a fade occurs when the radio signal decreases rapidly in strength, thus disrupting communication for seconds or even minutes. Cutting of a fiber optic cable is not a true fade, but it is deep and it is long!

Earthquakes and severe storms have knocked out fiber terminals of both the private network user and the telco or long-distance carrier. Breakage of a fiber link can cause an outage of many hours, or even days, because of the difficulty of properly repairing the fiber links or replacing the terminal equipment. This has motivated some organizations to implement backup and alternate communication paths, paralleling the route of the terrestrial fiber where it is potentially the most critical. Such implications of the growth of digital fiber capacity are reviewed extensively in this book. Clearly, the operators of a private telecommunication network will need to understand the capabilities of fiber, although they may never purchase it themselves.

Long-distance carriers are now using their networks in new and innovative ways. By increasing the processing power of their long-distance switches, developing proprietary software upgrades, and adding sophisticated automatic network control systems, the long-distance carriers are making their networks more effective and less expensive to use on a call-by-call (switched) basis. Under proprietary programs like AT&T's Megacom and MCI's Prism℠, bulk discounts for calling on their networks tend to undermine the attractiveness of full-time voice circuits between locations provided on private telecommunication networks. The types of DS1 and DS3 channels discussed previously must be examined carefully and compared with the costs of switched long-distance service. These developments are vital to the planning of private networks, so considerable information on the various offerings of the long-distance carriers is presented in Chapter 2. The competition for the corporate customer is so strong that new calling packages and bulk discounts are being introduced by the long-distance carriers almost on a monthly basis. Buyers of telephone services must maintain an awareness of new calling packages and perform economic evaluations when appropriate. (Chapter 6 provides information on how to perform such an investigation.)

1.2.2 Rapid Growth of Data Communication Applications

Data processing (DP), also called *information processing,* is the function carried out by computers, as opposed to data communication, which is the role of telecommunication in connecting computers to one another and to distant users. Essentially every strategic unit relies upon DP for some critical part of its operation. The task is simple for a company which is fortunate enough to have all its DP activities consolidated at one location. For most, however, several sites frequently must gain access to a central computer or to each other's computers. Data communication has consequently become a critical part of the performance of strategic units. As an indication of the pervasiveness of data communication, 78% of office workers in a Harris poll had computer terminals at their desks.

In traditional DP systems on mainframes, users submit jobs in the *batch mode,* consisting of programs and associated data for processing. Results in the batch mode are either printed on paper or recorded on magnetic media for storage and subsequent processing. Batch mode is used where processing is done once per evening, for example, by gathering up the appropriate data during normal working hours. Results may be electronically transmitted to distant locations over a private data communication network.

With real-time programming and computing, the computer responds almost immediately to the stimulus from a local or remote input device such as a terminal. The effectiveness of real-time computing for serving multiple users (numbering in the hundreds, thousands, and even millions) is causing it to become a central facet of many businesses. Another term often used is *on-line transaction processing* (OLTP), because users interact with the central computer as if dealing with a bank teller or file clerk. Examples include airline reservation systems, automated bank teller machines, inventory control systems, and credit verification. The communication links which support the real-time systems are typically part of a private data communication network. There are similarities between this type of network and that used for telephone or video, but there are also many critical differences. A comprehensive discussion of data communication networks can be found in Chapter 3, and integration of data with the other types of communication is covered in Chapter 4.

When the first data communication applications emerged in the 1950s, the only available communication channels were through the analog telephone network. Data are linked with the analog telephone circuit by using a modulator-demodulator (modem), which converts digital pulse information into a set of audible tones capable of passing through the voice frequency bandwidth of the analog network. The user has two choices when using analog circuits: the channels can be established as needed on a "dial-up" basis, or they may be leased on a long-term basis from the carriers. Because these circuits are designed to pass an intelligible voice signal and not for optimum data transmission, the performance of the channels may barely be adequate.

The telephone companies have improved the quality of data communication service in two ways. An interim solution was to alter the analog characteristics of leased telephone lines by using a technique called *equalization*. Conceptually, this makes the bandwidth of the channel appear wider to the data than it actually is. AT&T then introduced Dataphone Digital Service (DDS), which consists of digital leased lines implemented specifically to carry high speed digital signals without the use of analog equipment such as modems and equalizers. The interface between the computer terminal and the DDS is with a separate device called a *digital service unit* (DSU). These leased data communication facilities give reasonably good quality transmission and confidence that the line will be available when needed to carry data. Leased lines, however, whether analog or DDS, have always been expensive and difficult to obtain on short notice. DDS is often limited in outlying regions, particularly rural areas. A nationwide terrestrial data communication network must incorporate private analog telephone lines as well as higher quality DDS lines.

A new type of communication service was introduced by companies like Tymnet (now part of McDonnell Douglas) and Telenet (now part of US Sprint) to provide slow-speed DDS quality links on demand. As a general class, these offerings are called *public data networks* (PDNs). The forerunner to switched data service is Western Union's telex service, which allows subscribers to place calls and exchange character-oriented data at approximately 100 b/s. The more advanced offerings are referred to as *value added networks (VANs)*, since the network operator obtains leased lines and serves its customers by using sophisticated data switching equipment at nodes in its network. Most VANs are packet switched, wherein the customer's data are broken down into blocks of data called *packets*. Individual packets are "addressed" by the VAN's node equipment and routed over the appropriate link or series of links with packets being "handed off" from node to node. VANs have become so popular that the local telephone companies are entering the business, as we will discuss in Chapter 2. A public VAN is basically a PDN.

Data communication was once only a small piece of the telecommunication expense pie, representing approximately 10 to 20% of the annual business. The trend in growth of data communication *versus* voice is illustrated in Figure 1.7. Projecting the trend out to 1995, we expect the demand for data communication networking to increase more rapidly than that for voice. Note also that total telecommunication expenditures are rising at an annual rate of approximately 10%. Today, data communication expenses are growing at a much faster rate than those of the more traditional services. This demonstrates the importance of data communication to strategic units and operators of public telecommunication networks.

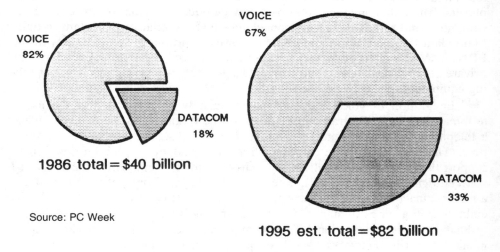

Figure 1.7 Growth in data communication service revenue in long haul networks.

1.2.3 Integration of Voice and Data

The digitization of voice networks and the growth of data communication lead to the obvious question: can voice and data be combined (integrated) so that common telecommunication network facilities are used effectively? The answer is yes, since the information in all cases is in digital or numeric form. We should not, however, underestimate the technical and organizational complexity of doing an effective job. Telephone communication networks are well structured, and users understand the capabilities of these networks. Nonetheless, new developments like voice messaging and speech recognition make telephone communication more complicated than in earlier networks. The structure of data communication, on the other hand, is dependent on the particular needs of the user as well as the type of DP equipment and software being used. Integrating voice and data will therefore involve a compromise on both sides of the equation.

In terms of equipment, the sophisticated types of multiplexers mentioned previously (described in detail in Chapter 4) provide an efficient means for integration of voice and data. The application of T1 multiplexers in an integrated network is indicated in Figure 1.8. When combined, the traffic stream is routed over T1 transmission links, which can be leased from telephone companies and long-distance carriers. In some cases, the user can implement its own T1 links by using microwave or satellite communication. Each link is terminated in a T1 multiplexer such as the one illustrated. Voice switching equipment of the private telecommunication network can be connected to this network through the multiplexers. The data communication part may be equally simple to introduce if the computers and terminals require connections that are relatively constant in time. Many data communication applications, however, are more compatible with the PDNs or VANs cited previously because data may only be sent on occasion. The private network developer could also build a private VAN by adding packet switching equipment to the T1 nodes.

Integration of voice and data may ultimately be performed at the user's terminal device, as illustrated in Figure 1.9. The terminal has the ability to place a telephone call and to send and receive data to a distant computer at the same time. The data part of the call might represent text or graphic information to support the voice communication. This scheme is the concept behind the *integrated services digital network* (ISDN), a new and evolving environment which will have a major impact on public and private telecommunication networks alike. ISDN is embodied in a series of standard technical specifications prepared under the coordination of the International Telecommunication Union (ITU). Several experimental and trial networks have been put together in Japan, Europe, and the United States, but the benefits of ISDN are still being identified and considered. Although its utility is uncertain, owners of large private telecommunication networks are now demanding of their equipment and service suppliers that ISDN features must be capable of being added to the telephone and data communication facilities in which they are investing today. Furthermore, public telecommunication network companies, notably the Regional Bell Holding Companies (RBHCs), discussed later in this chapter, and AT&T, are placing strong emphasis since ISDN offers a way to capture and hold the growing data communication business.

1.2.4 New Attractiveness of Satellite Networks

Geostationary satellites, which are capable of efficiently distributing information over a land area the size of a continent, have long been recognized as

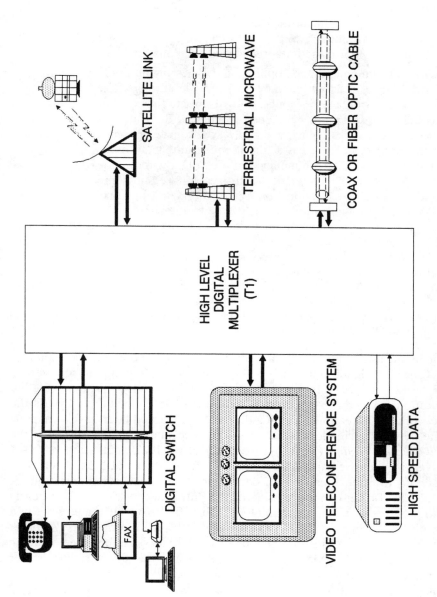

Figure 1.8 Integrated digital services using a high-level multiplexer with connections to satellite and terrestrial T1 transmission facilities.

Figure 1.9 Integrated voice-data terminal (model 513 Business Communications Terminal) for use with a Digital PBX. (Photograph courtesy of AT&T.)

useful for broadcasting TV signals and for international and emergency communication. Considerable information on satellite technology and applications can be found in the Elbert (1987). In the context of private telecommunication, satellite communication has a rather specialized role to play as a medium to bypass the public networks. The reasons for doing this will be reviewed later in this chapter. A satellite network technology which is gaining acceptance in this regard is the *very small aperture terminal* (VSAT), shown in Figure 1.10. By definition, a VSAT is a relatively inexpensive earth station with a rooftop antenna of 1.2 to 1.8 meters in diameter (4 to 6 feet), capable of data, voice, and video communication. A network of VSATs can operate outside of the public network, providing nearly total control by the user. Digital technology within the VSAT facilitates voice and data transmission, achieving some of the benefits anticipated from ISDN. Another attractive feature of VSATs is that high-quality color video can be received along with the voice and data service. In a terrestrial system, this would require either coaxial or fiber optic cable to the home or office.

The manner in which the VSAT is applied in a private network is illustrated in Figure 1.11. Two customer locations are shown connected to the public telephone and data networks. On the "roof" of each location, however, is a VSAT directed

Figure 1.10 A typical Ku-band Very Small Aperture Terminal (VSAT) with a transmit-receive capability. (Photograph courtesy of Hughes Communications, Inc.)

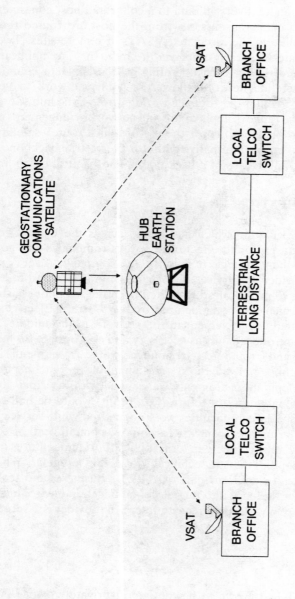

Figure 1.11 Terrestrial and satellite transmission for long distance telecommunication service.

toward a geostationary communication satellite. A hub earth station receives transmissions from VSATs and delivers all data to a collocated host computer system. Outbound control and data transmissions from the host are routed through the hub and over the satellite to the associated VSATs at remote sites. Two VSATs can in fact communicate with each other via the hub. In some applications, the hub is shared by several hosts, each operating a separate data communication network. While the satellite network allows the public network to be entirely bypassed, it is possible to enter the "outside world" through the hub. Considerably more information on the range of applications and network arrangements of VSATs is provided in Chapter 4. We also cover the use of "thick route" satellite links at T1 rates to provide backup and alternative paths for a terrestrial backbone network, which are potentially vital in case of a back hoe fade or central office fire.

1.3 AGE OF DEREGULATION

The wave of deregulation which has spread across the U.S. telecommunication scene is having a profound effect on the ways companies and government agencies perform out their strategic functions. One company, AT&T, previously was the principle provider of services and facilities. AT&T's end-to-end (some would say from cradle to grave) offering started from the telephone instrument in the home or office (which was manufactured by the Western Electric subsidiary of AT&T), and included every element in the network in the middle. For good or ill, technology was introduced only as necessary and appropriate for maintaining a universal, reliable, and profitable telephone network. Investments in outside plant (e.g., wire, telephone poles, radio systems, manholes), switching gear, buildings, vehicles, and telephone instruments were made with a virtual government guarantee that a reasonable rate of return (i.e., profit) would be had by AT&T. Furthermore, old or obsolescent facilities were not retired while service could still be provided. This is in marked contrast to how telecommunication companies, including AT&T, now must compete in terms of both cost and features of the most modern network capabilities that are technically feasible. Deregulation has allowed competitive entry into a significant part of AT&T's business, and technological change is introducing turbulence into the telecommunication environment. Except for local telcos, perhaps no telecommunication service provider is assured of profits or even revenues.

1.3.1 AT&T Divestiture

By 1984, AT&T had grown to be the largest privately owned company in the world, with more than one million employees. A legal suit by the U.S. Department of Justice was settled with AT&T agreeing to divest itself of the local telephone business. Judge Harold Greene of the Federal District Court of Washington, D.C., continues to monitor and direct implementation of his rulings. The corporate structure of AT&T, before and after divestiture, is illustrated in Figure 1.12. Under the Modified Final Judgement (MFJ), AT&T retained its well established long-distance business and was granted permission to move rapidly into the information processing and computer field. AT&T's impressive manufacturing business was injured by divestiture because the divested telcos are free to purchase equipment from any supplier, including those in foreign countries. Inroads from overseas suppliers and aggressive domestic companies quickly gobbled up large chunks of Western Electric's markets.

The MFJ spun-off local telephone service into seven RBHCs, indicated at the bottom of Figure 1.12. Judge Greene closely watches attempts by the RBHCs to expand their businesses into areas of activity for AT&T and others. The concern of the court is that the RBHCs can use their monolopy position to finance their entry into other markets. Furthermore, they probably have a guaranteed market in the form of their large, captive subscriber base. Judge Greene, however, has relieved some restrictions in the area of information services to allow the RBHCs to innovate where he believes other, more fragile suppliers will not be seriously harmed.

In contrast to manufacturing, AT&T's core long-distance business has continued to serve them well. The Long Lines division of the predivestiture AT&T was renamed AT&T Communications (ATT-COM). Their market share has dropped from 85% to 75% over the three-year period cited above; the amount of revenue has not decreased, however, due to the sheer increase in domestic and international calling. AT&T's profits have increased due to economy measures, such as staff reductions and increased use of automation. As the dominant long-distance carrier in the U.S., AT&T maintains its position as the clear leader. AT&T is investing in new plant and equipment, including the most advanced digital fiber optic transmission links and computer controlled switching. In 1987, the company announced a plan to invest $6 billion over a four-year period in these new facilities. They have been decommissioning obsolete plants, which have been written off more rapidly than prior to divestiture as a result of competition from other carriers. Another succssful area is AT&T's international operations which provide the bulk of international telephone service, employing satellite and cable transmission.

Manufacturing operations of AT&T have changed considerably since divestiture. Telephone instrument manufacture has been moved offshore and competition abounds. Sales of high-capacity digital switches, notably the Number 5 ESS, have been relatively good and AT&T is also involved in a venture on such projects in Europe. Smaller electronic telephone systems for businesses are selling very well and AT&T now dominates the supply of key telephone systems. These aspects are covered in Chapter 3. Long-haul transmission equipment has become a very competitive market, and AT&T is being forced into a lesser role than it

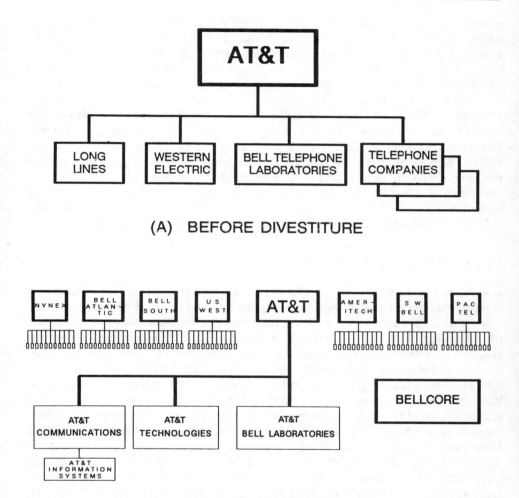

(A) BEFORE DIVESTITURE

(B) AFTER DIVESTITURE

Figure 1.12 The structure of AT&T and the Bell Telephone System (pre Divestiture).

once enjoyed in the area of microwave, cable, and multiplex equipment. The telecommunication equipment manufacturing part of AT&T is named AT&T Network Systems.

The aggressive move of AT&T into computer manufacturing has not met with particular success. AT&T first marketed its own minicomputer line, originally developed for telephone switch control, and teamed with Olivetti of Italy to offer personal computers. Their principal computer product is a line of personal computers which are 100% compatible with the IBM PC. All computer operations were put under AT&T Information Systems, which was subsequently combined with AT&T Communications. This was purportedly done to improve marketing by allowing a single sales force to handle computers and communication services.

AT&T correctly chose their competitor in computers: International Business Machines (IBM). IBM, however, has maintained its dominance as a provider of large or small business computer systems. Conversely, IBM's entry into telecommunication through its venture in Satellite Business Systems (SBS) was discouraging. The SBS experience is summarized in Chapter 2. History now suggests that the proper way to view AT&T is as a communication company which employs computers, while IBM can be viewed as a computer company which employs communication. Economy of scale, technology, and financial strength will make AT&T's long-distance network a credible force into the twenty-first century.

1.3.2 Local Telephone Service in the US

Divestiture resulted in a smaller AT&T and seven new regional Bell holding companies, identified at the bottom of Figure 1.12 and on the map in Figure 1.13. The RBHCs, in turn, own the local telephone companies that formerly were part of AT&T's Bell System. These are referred to as Bell Operating Companies (BOCs). For example, New York Telephone Company and New England Telephone and Telegraph are BOCs that became a part of the RBHC named NYNEX. The Modified Final Judgment allows the RBHCs to enter into some nonregulated businesses, but these are severely restricted by the court order and subsequent rulings of Judge Greene. Curiously, the RBHCs and local companies may retain the Bell logo, while AT&T is forced to use a new identification. The Bell Telephone Laboratories remained a part of AT&T and could continue to use the Bell moniker. To support the RBHCs in efforts to maintain common technical standards and equipment specifications, part of Bell Labs was spun off as Bell Communications Research (BELLCORE) and put under shared RBHC ownership. BELLCORE now maintains Bell System standards and provides technical assistance to the RBHCs and BOCs.

Having seven large telephone holding companies with common lineage introduces some interesting new elements into the US telecommunication picture. Employing the capabilities of BELLCORE in conjunction with their own individual resources (which are considerable), RBHCs are experimenting with new business offerings. One motivation for this is that with local regulation their revenue growth is generally limited, corresponding to population growth. This is adequate for Pacific Telesis and Bell South, both of which are located in rapidly growing areas. NYNEX and Ameritech, however, operate in the industrialized north, which is

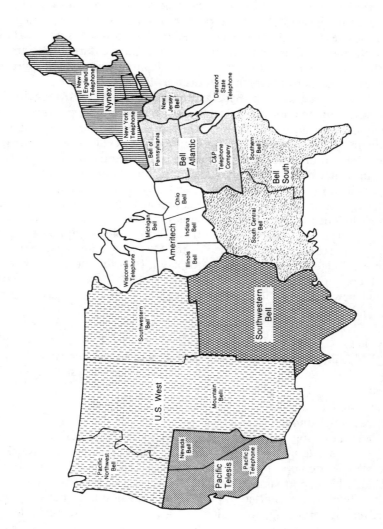

Figure 1.13 The seven regional Bell holding companies (post Divestiture).

not experiencing much population growth. When one RBHC opens up a particular business area, such as office equipment sales, the others watch closely. A successful new business thrust by one is adopted by the others. This provides a multiplying effect in terms of the growth of their respective businesses. An interesting result is that a given business activity may be on a nationwide basis, causing the RBHCs to compete with one another. For example, "yellow pages" directories are published in each other's areas, and cellular radio telephone systems are purchased from nonwireline carriers to compete with distant RBHCs.

Local telephone service is provided on a monopoly basis by several independent companies other than the RBHCs. The second largest provider is GTE Corporation, which formerly operated under the name of General Telephone. With approximately 12% of local telephone subscribers, GTE is now comparable in size to the RBHCs. GTE service areas are not contiguous like an RBHC, but can be found in nearly every corner of the country, including a major portion of the system in British Columbia. Other independent telephone companies include Contel (formerly Continental), United Telecommunications, Centel, and Pacific Telecom. These independents engage in other nonregulated business activities such as government contracting, long-distance service, cable television, and equipment manufacturing. The RBHCs are precluded by the MFJ from entering into the latter three categories. Judge Greene continues to play a forceful and dynamic role in overseeing adherence to the terms of the MFJ. On many occasions, the RBHCs have requested new authority for entry into restricted areas, such as equipment manufacture and provision of information services, but Judge Greene has steadfastly refused. Political pressure is being applied on the Congress and the President to enact new laws to allow such entry. Entities like AT&T, the cable television companies, IBM, and other existing equipment manufacturers, however, stand on the other side of the debate, so it is likely that removal of restraint will occur slowly if at all.

One contested area relates to the RBHCs entering the business of long-distance service, which Judge Greene refuses to allow. This prevents the RBHCs from using their monopoly positions and subscriber bases to gain an advantage over AT&T and other long-distance carriers in the long-distance market. Established by the MFJ were compact geographic regions called *local access and transport areas* (LATAs), which approximate the size of a metropolitan area or county. An RBHC is the only carrier permitted to provide transmission links between central offices (interexchange trunking) within its LATA franchise. As an example of this division, Figure 1.14 presents a map of the LATA boundaries for the state of California. An RBHC such as Pacific Telesis may be the sole provider of local telephone service in an entire region such as California but, it cannot provide trunking paths between LATAs (inter-LATA service). Within a LATA (intra-LATA service), the RBHC is the principal transmission provider. Service between LATAs (inter-LATA service), however, is by definition long-distance, and can

Figure 1.14 Local access and transport area (LATA) boundaries in California.

only be offered by a long-distance company. Within a state's boundaries, inter-LATA service still comes under the jurisdiction of the state's regulatory agency such as the Public Utilities Commission (PUC). Some PUCs restrict the number of carriers providing inter-LATA service, while in other states the market is open. The only long-distance company providing inter-LATA service within every state (and between states as well) is AT&T. Interstate long-distance services are regulated by the FCC.

1.3.3 Long-Distance Companies

The division of long-distance revenues among the major service providers is shown in Figure 1.15. There are three primary players in the U.S. long-distance market: AT&T, MCI Communications, and US Sprint. AT&T is clearly the leader with over 75% of the market, which includes much of the inter-LATA traffic within a given state as well as interstate traffic. The second largest long-distance carrier, MCI, began operation in 1968 and established a profitable alternative long-distance business using microwave transmission even before AT&T divestiture. Many observers expected MCI to benefit greatly by divestiture, but the market power of AT&T proved more difficult to defuse than had been expected.

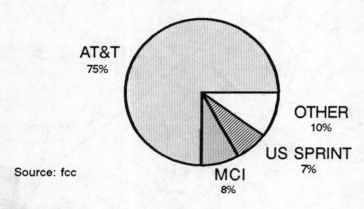

Figure 1.15 Market share of the three major long distance providers.

US Sprint was formed in 1986 by the combining of United Telecommunications' US Telecom operation with GTE Sprint. Both companies provided local telephone service as independents, but they believed that the cost and risk of developing a major long-distance network would be better managed and funded jointly. Previously, GTE purchased Sprint from Southern Pacific Railway. US Sprint is operated as a partnership, although United now dominates management. The cornerstone of Sprint's strategy is a fiber optic network being implemented between major cities in the U.S. (see Figure 1.6). Television commercials might imply that US Sprint had invented the technology and no other company used fiber optic transmission to a significant degree. In truth, both AT&T and MCI are installing fiber transmission at an equally rapid rate along many of the same routes. Moreover, the three carriers are converting from older analog facilities to digital. The benefits of this are many, including low unit cost of transmission, reduced circuit noise, and improved facility for data communication. Since 1987, the foundation of long-distance service is digital transmission on fiber optic cable and microwave radio.

A number of other investors and resellers have entered the long-distance market, locking up existing facilities and entering segments of the long-haul transmission business. The FCC has taken the position not to regulate carriers that use fiber, making it possible for foreign companies to gain a foothold in the U.S. market. Cable and Wireless PLC, a respected international carrier which was privatized in the United Kingdom, purchased a U.S. fiber optic carrier called TDX, and now is offering domestic and international digital transmission services. The types of services offered by investors and resellers may tend toward private line rather than switched, making these offerings more suitable for private networks.

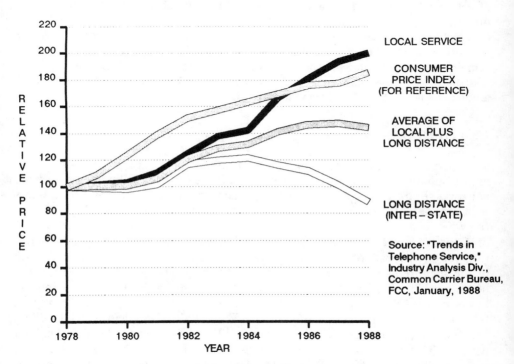

Figure 1.16 Price trends in local and inter-state long distance service.

The competition among rival long-distance firms is indeed fierce. Since divestiture, long-distance rates have been steadily declining, as illustrated in Figure 1.16. Actions by the FCC in the area of "equal access" have allowed MCI and US Sprint to offer service which acts almost exactly like AT&T's long-distance. Previously, to use a competing long-distance carrier's service, the subscriber needed to dial a local-seven digit telephone number to reach the carrier's switch. Then, the area code and seven-digit number of the called party would be entered, followed by the subscriber's account number. In exchange for their trouble, subscribers got long-distance rates which were 10 to 20% lower than AT&T's. The subscriber also needed to have a touch-tone phone to be able to communicate with the carrier's switching equipment. Under equal access, the subscriber preselects (or votes for) their long-distance carrier, which then would be reached by dialing the number 1 ahead of the normal area code and seven-digit number of the called party. (This voting process was probably the motivation for AT&T's advertising line, "The Right Choice.") Whichever carrier is able to convince the subscriber to choose it as the designated carrier will automatically be the subscriber's principal provider of long-distance service. (Another long-distance carrier can still be accessed by dialing a longer but still abbreviated code.) MCI pushed for the implementation of equal access because it theorized that simplified calling on a competing carrier's network would tend to increase its use. At the time of this writing, however, AT&T remains dominant with 81% selection, MCI achieves actually more than expected with 8%, US Sprint has approximately 6%, while the remaining 5% is divided among smaller national and regional long-distance carriers.

1.3.4 Private Network Suppliers

A type of prime contracting organization has gained acceptance in the U.S. as a response to the need of strategic units to design and to implement private networks. In the past, developers of private networks were government contractors with considerable experience in overseas projects, notably including Collins Radio (now part of Rockwell International), Page Communications Engineers (now part of Contel), Philco (now part of Ford Motor Company), and Hughes Aircraft Company (now part of General Motors). Collins gained an excellent reputation in the high-frequency (short-wave) radio and the microwave line-of-sight radio businesses. In addition to manufacturing the equipment, Collins engineered the links, installed the facilities, and provided maintenance support to the buyer. This became a model for others to follow. Page and Philco designed and installed a backbone telephone network throughout Southeast Asia during the Vietnam War for the U.S. armed forces, using microwave and tropospheric scatter radio links. An example of an over-the-horizon tropospheric scatter site featuring the prominent "billboard" reflector antenna is shown in Figure 1.17. Hughes Aircraft designed and participated in the construction of Indonesia's first domestic satellite network (the Palapa A system) during the mid-1970s.

Projects of this scope required that the contractor build a rather large team, which often worked in highly inhospitable surroundings with a very long supply line back to the manufacturers of the equipment. The facilities installed consisted of equipment produced by the prime contractor and many other manufacturers because each job was customized to a particular set of requirements. Operation of the system was either taken on by the integrator under separate contract or became the responsibility of the buyer. Each contract was unique, which made the

Figure 1.17 A tropospheric scatter antenna installation in southeast Asia (1967).

project both interesting and challenging. Technicians often worked a 12-hour day, seven days a week for months at a time. The final result, having a working communication system was seemingly worth the effort.

Today, many telecommunication companies including AT&T and the RBHCs participate in the private network business by taking on the network design and installation roles. Equipment manufacturers like IBM, Hughes, and Northern Telecom can also perform network integration functions, combining their own products with those of other companies. A key distinction between the pure integrators and other manufacturers is that the integrators do not incorporate their own proprietary products into the network; perhaps, they may be more objective in their source selection. Another approach is to employ a consulting company to design the network and then subcontract the major elements to the best available source (possibly the lowest bidder). The same consulting company might then monitor construction and oversee acceptance testing prior to the initiation of service. These important aspects of private network development are covered in Chapter 7.

1.3.5 International Services

Much of our emphasis thus far has been on domestic private telecommunication networks. Many strategic units, however, engage in international trade and operations for which reliable communication is often essential. Most experienced world travelers are familiar with record communication by telex, which is an established teletypewriter service available in almost every country on the planet. Telex messages can be sent at any time of day or night, allowing an office in New York to originate a written document, send it to an office in Tokyo, and have it waiting for personnel to arrive in the morning and act upon it that day (i.e., while the New York staff is asleep at home). This was a precursor of electronic mail. Telephone contact may be required for real-time conversation, but the lack of adequate or modern facilities in a distant country may restrict availability on demand.

The rapid growth of international operations during the 1970s and 1980s along with the familiarity of managers with more advanced information technologies have been forcing the international carriers to provide several capabilities found in domestic networks. Data communication links, facsimile transmission, electronic mail, and video teleconferencing are becoming popular among multinational corporations and diplomatic agencies. Some of this traffic can be carried over good quality analog telephone circuits. There has been a trend toward the use of dedicated private lines, however, because this tends to tie affiliates' operations more tightly together around the globe.

The two telecommunication technologies that have made possible international private (and public) networks are transoceanic cable and international satellite. The investment in a particular cable system is a joint undertaking of the countries being connected. Because the public telecommunication network within most countries is owned and operated by the government, the cables represent national assets. In the U.S., most cable investments have been made by AT&T. Currently operating systems use analog transmission over coaxial cables with repeating amplifiers spaced every several miles to overcome signal propagation loss through copper. Electrical power for repeaters is carried on conductors within the same cable sheath. These cables only support voice communication and low to medium speed data. High speed digital voice and data, and video communication are not possible over the existing cable systems. The newer fiber optic cables now being installed support the full range of digitized services like the domestic fiber and microwave systems in the U.S. Therefore, transoceanic fiber offers the capability to do almost anything, so long as the points to be reached are along one of the cable systems. In addition to AT&T, private companies in the U.S., the U.K., and Japan are investing in transoceanic fiber optic cable systems. As part of any cable operation, ships have to be kept available for installation and to repair the inevitable breaks induced by fishing trawlers and unknown natural causes.

The satellite networks of the International Telecommunications Satellite Organization (INTELSAT) represent the true international backbone, reaching every country and major city on the planet (including points in Eastern Europe and the People's Republic of China). Prominent within each country is one or more of the familiar 30-meter Standard A stations. These antennas, along with somewhat smaller ones permitted by the higher powered Intelsat V and VI satellites now used in the system, provide the full range of voice, video, and digital data that satellites are capable of delivering. Satellite transmission, unlike transoceanic coaxial cable, is inherently wideband. In fact, a single Intelsat VI satellite can deliver the same capacity at a unit cost comparable to that of a transatlantic fiber optic cable. The satellite has the added virtue of being able to establish many links throughout a hemisphere and to provide domestic services within countries.

Circuits on the cables and Standard A INTELSAT earth stations must be obtained on an end-to-end, common carrier basis from the telecommunication authority in the respective country. In the past, private network approaches often were difficult to implement on an international basis. To counter this, INTELSAT introduced a private-line type of service called INTELSAT Business Service (IBS). Capacity on certain of their satellites was reserved, and standards for more compact and less expensive earth stations were established. Countries that permit IBS operations at present include the U.S., Canada, the U.K., The Netherlands, the Federal Republic of Germany, France, Japan, and Hong Kong. Others can be expected to follow suit. In some cases, a private company can actually own its IBS antenna. INTELSAT stipulates that digital transmission be used over the satellite so that the quality of the signal can be properly monitored. Devices connected to each end of the circuit can communicate directly with each other. Nonetheless, interface with the foreign telephone or data network may be difficult due to differences in the local digital hierarchies or format, or may be prohibited by the telecommunication authorities for protective reasons.

A particular niche in international communication called *transborder service* is allowed with domestic satellites. Under the watchful eye of North American telecommunication administrations and INTELSAT, domestic satellite operators may use the incidental coverage of their satellites to provide links with neighboring countries. This is particularly popular for delivering video programming across the border from the U.S. into Mexico and *vice versa*. Two way communication links are commonplace between points in Canada and the U.S., where this mode is relatively easy to arrange with Telesat Canada or any of the established U.S. satellite carriers. Links between points on Caribbean islands and the continental

Figure 1.18(a) The U.S. INTELSAT earth station at Roaring Creek, PA, with three standard A antennas. (Photograph courtesy of COMSAT.)

Figure 1.18(b) A standard E (Ku-band) INTELSAT earth station antenna at Spring Creek, NY. (Photograph courtesy of Hughes Communications, Inc.)

United States (CONUS) will soon be implemented under the guise of transborder communication. This concept has been proposed for Latin America as a means to simplify communication between countries in this region of common heritage. A private company called Pan Am Sat initiated service of this type in 1988, through a specialized satellite launched that same year. In Europe, the Eutelsat satellite system is already used for international television transmission and other services on a limited basis.

1.3.6 Deregulation in Other Countries

Deregulation in the United States and the introduction of new capabilities around the world, like fiber optic cables and IBS, have provided impetus for telecommunication deregulation in other countries. The United Kingdom has moved to "privatize" the telephone system, with British Telecom (BT) being established from the old British Post Office. As a private company with stock-holders, BT is moving aggressively into the international marketplace by acquiring companies and offering services outside of the U.K. Meanwhile, back home in England, Mercury Communications, subsidiary of the established international carrier Cable and Wireless PLC (C&W), now provides businesses with local and long-distance telecommunication, in competition with BT. Further deregulation in the U.K. and more competition than is currently allowed in the U.S. would not be surprising. Deregulation has also occurred in Japan, now that Nippon Telegraph and Telephone (NTT) has been privatized, and other companies like Japan Communications Satellite Company (JCSat) are allowed to compete with it in certain areas. International service in Japan also has become competitive with two private companies being set up to compete with the only preexisting international carrier, Kokusai Denshin Denwah (KDD).

These developments in major industrialized countries bode well for the future of private telecommunication in an international context. While there are still barriers to private development in most countries, at least, progress has been made in leading countries like the U.K. and Japan. Once the trend is firmly established and multinational businesses begin to enjoy the benefits of efficient and modern digital communications, the trend will be difficult to stop.

1.4 GROWTH OF PRIVATE NETWORKS AND BYPASS

The last trend to be reviewed is the most dramatic, and it clearly provides motivation for writing this book in the first place. The breakup of AT&T has made internal administrative functions more difficult for many organizations (see Chapter 2). The preponderance of strategic units have either implemented private networks or they are evaluating how to do so. Digital technology has made possible the provision of private network features at reasonable cost, often using public network facilities. On the other hand, the time and cost of dealing with the local telco has encouraged many to purchase bypass facilities. These involve terrestrial links, switches, leased long-distance private lines, and satellite communication. The following paragraphs provide additional motivation for building private networks and bypass facilities.

1.4.1 Meeting Specialized Requirements

Many private telecommunication networks are put together for the simple reason that the public switched network is incapable of meeting a need. The leased private line business of telcos, long-distance carriers, and satellite operators is geared directly toward this aspect. Connecting users to computers for remote job entry or transaction processing usually requires special leased lines, such as Dataphone Digital Service (DDS) from AT&T. The particular network may be designed for a single purpose such as credit verification or inventory control. In the case of private telephone switching within an organization, the leased CENTREX and purchased PBX switching and transmission facilities are able to satisfy the needs at hand, but upgrading the network for another use such as telemarketing may not be possible.

A network implemented for one purpose can probably be depended upon only for that purpose. A single-purpose network, however, is usually difficult to modify once it has been ordered and installed. An approach for avoiding this network deadend is called *integration,* wherein several telecommunication needs are aggregated and carried over a single network. This topic is discussed in a subsequent paragraph and in detail in Chapter 4.

A special class of single-purpose network called the *strategic network* implements part of an organization's fundamental strategy. The classic example often cited is the computerized reservation system of an airline or car rental company. When the technology of the strategic network is adopted by other organizations in a competitive environment, it represents an entry barrier or new price of entry into an industry. Occasionally, the technology of the strategic network is so tightly controlled, due perhaps to protective patents, that a competitive advantage is gained over other firms in the particular industry. A more detailed discussion of strategic networks is found in Chapter 2 and further considered in Chapter 8.

The key to the current popularity of private digital networks is that the features and interfaces can be tailored to fit the strategic modes of the business. Sophisticated digital PBXs, programmable multiplexers, and packet switches efficiently manage and pack information for its trip from source to destination. A dimension now only beginning to appear in telecommunication is that of *artificial intelligence* (AI). High speed computers with tremendous memory capacity can increase the power and utility of private data networks, giving a new degree of intelligence not only to the hardware, but also to the people who use it.

1.4.2 Achieving Economy of Scale

When able to aggregate communication requirements for carriage on a private network, an organization has the potential to achieve an economy of scale. The principal U.S. government agency which buys telecommunication services is the General Services Administration (GSA). By aggregating most of the nonmilitary voice and data requirements of the federal government, the GSA has tremendous buying power and therefore is a powerful force in the U.S. telecommunication environment. This is a double-edged sword because dealing with such diverse needs among so many different agencies can well represent a *dis*economy of scale. On a smaller scale, a regional organization like a state government, university system, or electrical power utility is in a position to install a comprehensive network to serve its internal needs, customers, and clients. A good example is the private network installed by Contel for government services in San Diego county. This implementation approach was discussed in the previous section on private network suppliers. The network uses modern digital switches, private microwave owned by the county, and digital private lines (i.e., T1, as discussed in the section on public telecommunication networks) leased from Pacific Bell. San Diego decided to install a private network primarily for telephone services, allowing citizens throughout the county to have better access to its resources and to reduce telecommunication costs. By aggregating all public services requirements (e.g., health, public safety, and government), the requisite economy of scale was achieved. Anyone who has worked to aggregate the various and diverse needs of different agencies and suborganizations will attest to the fact that the planning process is involved and difficult. Assuming that aggregation yields enough traffic to justify a private network, the results can be very gratifying and financially rewarding.

The technical basis for achieving economy of scale in today's environment is the availability of the digital infrastructure throughout North America and much of the developed world. Considerable information about the range of possibilities for employing this infrastructure is provided in Chapter 4. A trend toward the digitization of services at the user device now makes private and public network interface convenient at appropriate levels of the standard digital hierarchy (DS1, DS3, *et cetera*). Of course, signaling and control systems must also be connected properly; otherwise, traffic will not flow or be routed correctly. Considerable attention by international standards organizations is now being directed toward simplifying this key aspect of utilizing digital networks and interfaces. Because of the high motivation and potential payoff to all participants, we can expect that this process will eventually result in "seamless" networks, wherein digital information in various forms can efficiently pass with complete accuracy from private points of origination into the public network and even to another country.

1.4.3 Single Vendor Solutions

"One stop shopping" for telecommunication networks is for many a figment of the past. In today's U.S. telecommunication environment, a network is typically pieced together from the offerings of three or more vendors. The primary reason for this is that AT&T does not own and operate local telephone facilities, which are the mandate of the RBHCs and independents like GTE and Contel. Network development under these circumstances is a major undertaking with a concomitant effect on the way business is conducted. For some organizations, this new environment represents opportunity and challenge, nurturing new technologies and cultivating the introduction of new concepts in strategic networks. For others, the breakup of AT&T has placed unmanageable demands on the organization's administration.

There are ways that management of technology and network facilities can be passed to a single vendor. This should lighten the difficulties of dealing with multiple vendors. There may be an added cost in using a single vendor or network supplier because the overhead and profit of the middlemen must be covered. The possibilities for single-vendor approaches to networks are reviewed in the following paragraphs.

In terms of local telephone service within a LATA, the BOC or independent telco is clearly in a very favorable position for being that single vendor. Such a private network implementation is classified as a *metropolitan area network* (MAN). The local telephone companies tend to be responsive to their customers because of the existence of competitors like equipment vendors, integrators, and alternative carriers. Offerings include a wide variety of local calling plans, CENTREX service, T1 and DS3 transmission, equipment leasing, and maintenance. Rather than losing business to alternative suppliers, the local companies are willing to bypass themselves by installing the very private facilities that remove traffic from their public systems. Another area of development is in data communication through the local version of DDS. Offerings are beginning to appear in the form of central office LANs (CO-LANs) wherein business customers can use the telco switch to interconnect their local terminals and computers. This is the data communication equivalent of CENTREX and is a forerunner to ISDN.

Another single-vendor approach for building a private MAN capability is by contracting with an integrator. The San Diego county government network cited earlier, provided by Contel, is a good example of this. The switches and campus installations are performed by the contractor, or integration vendor, as is training and some of the maintenance. Selection of the particular switch manufacturer can be done by the integration vendor if the using organization has adequately specified the requirements. Because the telco has an elaborate backbone system and established usage base, leasing transmission from the telco is usually attractive for the bulk of trunking requirements. In the case of San Diego county, most of the terrestrial links between sites are over T1 and private line facilities of Pacific Bell (PacBell). This MAN should be called a two-vendor solution, since equipment problems would be referred to Contel, while transmission problems would fall back on PacBell. Eventually, county government telecommunication staff would take over equipment maintenance responsibility from the integrator.

Moving outside of the LATA establishes what is called a *wide area network* (WAN) and introduces a long-distance carrier (also called an *interexchange carrier* or *IXC*). In many cases, the strategic unit also must deal with two different local telephone companies. This is made even more difficult when the telcos on each end are owned by different RBHCs or are independents. Adding more sites in the network increases the number of telcos. The private WAN involves a minimum of three transmission carriers plus the equipment supplier or integrator. Perhaps this situation is the most fundamental difficulty produced by the breakup of AT&T. Having a single-vendor WAN using all-terrestrial facilities is impossible. At best we can hope that all facilities are designed to the same set of basic interface specifications and a good network management system is in place. The importance of network management cannot be overstated because the best refuge under problem situations is to be able to pinpoint where the problem in the network is so that the right party can be called in to effect repairs. Chapter 5 provides an in-depth discussion of network management principles and current implementations.

The last approach for achieving a single-vendor solution is the role of satellite communication. As we emphasize in Chapter 4, satellite communication offers total bypass of the terrestrial transmission network. An attractive approach is that of the VSAT at each user location. This type of network can satisfy a wide range of needs for internal communication within a dispersed organization. A satellite network does not replace terrestrial communication because it is doubtful that universal coverage of the public directly via satellite will ever be provided. The RBHCs and IXCs offer services and facilities that will always be important to doing business. We are talking about a role for satellite communication in delivering an aspect of a single-vendor solution, which can be attractive for certain applications. The strategic network is one of the best candidates for satellite communication because total control is generally possible. We discuss this aspect in the following section.

1.4.4 Regaining Control

The strategic unit must dominate its telecommunication resources, rather than *vice versa*. The concept of regaining control in this age of divestiture and deregulation is similar to that of the single-vendor solution discussed in the previous section. This ties into the breakup of the Bell System and the corresponding difficulty that organizations face in managing (i.e., controlling) their networks. In the past, strategic units might have found it adequate to obtain telecommunication services from AT&T in the same way that office supplies and equipment maintenance services were purchased. The unit's buyer of telecommunication services could order the requisite services and equipment, and demand help from the same supplier (e.g., AT&T) in the event of difficulty. Organizations now recognize that it is not that simple anymore; telecommunication is more complex (and more capable) than ever.

By definition, control of telecommunication is telecommunication management. Proper management of telecommunication demands that special people with experience and training in telecommunication must be a key part of the organization's structure. Control can be had in today's environment by establishing within the strategic unit a single point of responsibility, for example, the telecommunication manager. There has even been discussion of the need for a senior executive level position with a title of chief information officer (CIO) with total responsibility for telecommunication and DP.

Telecommunication management is a complex function, involving business, engineering, operations, maintenance, and contract negotiation. It represents one of the most challenging and rewarding occupations. The integration of services (voice, data, video) means that multiple disciplines are involved in planning, implementation, and operation. Many universities have implemented, or are implementing, degree programs in telecommunication management. The complexity of user requirements alone is enough to keep a telecommunication manager in school full time, and associations like the International Communications Association (ICA) and the Tele-Communications Association (TCA) provide courses and seminars for that purpose. Furthermore, the variety of vendor equipment and services is mind boggling. This situation is exacerbated by the appearance of new manufacturers and new products with capabilities needing to be investigated and understood. Likewise, the telcos and long-distance companies continue to add and to modify their service offerings, responding to each other's competitive inroads. Some good sources of current information about the technological innovations and new services in telecommunication and DP are listed in the bibliography. Also, an overview of the need for telecommunication management education is provided in Chapter 9.

Physical control of a private telecommunication network is exercised through the network management function. In the past, network management was done manually, using very crude test and monitor capabilities. People were required to go to points in the network where any kind of thorough investigation was needed. Automated systems are now appearing, and these continuously monitor network performance and can respond almost without human intervention to problems that will otherwise block traffic flow. This, of course, is an ideal, but it is one which is becoming realizable with modern computers and network control software. This topic is discussed in Chapter 5.

Building and running a network is not for the faint-hearted. The risks are great. New capabilities are needed, but if the facilities cannot do the job or are unreliable, the consequences can be career-limiting! On the other hand, the rewards are great. The person who can grasp meaning and value from the apparent disorder in the telecommunication environment is a special individual, one who will be valuable to strategic units in years to come.

1.5 EXAMPLE OF A BUSINESS NETWORK

The concepts introduced in the previous paragraphs are brought together in the following example of an actual corporate private telecommunication network. This network services the bulk of the telecommunication needs of a major diversified U.S. industrial company involved in engineering, manufacturing, financial, and data processing services. We have taken the liberty of using much of the terminology and nomenclature of telecommunication management, which, while confusing for the newcomer, will possibly whet one's appetite for what follows in later chapters. The name of the corporation is not mentioned in this description for proprietary reasons. The reader, however, may be assured that the example is basically factual.

The management of this corporation, the leader in its industries and one of the largest private enterprises in the world, thought that it was spending too much money on leased and tariffed telecommunication services. Aside from relatively small PBXs, all service was provided primarily by AT&T and, after divestiture, by the RBHCs as well. The corporation had experimented with a turn-key satellite network, which was supposed to have supplied the bulk of its voice and data communication needs. The experience with that carrier, however, was particularly unsatisfactory, and so the service arrangement was terminated after less than a year of operation. The reasons for the termination had to do with improper network design and software, rather than with the use of satellite links.

Responsibility for telecommunication and DP was changed from internal corporate staff to that of an information services company. The new telecommunication organization proceeded aggressively with the implementation of a private network of impressive proportions, relying on the use of a terrestrial digital backbone and an overlay of thin route links via satellite. This is the basic concept of a hybrid network, taking advantage of the inherent strengths of terrestrial and satellite technologies. Much of the research for the project was done by the previous telecommunication staff before the function was taken over by the information services company.

Reviewing the basic structure of the network, a terrestrial T1 backbone between major locations of the parent company provides for 80% of voice communication and host-to-host data transmission. The basic arrangement or *topology* of the private network is provided in Figure 1.19, where only major locations or *nodes* on the backbone network are shown. The PBXs are installed at corporate sites, routing telephone calls over the backbone to other sites and into the public telephone network. The calling plan allows company employees to use the backbone as if it were the network of a long-distance company, providing additional savings. Alternate routing is possible because the largest PBXs have tandem routing capabilities, meaning that calls can be connected through an intermediary PBX when a direct link is blocked or unavailable.

The computer environment is IBM Systems Network Architecture (SNA), with large data processing centers around the country connected by terrestrial T1s for load sharing and access to specific software applications which reside on various mainframes at different sites. This data network uses high-level T1 multiplexers to aggregate traffic from lower data rates in the range of 56 kb/s to 256 kb/s emanating from ports of the IBM model 3725 *front-end processors* (FEPs). The purpose of the FEP is to manage the data communication links, freeing the mainframe to process the information and respond to user demands. A benefit of using these T1 multiplexers, which are programmable, is that they allow network operations personnel to control and to reconfigure the network in response to traffic demands and terrestrial circuit failures.

The T1 backbone network provides bulk transmission of digitized traffic between the major telephone switching nodes and DP centers. From an economic standpoint, the use of the backbone and its associated private switching have reduced telephone expenses by 10 to 30%, and the specific reduction per strategic unit served depends on the amount of traffic and the location. Some of the data communication T1 links are backed up with C-band satellite paths for alternate routing and redundancy. These heavy route satellite links have been vital in case of breakage of the telco cables to the DP centers or failure of the terrestrial paths.

Integration of voice and data has not yet been implemented throughout the network, but can be accomplished in the near future with the telephone switches and T1 multiplexers. The management of the information services company intends to add ISDN features to the PBXs when such capability is economically available from the switch vendors. As attractive as it may be for network integration purposes, smaller corporate sites and subsidiary branch offices are not brought directly into the T1 backbone network due to the cost of equipment and T1 transmission service.

This high capacity mesh is overlaid by a satellite network of very small aperture terminals, providing thin route data communication links. The lower economic hurdle or prove-in cost of the VSAT makes it an attractive way to add

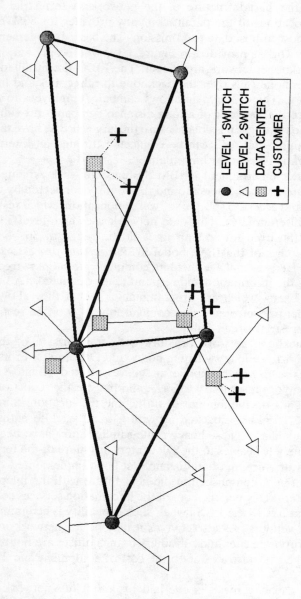

Figure 1.19 Topology of a nationwide private telecommunication network.

branch offices to a private telecommunication network (i.e., without going to a full T1 connection into the backbone). The first of two VSAT subnetworks consists of hundreds of television receive-only (TVRO) antennas (1.2 meters in diameter) at corporate and customer locations. The corporate staff uses the terminals for video teleconferencing, while other satellite video links deliver specialized programming to customers on product announcements and training. A one-way digital channel superimposed on the vertical blanking interval of the video signal has been added to carry a 56-kb/s data broadcast for use by customers. Applications for this channel include data downloading to minicomputers and delivery of bulletins.

The financial services subsidiary is served by the second thin route network arrangement with two-way VSATs and a hub earth station. In the network, IBM 3270 terminals are used by customer service personnel to determine credit ratings, to maintain customer histories, and to process money transfers. These applications existed before the satellite network and were neither altered nor affected when moving from terrestrial to satellite communication. The terrestrial network arrangement consisted of leased DDS lines between the 3725 FEP at the IBM host and model 3274 cluster controllers at the branch offices. In the satellite implementation, the cluster controllers are connected to a VSAT, which is part of an interactive "star" network. The 1.8-meter VSAT with transmit-receive capability also includes an indoor digital controller unit about the size of a personal computer. The hub earth station is shared with other networks, and hence is not at the same site as the host computer. Connection between hub station and the FEP of the host computer is by dual redundant 56-kb/s lines. The software and protocols of the VSAT network cause the FEP to continue to "see" the remotes as 3274 cluster controllers on polled multidrop lines. The SNA protocols are adapted in such a way that the additional time delay over the satellite path is not a problem for the transaction-based applications.

Activation of the VSAT network occurred smoothly by employing an A/B switch at each site to transfer the cluster controller from the leased line to the VSAT. Further, the host computer is connected to the backbone as discussed previously so that the satellite and backbone networks are logically tied together. A 3270 terminal at a VSAT-served branch office can log onto the nationwide data network and initiate an SNA session as if it were directly connected to the host. The real beauty of this transition is that all applications are running as they were before the cutover.

The two VSAT networks previously described are undergoing further development, and it is likely that many thousands of VSATs will be put to use. The low cost of the VSAT makes it feasible in many cases where DDS or T1 transmission to the branch office is not economically justified. The reliability of transmission has improved over the more expensive DDS circuits leased from AT&T. The time delay of the satellite link does not affect the performance of these applications. In the case of the interactive VSAT, end-to-end delay was actually reduced because data rates were increased substantially (i.e., from 9.6 kb/s to 56 kb/s). Since intermediary SNA nodes are not traversed, response time tends to be further reduced. Consideration is being given to adding voice communication to the VSAT to increase the economic benefit of the private network.

This hybrid network does not stand as the single example of such an approach. In 1988, Shearson Lehman Hutton, a subsidiary of American Express, disclosed plans to add a VSAT network overlay to its already well developed terrestrial T1 backbone. As is emphasized throughout this book, the backbone connects major locations in the northeastern U.S. with each other and with locations in other states across the country. The key to efficient loading of traffic and integration of services is the high level multiplexers which Shearson has deployed at each major node in the network. By contrast, VSATs will be installed at over 600 branches to take advantage of satellite communication previously reviewed. Through digital integration, communication channels of the highest quality are made available throughout Shearson's business units, which, in turn, facilitates extension of the network to offshore locations.

1.6 CROSS-REFERENCE OF FACILITIES AND APPLICATIONS

The information contained in this book covers a broad range of subjects. Quite often, and unavoidably, there is repetition, such as when first we discuss the service offerings of the long-distance carriers and then later we restate how these offerings can be blended into a private telecommunication network. A structure for viewing the topics in this book is the cross-reference matrix illustrated in Table 1.3. All telecommunication capabilities are provided by physical facilities installed on the ground and in space (in the case of satellite communication). The general classes of facilities are shown across the top of the matrix, ranging from telephone networks to data communication networks, satellite networks and, finally, integrated voice-data networks. The networks can provide various kinds of connections between users. The variety of demands placed on telecommunication networks for applications to serve the needs of organizations now forces telecommunication managers to blend the capabilities of these classes of networks. Discussions in Chapters 2 and 4 are basically centered on the facilities side of the cross-reference matrix, which is technology driven. Therefore, some capabilities of networks, while interesting to users and inspiring to engineers, may not prove useful in a business sense.

Applications and services appear along the vertical dimension of the cross-reference matrix. Ranging from voice (e.g., telephone) and messaging to data communication and business video, these applications and services represent capabilities of networks once they are operational. Detailed discussion of these is contained in Chapter 3. Perhaps, we can say that services and applications are

Table 1.3 Cross Reference Matrix of Telecommunication Services and Facilities

APPLICATIONS	TELECOMMUNICATION FACILITIES			
	TELEPHONE NETWORKS	DATACOM NETWORKS	SATELLITE NETWORKS	INTEGRATED VOICE AND DATA NETWORKS
VOICE				
MES–SAG–ING				
DATA–COM				
BUS–INESS VIDEO				

market driven, but some well promoted services have not taken off (e.g., ZapMail™ and Picturephone™). We can obtain these services from a telecommunication service provider, or we can implement owned facilities to provide them internally.

We have now introduced the concept of a private telecommunication network in the modern context. Clearly, there is a wide range of technology and applications. The field is both incredibly rich and challenging. Private telecommunication networks are needed for a very wide range of purposes, and those purposes change from day to day. At the same time, the available technologies are continually evolving. The regulatory environment, which effects what is possible (or permissible) in a given country, either increases or diminishes the opportunities for meeting these needs and employing new technology as it becomes available in the marketplace.

Private telecommunication is anything but static, so the reader should understand that many of the specifics will change over the course of months since the writing of this book. Keeping up to date in the field will require a lot of investigation by telecommunication professionals, for which the organizations and publications listed in the bibliography will be useful. One telecommunication manager of a major corporation was recently quoted as saying that he read or scanned twenty publications a month to keep up with new developments. The remainder of this book should put some more meat on the skeleton just outlined.

Chapter 2
General Classes of Public Telecommunication Networks

There is a well established telecommunication infrastructure within North America, which provides 80 to 90% of the resources needed to build most private telecommunication networks. The purpose of this chapter is to review briefly the functions of that infrastructure. The greatest breath of capability comes from the public telephone network, which, as discussed in Chapter 1, was formerly the domain of AT&T. Today, the RBHCs and the principal long-distance carriers are modernizing and expanding the public telephone network so that the range of possibilities is beginning to exceed the current definitions of user needs. There are other public networks, which are not designed for telephone service, that play an important role in private telecommunication. Included are the Telex network, information delivery networks, public data networks (PDNs), and a variety of other offerings such as cellular radio (mobile telephone), and electronic mail (E-mail). These networks are reviewed in the following paragraphs. Satellite communication networks, also reviewed below, do not offer the full range of telephone and data communication services, but are found useful in important segments of the overall telecommunication picture.

2.1 PUBLIC TELEPHONE NETWORKS

At the risk of sounding inconsistent, a dedicated telecommunication network can be implemented and operated with the primary purpose of providing telecommunication services to the public. In the U.S., any private network can provide service to outside users under the generally recognized principle of sharing. The ultimate case occurs where the implementer has no intention of using the network himself in a significant way, and consequently the network is truly a public network.

The strategic unit that provides these services therefore has this as its primary line of business, and thus its strategy deals with how it meets the telecommunication service needs of others.

2.1.1 Local Telephone Service

A local exchange network provides basic telephone service to a particular area. In the U.S., the local telephone company performs this function. Services are offered on a *common carrier basis,* meaning that rates charged will be the same for all and users will neither be discriminated against nor given special preference. Throughout the U.S., the telco is granted a local monopoly by the federal government and the state in which it operates. Essentially all telcos in the U.S. are privately owned and their stock is publicly traded. This is not the situation in most other countries where telephone service is often a part of the same ministry that provides postal service, and the resulting post, telephone, and telegraph (PTT) agency has a total monopoly throughout the country. As we mentioned previously, privatization and deregulation are gaining momentum now that Japan and the U.K. have taken major steps in these directions.

As a common carrier, the telco accepts information in electrical form from "third parties" (i.e., individuals, businesses, and government agencies), carries it over distances, and delivers it to the distant "subscriber." The obvious advantage of having a single network for the telco is that essentially everyone in the area can receive and send messages, not unlike the situation with the post office and mailing addresses. This analogy actually becomes more apropos as we move from analog information to digital. During carriage, there should be protections on the information. For example, to monitor telephone conversations through a wire tap or by radio means without a court order is illegal. To monitor telephone and digital transmissions is technically easy, and therefore many organizations encrypt their proprietary data at the source so that monitoring and intrusion are made difficult.

Local telco service is basically a profitable business, particularly in urban areas. Foreign governments which operate through the PTT achieve attractive return on investment from their telecommunication networks, which is perhaps why they prefer to retain these activities. Rural areas and remote regions tend to be unattractive for telco service, whether provided by a private company or a PTT. This is because the cost of serving a customer in such a region is many times greater than in urban areas, in terms of initial investment, and can be difficult to reach in the event of outage. Universal service can be both a blessing and a curse. Some telcos are exploiting technologies such as thin route radio and satellite communication, which offer a relatively inexpensive means of reaching rural and remote subscribers.

2.1.1.1 Basic Local Telephone Service

The telephone line obtained from the telco gives one access to the public network. That line is said to provide *plain old telephone service* (POTS). We are all very familiar with POTS because it is basically the same kind of service we have at home. When you lift the receiver, the instrument indicates by the dial tone that it is ready to accept the number you are calling. You may then dial a local number, or you can access a preselected long-distance company by dialing a 1 followed by the three-digit area code and then the seven-digit number. Your bill at the end of the month consists of a flat rate charge for the line and access to it. Additionally, toll call charges are itemized. To initiate service at a new address, you establish credit with the telco and pay a nominal connect fee with your first bill.

With the advent of digital switching in the *central office* (CO), other features are now being offered along with the basic POTS. In *call waiting*, a subscriber may be talking to one party when a second party calls on the same line, and then the subscriber can switch between them, obviating the need for a second line and key telephone. This author, having a teenager at home, finds call waiting to be particularly effective. Speed calling (i.e., storing several numbers in the telco switch and being able to key them with a single digit code, also known as *speed dialing*) and *call forwarding* are two other common additions. As more of these "bells and whistles" are added, POTS will cease to be so plain! To sell us more services over the same line and to encourage our use of the network is obviously to the telco's advantage.

When a business arranges for local service, it typically pays a higher rate per line than do home customers, even though the service may be basically the same. Multiple lines are required, however, when CENTREX service or a PBX is employed. The lines typically have numbers within some group. All telephone lines at one company location probably have the same three-number "exchange" designation. Individual station lines are assigned a different four-digit "subscriber" number. For example, local numbers in the range 540–4000 to 540–4199 might be assigned by the telco to the headquarters of a fictitious company called Enterprise Industries, located in Torrance, California. This would imply that a maximum of 200 stations could be connected to the CO switch (either directly for CENTREX or through a PBX). Typically, the 4000 line would reach the main operator "attendant" position, while numbers 4001 to 4199 would be used by the actual office telephones. Dialing within the building would use the last three or four digits.

The principal advantage of POTS is that it is universally available and understood. It is the first thing a child learns how to use after the television set. Local exchange service is being upgraded to include the features previously described.

Also, the introduction of data transmission on the local loop will change the texture of POTS. We can then expect the bill for local exchange service to increase, due to inflation and to pay for the added services.

The basic approach of providing local telephone service is by connecting subscribers through electronic switches, as illustrated in Figure 2.1. Each subscriber has his or her own line, but the links between switches are accessed only on demand. The local loop connects from subscriber to local switch, while the lines between switches are called *trunks*. The ratio between the number of local loops to trunks is much greater than one, but the particular value depends on the number of switches and loops. To facilitate making physical connections which are semipermanent, a large wiring panel called a *distribution frame* is provided on the loop and trunk sides of the actual switch. The frames are useful when making changes and for connecting private lines (which are not switched in the CO, as we will discuss in a subsequent section).

A single CO cannot possibly handle the switching requirements of a large metropolitan area; consequently, several telephone exchanges are dispersed around any such region. Generally, a hierarchy is followed in connecting switches, where level 5 is used in reference to the local CO in closest proximity to subscribers. Calls are routed up the hierarchy to a level 4 toll office to span the distance to another level 5 CO. This arrangement is depicted in Figure 2.2. Since level 4 offices can only connect a specific number of COs, another stage of the hierarchy is often required. The level 3 tandem office employs the principle of *tandem switching*, wherein calls are connected through intermediary exchanges. Together these levels of the hierarchy are sufficient to serve the switching needs of a highly populated region. Levels 1 and 2 are reserved for long-distance service.

A standard numbering plan was developed for the Bell System and is now used throughout North America. The structure of the plan along with an example are shown in Figure 2.3. Every telephone number has three components: the three-digit area code; the three-digit exchange number; and the final four-digit subscriber number. Telephone switching equipment must decipher these numbers so that the call can be routed properly. If a subscriber first dials a 1, the CO equipment will know that this is a long-distance call. With regard to the area code, the possibilities for the first two digits are restricted (which allows the switching equipment to identify this as an area code). The numbering plan specifies that an area code will not begin with either a 0 or a 1 (N is defined to represent any decimal digit except 0 or 1). In the case of the area code's second digit, only a 0 or a 1 is permitted. Local exchange numbers have the form NXX, meaning that the first digit (N) cannot be either 1 or 0 while the remaining two digits (X and X) can take the form of any decimal digit between 0 and 9. Special services numbers, such as 411 for directory assistance, 611 for repair service, and 911 for emergency help, are also recognized by the telco switches.

The form shown in Figure 2.3 looks complicated, but in actuality the system is quite simple to comprehend. In the future, the rules outlined above are being

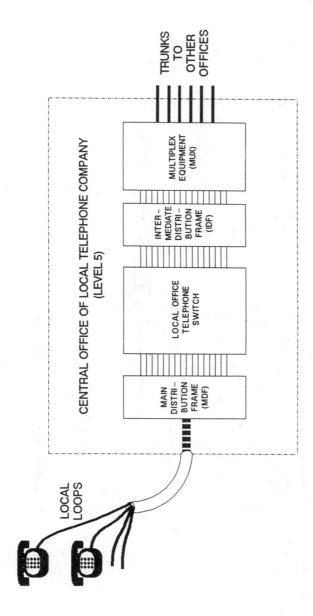

Figure 2.1 General arrangement of a local telephone switching office (level 5).

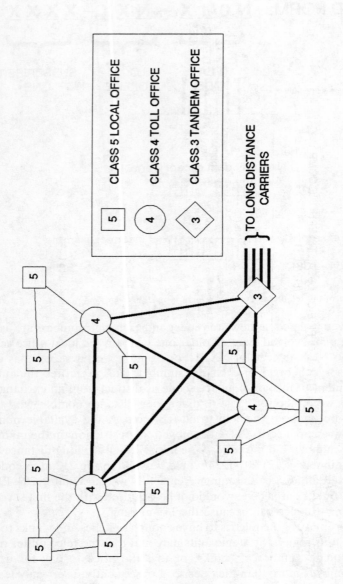

Figure 2.2 Conceptual topology of a local telephone network on the trunk side.

STANDARD FORM: **N 0/1 X — N X X— X X X X**

AREA   LOCAL SUBSCRIBER
CODE   OFFICE   LINE

Where
 N = any digit except 0 and 1
 X = any digit
 0/1 = 0 or 1

Correct example   Incorrect example

619—454—0381   146—037—5278

Figure 2.3 Example of the telephone numbering plan in use in North America.

modified, and even violated, as new area codes and exchange numbers are assigned. This numbering plan evolved from an older one wherein the local office sequence was restricted to NNX instead of NXX. For example, 201 was valid as an area code, but not as a local office or exchange number. This permitted the automatic switching equipment to recognize an area code as distinct from an exchange number. Subscribers were therefore able to dial a long-distance number with the ten-digit sequence, but without the 1 in front. To increase the available number of exchanges in a given area code, AT&T removed the restriction on the middle digit of the exchange number and introduced the initial 1 digit for long-distance calling. Therefore, the number (212) 201-1234 would be valid in the 212 area code, even though 201 is itself an area code sequence currently in use. Dialing 201-1234 as a local call within the 212 area code would not be confused with the first seven digits of a ten-digit long-distance call because the 1 is missing.

 The responsibility of the telco is to have enough switches and trunks to service the subscribers in its region. Strategic units may find that the telco either does not provide an adequate number of trunks, or that the price to pay for using the switches and trunks is more than they desire. Consequently, users may lease full-time lines between specific locations. These leased lines (also called *private lines*) follow the route of the local loop and trunk, but do not use the switching capability of the telco. To complete the circuit, the telco makes a fixed (hard-wired) connection from local loop to trunk by using the distribution frames in the switching office. Essentially all of the wires coming into the office are terminated on a distribution frame allowing connection to or around the electronic switching gear.

The local loop and switching facilities are engineered for voice frequency signals in the range of 300 to 3400 Hz. Human speech can be carried with adequate fidelity for the purpose of normal conversation, allowing the typical caller to recognize the voice of the other party. Distortion and noise are held to levels adequate for these purposes, consistent with allowing calls to be connected over short and long distances. Other types of communicating devices can also employ the voice frequency bandwidth of the telephone plant. Identified in Table 2.1 are typical examples, where separate columns are provided for switched (dial-up) and nonswitched (private line) connections. High fidelity audio for broadcasting purposes cannot be carried over the dial-up network. Innovations in modem technology are pushing throughput toward the theoretical limit of the voice frequency bandwidth of the telephone circuit. For example, a modem manufactured by Telebit can transfer data at a rate of 19,200 bits per second over a good quality line. Remember that the switched telephone network is not designed for nonvoice applications, and the telcos typically make no guarantees as to such performance. Introduction of ISDN would open up a separate pathway for high speed data communication without the use of modems, as discussed later in this chapter.

Table 2.1 Uses of the Local Loop

Use	Switched	Nonswitched
Voice	Public network dial-up	Private line, trunks
Data-Voiceband	Dial-up, typically 2400 b/s	Private line, typically 9600 b/s
Facsimile (fax)	Dial-up	—

2.1.1.2 Other Telco Offerings

Leased lines are used extensively by strategic units to implement private network capabilities. (This is discussed further in Chapter 3.) Pricing of leased lines is dependent on many factors, and often can run counter to the business objectives of the telco. In the U.S., the RBHCs have generally made leased lines less attractive than switched service, probably because of the difficulty of maintaining the quality of fixed connections. In addition, an effectively used trunk would earn more when in the telco's inventory. For a period in the U.S. there was a lot of competition for the leased line business with terrestrial and satellite carriers offering very reasonable rates. The new environment could favor switched telephone service over leased lines as private trunks.

Telecommmunication service companies compete with their unregulated counterparts by offering equipment that allows users to customize their private networks. Usually the customer decides whether the equipment is purchased or leased (e.g., paid on a monthly basis). Deciding whether to purchase or lease is based on the same "lease *versus* buy" criterion generally employed by business organizations, as discussed in Chapter 6. Therefore, the real issues here deal with whether the user ought to commit to the equipment for the long term, or whether employing the equipment as a service is more opportune. In the latter case, the telecommunication service company maintains ownership and responsibility for the quality and operation of the equipment. The user may even be able to cease use of the equipment under the service agreement or tariff. The decision becomes simply a question of who retains the risk that the equipment will not work effectively for the intended purpose, or that the user will no longer need the equipment due to business or other reasons. (The service agreement may have up-front "construction" charges and early termination payments, which effectively transfer most or all of the business risk from the telecommunication service provider back to the user.)

The types of equipment that telcos offer include PBXs, multiplexers, telephones, terminals, and almost any kind of telecommunication device which a user may dedicate to the private telecommunication network. (As discussed in other chapters, telecommunication users may purchase these devices directly from the manufacturer or from system integrators.) Maintenance of the equipment could be contracted to the supplier, which may be either the telecommunication service company, the manufacturer, or a specialized maintenance company. The maintenance agreement is an important document, providing for guarantees as to how quickly the vendor will respond and for what duration the maintenance service will be rendered. One of the thorny problems of maintenance contracting is ensuring that a manufacturer or supplier will support the particular equipment in the future, possibly after the product line is discontinued. Specialized companies can provide third-party maintenance service (i.e., maintenance on someone else's equipment). This requires an inventory of spare parts obtained from the original source. If the manufacturer does not maintain an inventory of spare parts, then cannibalizing salvaged equipment of the same design may be necessary.

Local telcos usually manage their own networks very effectively because network failures result directly in lost revenues. In the case of private telecommunication network facilities supplied by the telco, it may not have the same motivation. The strategic unit could possibly obtain network management services from the telco as a subset of the overall local system. One of the most common telco-provided network management services is *station message detail recording* (SMDR), wherein the customer is provided with billing information for all calls on the network. Individual responsible areas within the organization can then be billed for the telephone expenses applicable to them. Network management also takes the form of equipment control and reconfiguration. This is particularly important in T1 metropolitan area networks. Part of the advantage of using T1 is that systems are able to alter the traffic routing in response to changes in user locations and requirements. Leased telco facilities may be controlled by customer terminals connected to the network. Otherwise, special orders would need to be placed with the telco, resulting in additional charges and time delays. (A detailed discussion of network management can be found in Chapter 5.)

2.1.1.3 Local ISDN Services

That the RBHCs and major independent telcos are aggressively pursuing ISDN is clear from reading the telecommunication trade press. Many private network operators are still waiting to see the benefits that ISDN will bring. If ISDN becomes a widespread phenomenon, the telcos will be in the middle of the action. The area which may show the most results is the local loop, which today is constrained to carry relatively narrow bandwidths due to its composition of twisted wire pairs. Narrow-band ISDN, also called the *basic rate interface* (BRI), would increase the capacity of a twisted pair from one 3000-Hz analog telephone channel to two 64-kb/s equivalent voice channels and a 16-kb/s data channel. The abbreviation 2B + D is used in reference to the composition of the basic rate interface. Digital communication services of this order are possible using the latest generation of high speed, high capacity digital switches being installed by the leading telephone organizations throughout the world.

To date, the approach taken for using basic rate ISDN service is for simultaneous voice and data in appriopriate applications. Circuit-switched services are provided by the B channels. Each B channel can carry a standard voice channel which has been digitized at the source. In 2B + D, one B channel carries voice and the other is used for 56 or 64-kb/s data. This relatively high speed is currently beyond the typical data communication application at the subscriber level. In the future, this rate would allow compressed video teleconferencing in the format of the old Picturephone™ promoted by AT&T in the 1960s and 1970s. The 64-kb/s data channel could be used for high speed bulk transfers, such as downloading databases. Interactive services for reservation systems and order processing could benefit from higher transmission speeds. Regarding the D channel, which connects to a packet-switched network at 16 kb/s, most of the data communication service will be associated with setting up and monitoring the calling process. The services identified in the next section under *open network architecture* (ONA) could be activated over the D channel. There has been some discussion of allowing users to employ the D channel for functions outside of the telephone network, such as on-line transaction processing.

Basic rate ISDN service is being offered in the U.S., Japan, and certain European countries to a select group of corporations and government agencies.

To entice users to experiment with ISDN, Soutwestern Bell has opened its Advanced Technology Laboratory. This permits the testing of ISDN devices and services before a commitment is made to the technology. The expanded service possibilities of the basic rate interface (in comparison with the current analog voice frequency interface) will not come free of cost. Preliminary tariffs filed by telcos that are offering ISDN service indicate line charges which will be in the range of 40 to 100% higher than normal. Of course, the user has three separate channels available per line instead of only one.

A class of local loop which shows potential is referred to as *broadband ISDN*, or B-ISDN. While standards have not been worked out for this mode, the basic concept is to run fiber optic lines into the customer's premises or home. Planning in Europe for B-ISDN is focusing on sending video signals over the local loop, substituting fiber for coaxial cable. Switching of B-ISDN is still under experiment, as it involves completely different technology than narrowband ISDN. The obvious application for this is cable television. In the U.S., however, the RBHCs cannot overtly engage in video distribution due to the restrictions of the AT&T divestiture. Almost 80% of all homes already have access to cable, and 50% now take the service from their local cable television company. B-ISDN may find a warm reception in the business environment, where wide bandwidths are in demand for network capabilities for commercial video links and host-to-host computer data transmission.

2.1.1.4 Open Network Architecture

Another developing area for local telcos to pursue as an adjunct to private networks is providing data communication and videotex services as a common carrier. The data would be supplied by a specialist, as now done on videotex offerings discussed later in this chapter. During the process of defining ONA, the FCC required that the RBHCs and AT&T establish standards and interfaces allowing third party vendors to offer information services via the public network. The term "intelligent network" is also used by the telcos and AT&T. Currently, the only service in this class is protocol conversion, which is one of the offerings of the public data networks of Tymnet and Telenet. The RBHCs will only be able to serve customers within LATAs, and reaching other LATAs will involve the long-distance carriers.

Using the term "enhanced services" to refer to the information service offerings of third party information providers and potentially the RBHCs and AT&T, ONA would establish a set of standard capabilities via the telco switch and network. Plans for ONA were submitted to the FCC by the RBHCs in February 1988. The *basic service elements* (BSEs), summarized in Table 2.2, permit the information providers to design their services with the knowledge that the "hooks" will be available for direct service, tracking, and billing of their subscribers. The actual

Table 2.2 Open Network Architecture: Typical Examples of Basic Service Elements

Automatic number identification
Central office announcements
Answer-disconnect supervision
Call distribution
Trunk busy
Call forwarding
Billing services
Call transfer
Conference calling
Call screening
Interoffice channels
Distinctive ringing
Abbreviated dialing
Call waiting
Traffic usage
Multiplexing
Network management
Speed calling

interface between service provider and telco central office would be from a collection of options called *basic service arrangements* (BSAs). Examples include circuit switching, packet switching, private lines, and other special access lines. The BSAs and BSEs are comprehensive enough that many large private telecommunication network users are finding ONA attractive for direct access to the telco network. Also, the tariffs filed by the telcos for ONA could offer substantial savings to these users. This possibility for direct access to ONA, however, would bypass the providers of enhanced services.

An example of direct-access enhanced service is the pay telephone service (not pay phones) available in many U.S. cities. A subscriber dials the number of the service, such as a ski report or stock market summary, which has 976 as its three-digit exchange number. Billing for the call is significantly greater than for a toll call, and the telco passes a portion of the charge on to the information service provider. In the case of ONA, revenue could be passed through the telco or be handled directly by the provider, as is currently the case with videotex services like CompuServe® and Dow Jones News-Retrieval.

At the time of this writing, ONA is a concept to evaluate the regulatory framework for controlling the entry of the RBHCs into the information services business. The term "unbundling" has been used to describe ONA's objective for defining the RBHCs as mere conduits, thereby facilitating the growth of new information industries. The modern digital switching facilities can simplify the business structure by being used for mundane problems such as protocol and speed conversion, billing, and packet routing. Data communication service on demand is currently the business area of public data networks like Tymnet and Telenet.

When data communication is universally integrated with voice via the local loop (e.g., ISDN), the telcos will be in a powerful position to offer similar or better capability than the public data network.

The RBHCs and large independents, however, have the financial resources to create packet network capabilities which are much more elaborate and powerful than those offered by the existing data networks. The telcos would still need to find the means to interconnect between one another, which is analogous to the situation with telephone service. Logically, the long-distance carriers will provide essentially transparent transmission for the data network services. Another approach being taken by Bell South allows Telenet to connect directly into the telco switch. This approach will use the ISDN features where they are extended to business customer locations. The first direct offerings of Bell South are through their Transtext Universal Gateway, which users can access with modems over the telephone nework. Some information services available at the Gateways include Comp-u-store OnLine, Market Watch, News Net, CompuServe, Dialog, EAASY SABRE, and The Source.

Another important question regarding ONA is the availability of attractive information services which customers would want to access. This situation is akin the that of direct broadcast satellite (DBS) networks, which have had a rough beginning in the U.S. The technology can deliver a television signal of high quality to a home receiving dish of less than two feet in diameter. The obstacle, however, has always been the availability of programming which is attractive enough to consumers that they will purchase the receiving equipment and pay for the programs on a subscription basis. One serious problem is that the consumer is already inundated with programming in the form of television shows broadcast over the air and movies available through cable television and home video cassettes. Such alternatives to ONA information services are currently available through public data networks, floppy disks, data broadcasting by satellite, and, of course, print media.

2.1.2 Long-Distance Services

Introduction of intercity telephone service in the U.S. in the early 1900s was perhaps the second greatest innovation in voice communication (after invention of the telephone instrument itself). Patents on key devices that made long-distance connections possible were under the control of AT&T. This permitted AT&T to grow as both the holding company for local telcos and the sole provider of long-distance service. Government regulation helped keep AT&T in control of its monopoly until serious inroads were made by MCI in the 1970s. Microwave radio provided MCI with the vehicle to implement long-distance links relatively inexpensively and quickly. The rates charged by MCI were less than those which companies had been paying, causing AT&T to accuse MCI of "cream skimming."

This idea is based on the notion that AT&T needed to earn the highest possible long-distance revenues from corporations to compensate for the expense of providing universal service (AT&T's mandate), even to unattractive locations in remote areas. Since MCI was not so "burdened," it could charge lower rates and thereby easily garner customers. The growth of MCI, formed in 1968, into a major communication company in the 1980s has been phenomenal.

2.1.2.1 Evaluation of Long-Distance Services

In the days of manual service with human operators, placing a long-distance call was somewhat inconvenient. You would first lift the telephone receiver and request that the local operator connect you with long-distance. The long-distance operator would take the number you wanted to reach and then set up the call through the network. When automatic switching was introduced, the operators initially dialed for you. The big innovation came in the form of "direct distance dialing," wherein the subscriber could initiate and complete the entire long-distance connection with the dial on his or her telephone instrument. Obviously, AT&T had a lock on the subscribers, their instruments, the switches, and the long-distance transmission system, making it relatively easy to upgrade the network as the technology matured and as requirements arose. The need for automatic control, billing, and monitoring was sufficient to drive AT&T to evolve into a capable computer hardware and software developer. Many technical innovations such as the transistor, bubble memories, and the Unix computer operating system software were developed by researchers at Bell Labs.

Whenever one company finds an attractive business with exceptional profit margins, imitators are generated [Porter, 1980]. The private line business is one that attracted companies like Western Union and RCA. These circuits were provided primarily on satellite networks of the 1970s. Private lines are full-time leased circuits which organizations use as pathways within private telecommunication networks. Switched long-distance service, however, is considerably different, even though similar circuits are used as components. One of the biggest markets for private lines is their use as components of the switched long-distance networks because the long-distance carriers, particularly those other than AT&T, cannot afford to install tie lines and trunks at every location in the country.

Southern Pacific Communications Company (SPCC) was set up by the parent railway company to resell capacity on their private transcontinental microwave network which paralleled the railroad's right of way. Investing many millions of dollars in microwave and switching equipment in the pursuit of MCI, SPCC elevated itself to become a contender in the long-distance business. SPCC was subsequently purchased by GTE. Switched long-distance service became a new competitive arena where AT&T was met by MCI, SPCC, Western Union, and

IBM through its Satellite Business Systems (SBS) partnership with Communications Satellite Corporation (COMSAT). More information on SBS can be found in the next section.

Technological advances in fiber optic transmission and the attendant reduction in the cost of installing such networks opened up opportunities for new entrants in the long-distance business. In the mid-1980s, several companies and joint ventures established themselves and began obtaining the physical rights of way needed to install the fiber cables. Some of these enterprises never made it past the planning stage. United Telecommunications, Inc., however, was a particular fiber optic developer which aggressively entered the business. With a strong financial base in local telephone service as an independent, United Telecommunications moved quickly to implement its network and acquired a relatively small long-distance service company called US Tel to be its marketing agent. Eventually, US Tel and the fiber optic network were merged with the long-distance operation of SPCC to form US Sprint. GTE and United Telecommunications started as partners in supporting the massive capital requirements of US Sprint as it develops its network and customer base. GTE, however, reduced its participation by selling a controlling interest in US Sprint back to United Telecommunications in 1988. As shown in Chapter 1, the US Sprint fiber optic network is one of the most complete digital backbones in the U.S.

2.1.2.2 Service Arrangements

Prior to divestiture, telephone subscribers accessed AT&T's long-distance service by dialing 1 and then the ten-digit number. (The numbering plan has been reviewed and is summarized in Figure 2.3.) Subscribers could access another long-distance network (e.g., MCI) by dialing a local seven-digit number to connect from the telco tandem office to the appropriate carrier's switch. Subscribers who were a considerable distance from the closest access number could use a toll-free 800 number supplied by the carrier. While the cost of access was relatively modest, the competing carriers thought that the complication of dialing multiple numbers was a burden which put them at a severe disadvantage relative to AT&T. This problem was rectified by the federal court under Judge Greene, directing the RBHCs to implement equal access, under which local subscribers first dial 1 to connect to the particular long-distance carrier that they had preselected through a mail-in balloting process. Equal access may be a blessing to the other long-distance carriers because customer access is simplified, including the process of submitting and collecting bills. Still, AT&T has been able to retain its market share.

The various access arrangements in effect today for reaching the long-distance carriers are illustrated in Figure 2.4. Configurations (a) and (b) were in effect prior

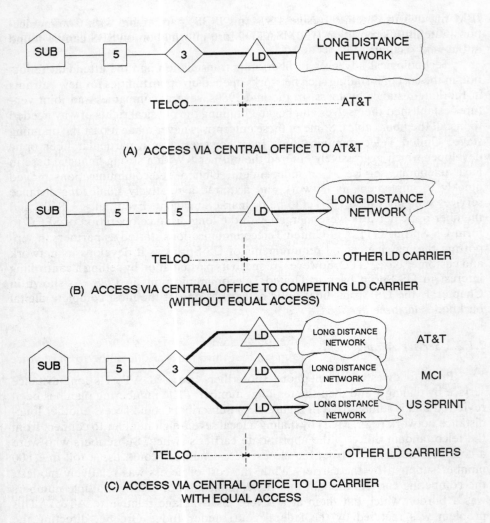

Figure 2.4 Customer local access arrangements for long distances service.

to divestiture, as discussed at the beginning of the previous paragraph. Equal access is illustrated in (c), where the preselected carrier can be reached by dialing 1. Direct access to a carrier for dedicated long-distance service (e.g., WATS, Megacom, and Prism™) employs a permanent connection by the telco.

Of vital importance to businesses in the U.S. is the availability of bulk purchase offerings of long-distance voice calling. The three leading long-distance companies are vigorously competing for the business customer. Bulk calling packages are so attractive that some major corporations have opted for using them in lieu of building up a private network capability for telephone. Bulk calling schemes are reviewed later in this section.

2.1.2.3 The Long-Distance Marketplace

The following discussion centers on the service and pricing schemes used by the carriers to entice business customers. AT&T has about three-quarters of the U.S. long-distance market, with MCI, US Sprint, and a few others such as ITT and Allnet fighting over the remaining quarter. This competitive environment has helped to reduce the cost of long-distance service and has encouraged the leading long-distance carriers to become very innovative. Because of continually changing offerings and rate structures, describing with great clarity how each offering works and how it compares with that of the competitors is difficult. Nevertheless, we review the current picture to help readers comprehend the possibilities for buying and using long-distance services. The services discussed are all provided on a dial-up basis through the carrier's switching and transmission facilities.

Message Telephone Service (MTS)

The most basic service of the long-distance companies is *message telephone service* (MTS). By "message" we mean that third party traffic in the form of voice information is being carried by the network (we are not referring to recorded messages like telegrams and telex printouts). Customers access MTS through the public switched network and the telco's local loop. The time at which the distant called party answers to that when the call terminates is measured (metered) for billing purposes. The billing rate per minute is based on the distance covered. Charges are also affected by the time of day, with the highest rates applying during weekday business hours. Additional features of MTS include operator assistance, credit card calling, and collect calling. All charges are accumulated and then accounted at the end of the month. In the case of AT&T or another carrier selected through equal access, the billing is included in the monthly statement from the local telco.

Induced by competition for the business customer, the carriers offer a variety of optional pricing packages, which basically give discounts for volume usage. For example, AT&T's PRO America packages are aimed at small to medium sized businesses with normal MTS bills in excess of approximately $150 per month in 1988.

Wide Area Telecommunications Service (WATS)

Business users of long-distance service can employ the long-distance switching and transmission facilities without going through the local telco switch. The WATS offerings allow the customer to connect to the particular long-distance carrier's switch by a dedicated leased line. This leased line is usually provided by the local telco, but the user pays for it as part of the WATS service. This particular definition of WATS is changing as the carriers make arrangements with the telcos to permit the use of dial-up lines. Using existing access lines and avoiding the dedicated line can significantly reduce cost, particularly for the relatively small user. (Readyline™ is an offering of AT&T which employs this feature.) When this is done on a wide scale, the dividing line will blur between bulk discounted MTS and WATS.

WATS was originated by AT&T in the early 1960s and is used frequently as a toll-free access service for telemarketing (discussed in detail in Chapter 3). Inward dialing WATS numbers typically begin with the familiar 800 area code. The caller is not billed for the call, but charges are automatically made to the account of the called subscriber. Service areas, called "bands," define geographical regions into which calling is permitted. Therefore, a given WATS line would be tied to a specific band for which the billing rate (cents per minute) is constant. Organizations use outward WATS to save on long-distance calling because the rate per call is lower than for MTS. Recently, the carriers are allowing multiple bands to be served by the same WATS line. Other services like call diversion, dialed number identification (both discussed in Chapter 3 under telemarketing), and automatic route selection are available as options. In comparison with the other discounted service offerings, WATS has the advantage of being universal throughout the U.S. Both MCI and US Sprint now offer their own version of WATS. This means that these carriers must negotiate with every individual telco in the U.S. so that 800 numbers are properly translated into standard numbers with area codes and seven-digit local exchange lines.

Advanced Bulk Long-Distance Services

This special classification of long-distance service continues to evolve as the carriers compete for the long-distance business market, particularly that comprising the largest users such as the *Fortune 500* companies. The WATS offerings of the three major long-distance providers represent the starting point in this part of the market. AT&T's Megacom and MCI's Spectrum, however, are designed to take discounted long-distance service to a much higher level, one not affordable for small to medium sized businesses.

AT&T's Megacom offering is targeted at the largest corporations in the U.S. which have monthly long-distance calling that aggregates in excess of 1000 hours per business location. The customer pays a rather hefty fixed monthly fee plus other charges per call. Unlike standard MTS and Readyline™, which are accessed through the telco switch, Megacom requires a dedicated access line which will be the user's responsibility. This line is normally provided by the telco and charged separately as a private line. Local bypass service may be available from other common carriers, or can be implemented expressly for connection to the long-distance carrier. A trend is toward the use of digital T1 access lines to carry bulk quantities of simultaneous calls over the local loop. These bulk services are somewhat like WATS, and resolving the differences can be difficult. A rather complete investigation of initial costs and monthly usage rates would be needed before committing to a particular bulk offering.

MCI offers an advanced bulk long-distance calling service called Prism™. The principal advantage of Prism™ over conventional WATS appears to be the flexibility that it offers to the medium sized customer. For example, the service area limitations of WATS do not apply to Prism™. The three classes of Prism™ cater to different sized customers, the smallest and the largest potentially benefiting.

The detailed nature of bulk discounted services as well as their names are in constant flux. The carriers are able to restructure their services because they use the most modern switching facilities and sophisticated software. New pricing packages are introduced to lure customers away from competitors (or, conversely, to prevent customers from defecting). For example, AT&T is introducing Hospitality Network Service to hold the long-distance business of large hotel chains. This service would not be appropriate for the home since the aggregate usage should exceed four million minutes per month. In addition to cost reduction, switching systems also increase the ways in which the customer can control his usage of long-distance services, and provide more accounting and traffic information for network management purposes. As the IXCs make their networks more effective for business use, private dedicated facilities will tend to be less attractive. As discussed in the following sections, however, dedicated facilities offer other advantages which go beyond strict economics.

2.1.3 Private Lines

The private line network business is actually an adjunct of the public telephone network. Any dedicated telephone circuit between two points is called a private line. To implement these capabilities, the resources of a long-distance network and the local telcos at each end are necessary. The routing of private lines is through the cabling and long-haul transmission systems of the carriers. Switching offices are traversed, but the switches themselves are bypassed by making appropriate connections at the distribution frames where all lines are physically terminated.

Private lines are offered by the telcos and IXCs in either analog or digital form. Normal telephone voice frequency channels are analog in nature, and they are suitable for tie lines and private trunks, as discussed in Chapter 3. Data communication can be carried on analog private lines by using modems to perform the conversion between digital and analog transmission. Wider analog bandwidths for high fidelity audio are provided on a special-order basis (these are used by radio and television networks). Digital private lines were introduced by AT&T prior to divestiture under an offering called Dataphone Digital Service (DDS). This was expanded with Accunet 56 and Accunet 1.5, which are respectively capable of 56 kb/s and 1.544 Mb/s and allow connections on demand. The RBHCs provide similar capabilities in their service regions.

2.1.4 Virtual Private Networks

The features of a private backbone network for internal telephone services are now available through the switched long-distance networks of the three major long-distance carriers. Designated as either *virtual private network* (VPN) or *software defined network* (SDN), the service relies on the carrier's ability to translate a short seven-digit internal line number into the conventional ten digits of a long-distance call. The calls are routed over dedicated access lines (as in WATS and Megacom) from each customer location to the carrier's closest switch. Customer personnel would place calls between fixed locations as if the virtual network consisted of dedicated trunks between the switches. The user can also place long-distance calls to off-network locations, although these will automatically be routed over the particular carrier's network. The features of a VPN typically include seven-digit on-network calling and ten-digit off-network calling (as explained above), station-to-station calling within the same area, speed calling, and authentication codes to prevent misuse of services.

The VPN services of AT&T, MCI, and US Sprint are called SDN, Vnet, and VPN, respectively. These offerings have attracted a limited number of corporate customers and the associated revenues being collected by the carriers are modest. Virtual private networks appear to be attractive to dispersed organizations that need simplified dialing but cannot justify the cost of private lines. Larger organizations which employ dedicated private lines for heavy trunking applications can still take advantage of VPN services to provide smaller locations with some of the conveniences of the corporate backbone.

2.2 INFORMATION NETWORKS

Many networks have been implemented for delivering information products to businesses or the general public. One well known class is the broadcast network for television and radio. Our interest, however, is in networks used for business purposes. The types of information offered are primarily textual in nature and require a data delivery vehicle of some type. This can employ the telephone network, a terrestrial radio broadcast medium, or satellite communication. The information may be part of an interactive network, whereby the user can make requests for specific information, and he or she may use the information base as if it were a very large library. A one-way broadcast medium can also supply information, but in this case the information should probably be collected in a local computer disk file for off-line examination.

2.2.1 Videotex and Teletext

The availability of inexpensive video display terminals and personal computers has stimulated the introduction of information services derived from text and graphic sources. Around 1980, home computer enthusiasts began dialing into a computer time-sharing service called CompuServe® to reach on-line wire services and several different statistical databases. Stock market quotations and a facility for making stock purchases from a discount broker were added. The generic name for connecting to a central time-sharing computer over the telephone lines is *videotex*. One of the most popular early applications for videotex was a simulation game called "Adventure." (This author spent too much time and money pursuing "digital" treasures while avoiding nasty gnomes and dragons.) At almost the same time, an experimental form of service was introduced wherein a home terminal would display information taken from the incoming video signal over the television set. This type of one-way information distribution service through the television set is called *teletext*. The original distinction between the two approaches is that videotex is a two-way service allowing subscribers to interact with a central computer, while teletext uses a one-way broadcast data channel delivering a continuous stream of information, most of which is not concurrently displayed. The distinction has all but disappeared, however, making it necessary to research the particulars of the offering.

The most popular videotex service in the U.S. is CompuServe®. With approximately 300,000 subscribers as of 1988, this offering is considerably ahead of the next most popular service called The Source. The Dow Jones News-Retrieval Service has also been in operation for several years, offering an interesting mix of financial information and current news. All provide stock market data services with the ability to store one's portfolio, to obtain quotes, and actually to make purchases on national exchanges through a discount brokerage house. Wire services such as Dow Jones, Associated Press, and Reuters are available via both teletext and videotex. Home banking is an idea that has not attracted enough customers to make the offering attractive to the banks themselves. The current videotex services offer electronic mail for use by subscribers. In a kind of a ham radio mode for sheer enjoyment, party line connections allow simultaneous messaging among subscribers who care to chitchat.

The videotex services of the Knight-Ridder newspaper chain provide on-line news service and quotations for securities traders and analysts. Their capabilities will expand even further through the acquisition of Dialog Information Services, Inc., formerly owned by Lockheed. Dialog, which is actually an electronic library service, is currently the world's largest provider of database information services. Knight-Ridder will now have a substantial share of the videotex market and should represent significant competition for Dow Jones, CompuServe®, and others.

Videotex is still experiencing user acceptance problems in the U.S. Approximately two million homes have personal computers capable of accessing the service, but less than 20% are equipped with the modem needed to interface with the telephone network. Consequently, there is a hurdle for consumers to try a service of this type. Overcoming this hurdle would probably require the videotex provider to offer "free" modems and communication software.

The most ambitious example of videotex can be found in France, the Teletel offering of the French Post, Telephone, and Telegraph (PTT). The French government has promoted the development of advanced information technologies, including videotex. As the nation's public data network, Teletel is used extensively by businesses to connect to information services offered by third parties. The PTT also offers a telephone directory service via Teletel. Further stimulation for the growth of videotex was provided by installing almost two million "Minitel" display terminals free to households with telephones. This type of terminal, shown in Figure 2.5, is compact and relatively inexpensive to manufacture, since its only function is to display information and to enter inquiries through the keyboard. Further, users are obliged to use the system because it is the only way to obtain directory assistance. Information services can be accessed through the network from third parties to which users can subscribe through the PTT. Home usage of Teletel is still evolving, while business usage has already made Teletel a success for the PTT. A subsidiary of the PTT is working toward exporting the technology and service to foreign countries, notably the U.S. A subset of Minitel is available in the U.S. through a gateway to the French network. Demonstrated at the 1988 ICA Convention in Anaheim, CA, the Minitel videotex system is quite advanced, and it presents an astounding variety of information and even shopping services for the average consumer.

Teletext, the receive-only information service, is being applied by cable television networks and local systems for a variety of purposes. One type of application is similar to "closed captioned" textual transmission, which provides subtitles on broadcast television for the hearing impaired. One such service familiar to many in the U.S. is the way in which the Weather Channel provides local weather forecasts. In the regular live video broadcast, a weather man or woman stands in front of the familiar weather map and provides the national and regional forecast. At the conclusion, the weather person indicates that the local forecast will be displayed on the screen. The signal switches to a flat blue "color key," and a

Figure 2.5 The Minitel Terminal, developed for the French videotex market, is available in North America. (Photograph courtesy of France Telecom.)

character display is substituted in the video channel by the local cable system. The teletext aspect arises because the actual local forecast is sent via a data stream over the satellite which contains forecasts for each locality in the nation. The character generator at each cable system is addressed by the teletext information.

An innovative teletext service carried by some cable systems is called X-Press. Again, the information is in the form of a data stream carried alongside one of the satellite video channels. The intended subscriber would have a home computer and software for selecting and displaying particular subsets of the data. A modem is not required because the service provides a digital interface unit to each subscriber. The kinds of information offered mirror videotex services like CompuServe® and The Source. In this case, however, the subscriber only interacts with his computer and the digital interface unit, while the information stream continues in seemingly endless procession. Offerings include news wire services from the U.S. and other countries, and stock market "tickers" such as the New York and American Stock Exchanges and indexes. Historical stock data would be loaded in the subscriber's personal computer from previous sessions on disk. The cost of using X-Press is typically lower than currently available videotex because the signal is broadcast on a point-to-multipoint basis over a satellite and no return communication path is established.

Two possible explanations are offered as to why teletext and videotex have yet to succeed as consumer marketing businesses. First, there is the critical mass theory, wherein it is argued that the networks are not attractive to information providers when an insufficient number of users are connected. Also, some of the use of the networks is to allow users to communicate with each other, and if a critical mass of users is not connected, the benefits of communicating between users are diminished. (This last point is more appropriate for E-mail applications.)

The second theory states that without enough attractive information services, the public will not make the necessary investment in terminals or home computers; hence, critical mass is discouraged. The Minitel experience may provide the answer (i.e., free terminals), or perhaps some company with "deep pockets" can finance the creation of interesting information products to offer on the networks. A recent attempt to enter the market in this regard is a videotex service called Prodigy, a joint venture of IBM and Sears. Consequently, we can expect continued investment and experimentation in videotex and teletext as companies seek the magic formula.

2.2.2 Electronic Library Services

The versatility and effectiveness of libraries have been improved through a variety of electronic services. These employ large on-line databases, which are accessed over public telecommunication networks. Professional librarians once were only concerned with books and card catalogs, but the appearance of electronic services has opened up opportunities for them in research. Performing an automated search is now possible in a computer database, rapidly scanning thousands of titles and abstracts of articles, books, and other publications. In some cases, the entire document is available on-line, so that it can be transmitted electronically and printed locally. Using conventional telecommunication network facilities, the cost of obtaining the electronic document in this manner is still too high for everyday purposes. Being able to find valid references in the literature, however, can be of enormous value to high school or college students and working professionals. Increasing transmission rates on the public network or delivering data directly by satellite offer the prospect of efficient electronic document delivery.

Communicating with an on-line database is much like videotex, since you use a terminal or personal computer and a telephone connection. In fact, the existing videotex services offer limited on-line databases, such as past issues of national newspapers and financial research reports. To be useful in research, however, the database is contained on a large mainframe, which has direct access to vast stores of information. Until economies of scale are realized, the expense of using on-line libraries makes them more appropriate for professional or industrial purposes. The Dialog services previously mentioned support professions such as engineering, medicine, and the law.

2.3 SATELLITE NETWORKS

The proper place for satellite communication in the private telecommunication network picture has been a subject of considerable debate in the U.S. over the past several years. Many major industrial companies experimented with the medium in the early 1980s and were not pleased with the results. This is primarily the story of SBS, told elsewhere in this chapter, but many organizations have found that satellite communication can be a vital lifeline in their business because of the advantages cited in Chapter 1 and below. Currently, the VSAT and Ku-band satellites are gaining in popularity, even among the medium sized firms which would otherwise use private lines and the public telephone network. The following discussion is brief, and more detail can be found in our earlier work [Elbert, 1987].

The basic building block of satellite capacity is the *transponder,* which is a physical channel for transmission of one or more microwave signals through the satellite's communication repeater. A transponder typically relays one standard color television channel, or up to 60 Mb/s of high speed data. The design of a modern satellite has 16 or more transponders, each capable of meeting the needs of a particular network of earth stations. Transponder power levels and bandwidths differ among satellites, depending on the frequency band and the particular requirements of the overall system. Two frequency bands, designated C and Ku, are used extensively for commercial satellite communication. C-band (3.7 to 4.2 GHz in the downlink, 5.925 to 6.425 GHz in the uplink) was the first to be developed in the early 1960s because it has favorable radio link characteristics due to low environmental noise and atmospheric absorption. A typical C-band satellite provides 24 transponders, each delivering 8 to 16 watts of power in an RF bandwidth of 36 MHz. The majority of television programming is transmitted to television stations and cable television systems at C-band.

The main difficulty with C-band is that the same satellite frequencies are used extensively by terrestrial microwave systems, forcing these media to share this band on an equal basis. In particular, the International Telecommunication Union and the Federal Communications Commission require that new earth station licenses undergo the process of frequency coordination to make sure that existing terrestrial radio links are not disturbed by earth station transmissions and *vice versa.*

Because of the problems with C-band sharing, a dedicated frequency allocation was made for satellite communication in the higher Ku-band range (11.7 to 12.2 GHz in the downlink, 14.0 to 14.5 GHz in the uplink). On a typical medium power Ku-band satellite, there are 16 transponders, each employing a power amplifier of 20 to 50 watts in a bandwidth of 54 MHz. With terrestrial microwave transmission precluded, Ku-band is ideal for use with small transmit-receive earth stations.

The view of satellite communication from the 1970s is that of a competitor to terrestrial microwave systems for the telephone trunking business. To gain maximum economic advantage, earlier satellite voice networks attempted to use the bandwidth of the transponder as efficiently as possible. This minimizes the cost per voice channel of the satellite. The earth station investment, however, would generally be high by current day standards (i.e., around $2 million for a major trunking terminal *versus* $15 thousand for a VSAT). Such large sums were needed to provide and support the multiplexing and transmission equipment needed to bring voice traffic together and link it efficiently through the satellite. A trunking earth station may have had a capacity of 500 to 5000 voice channels as compared to today's VSAT with the equivalent capacity of one or two 64-kb/s voice-data channels.

The old philosophy of maximum transponder utilization was disregarded by the Equatorial Communications Company, founded in 1979. Bandwidth was sacrificed with spread-spectrum modulation to allow the use of a very inexpensive class of VSAT called the Micro Earth Station. At a cost of under $3000 per installation, this technology made one-way data broadcasting feasible for delivering low and medium speed data to thousands of locations. Over 30,000 of these simple receiving terminals were in operation in 1988 at businesses and even homes. C-band was used because spread-spectrum modulation is quite insensitive to various forms of interference (and itself causes little interference), allowing reception with antennas of two feet in diameter. Transponder occupancy, however, drops from an efficient maximum of 60 Mb/s to only 100 or 200 kb/s with spread spectrum.

In satellite data broadcasting at C or Ku band, a source of data such as a host computer is connected to an uplink earth station, which transmits data packets to the satellite. The radio frequency signal occupies a portion of the power and bandwidth of one transponder, and is radiated toward ground to be picked up by receive-only dishes. Therefore, one transponder supports several simultaneous broadcast channels by dividing up the bandwidth and power. Requests from a user at a remote receive-only dish for specific packet transmission can be sent to the host computer over the terrestrial network, possibly on a dial-up basis. Data broadcasting is satisfactory for applications such as computer file download, where the ratio of outbound to inbound data traffic is very large (in true broadcasting, the ratio is infinity).

One-way (or broadcast) data networking is a natural fit to satellite communication, and many such networks have been implemented. A means to share the broadcast transmission with video was developed by the Public Broadcasting Service (PBS) and has been adopted by Visa International, the credit card clearing organization. The purpose of the broadcast, which follows the video from the PBS earth station to local television transmitters, is to provide consumers with a fast and convenient way to use Visa cards for smaller purchases. The data broadcast contains the account numbers of lost, stolen, or bogus credit cards and keeps the information in merchants' credit authorization terminals updated. The satellite segment operates at C-band because it employs PBS's established video distribution network.

Data broadcasting is taken several steps further in interactive VSAT networks by adding the transmission function to each earth station. A transmitting C-band VSAT using the level of power needed to reach a geostationary orbit satellite (in the range of one to ten watts) can easily interfere with reception at a terrestrial microwave station, making difficult the performance of frequency coordination for new earth stations in densely populated areas in and around major cities. The location of transmitting earth stations at Ku-band is greatly simplified; in fact, an entire network can be licensed by one application to the FCC (referred to as *blanket licensing*). The first Ku-band VSAT product, called the Personal Earth Station, was produced by M/A-Com Telecommunications (now Hughes Network Systems). As in data broadcasting, a given network occupies a single channel and can be placed in a common transponder along with several other interactive or broadcast data networks. The hub earth station acts as the gathering point for the VSAT network, giving access to a host computer and controlling transmissions from the VSATs as appropriate.

Since the power of a Ku-band transponder is four to five times greater than that of a typical C-band transponder, a relatively small antenna can be used to receive a video broadcast. Heavy rainfall has more of an attenuating effect on Ku-band transmissions from space than C-band; hence, additional power margin is recommended. After accounting for Ku-band's greater atmospheric absorption due to rain, the ground receiving antenna for good quality video reception can be as small as four feet (1.2 meters) in diameter. This is ideal for corporate video teleconferencing applications, and can be accomplished from the same VSAT used for one-way or two-way data.

2.3.1 International Satellite Communication

International satellite communication continues to have strength and vitality, even as fiber optic cables begin to replace the older coaxial cable systems. One satellite positioned over the Atlantic, Pacific, or Indian Ocean provides complete connectivity among dozens of countries in the region. The overall coverage regions for these ocean basins are illustrated in Figure 2.6. Dots represent nominal satellite positions on the equator, and the curves define the outer edge of visibility to the satellite from the earth's surface. Many of the routes between countries are too thin to justify cable links. Satellite links are inexpensive and quick to establish from and between existing earth stations.

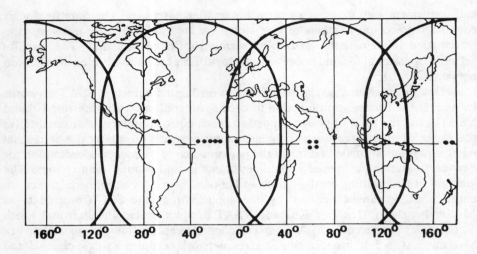

160°   120°   80°   40°   0°   40°   80°   120°   160°

Figure 2.6 INTELSAT orbital positions and geographic coverage.

Many countries have leased or purchased capacity on INTELSAT's satellites for domestic communication. Whole or partial transponders can be leased by members of the organization. The earth station networks are typically designed and installed by system integrators, such as those discussed in Chapter 1. These systems can be operated and maintained by indigenous personnel who have been trained by the installing contractors. Consulting and training services are often available from the UN and the ITU as well as part of foreign aid programs from the U.S., Japan, and many European countries.

INTELSAT, having taken over complete responsibility from COMSAT for construction and operation of its satellites, manages the largest fleet of commercial satellites in the world. In terms of the business, INTELSAT only provides the satellite end of the link, referred to as the *space segment.* Control of satellites (e.g., the tracking, telemetry, and command function) requires that INTELSAT have access to earth stations at strategic positions around the globe so that radio control links to the satellite can be maintained. The communication earth stations themselves, however, are the users' responsibility.

INTELSAT users are the same PTTs that employ terrestrial means such as cable and microwave. The U.S. was one of the few countries to set up a commercial organization with private ownership, COMSAT, but most countries kept satellite communication under government authority. In the late 1960s and early 1970s, using satellite communication on INTELSAT was actually more expensive than using the transatlantic cable between the U.S. and the U.K. This resulted from two factors. First, the channel capacity of the earlier generation of satellite was somewhat restricted, while construction and launch costs were relatively high.

Second, pricing of INTELSAT satellite capacity is averaged over the entire system, which causes heavily used paths, for example, between the U.S. and the U.K., to subsidize the thin routes to lesser developed countries. Eventually, INTELSAT drastically reduced the charge for space segment to be competitive with the highest capacity cable systems.

The trend in INTELSAT's charge for space segment is shown in Figure 2.7, which presents the monthly lease rate per half circuit. A half circuit is an imaginary two-way (duplex) telephone link between one earth station and satellite. The circuit is imaginary because it consists of bandwidth on a radio link without any physical connection to terrestrial facilities. Therefore, two of these half circuits are required to achieve a link from earth station to earth station. Because of the way in which all links in the INTELSAT system are aggregated, the charge is expressed in this manner, regardless of the path or destination. The composition of a typical end-to-end international satellite circuit is shown in Figure 2.8. To be added to the space segment cost is the charge for using the earth stations and the terrestrial "tail" links back to the international gateway telephone exchanges. These additional charges are levied by the operator of the ground facilities and vary among countries. As is usually the case in transmission, the cost of the satellite link itself is a relatively small fraction of the cost of the end-to-end circuit. Using larger and more capable satellites, INTELSAT has been able to reduce the cost per channel on a yearly basis.

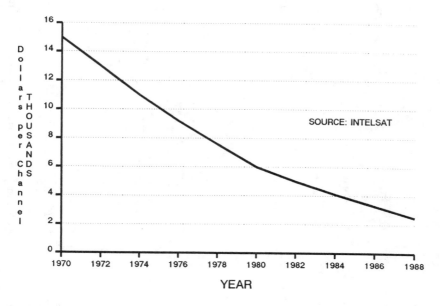

Figure 2.7 Annual cost of a typical international half circuit over the INTELSAT satellite system (space segment only).

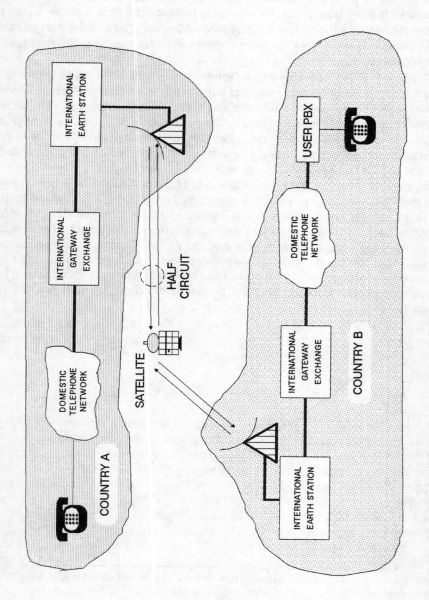

Figure 2.8 Definition of an international satellite half circuit.

Historically, users gained access to INTELSAT through their domestic tele-communication networks. Thus AT&T simply employs satellite links as a part of their overall international network connections. Recently, INTELSAT introduced the INTELSAT Business Service (IBS) to give builders of private telecommuni-cation networks direct access to the INTELSAT space segment. In the U.S., a strategic unit can install its own earth stations and obtain space segment through COMSAT. The complicated part, however, is getting the cooperation of the foreign PTT. PTTs generally exercise tight control over their networks and can impede the development of private networks.

2.3.2 Early Domestic Satellite Systems

Satellites used for domestic telecommunication appeared on the scene in the early 1970s. The benefits of satellite transmission of telephone and video signals were understood and debated well before the technology was reduced to business practice. As discussed below, Canada was the first to proceed, and the U.S. came later due to regulatory hurdles. From a small beginning, domestic satellite com-munication has become a mainstay in the U.S. telecommunication picture. For example, Figure 2.9 displays the section of the geostationary orbit containing U.S. domestic satellites. This large capacity in orbit is composed mostly of C-band satellites used primarily for video transmission and Ku-band satellites which are applied to a variety of purposes including video and data communication. As discussed throughout this section, the Ku-band satellites can perform important functions in private telecommunication networks due to the versatility of the VSAT.

2.3.2.1 Canada

Unlike the INTELSAT system, the implementers of domestic satellite systems built earth stations simultaneously with the launch of satellites. Canada was the first to establish a domestic telecommunication service company, Telesat, to exploit satellite technology and to improve rural telephone and television service. These objectives were satisfied by using a first generation of C-band domestic satellite with twelve transponders. Each transponder could relay one analog television channel, or up to 500 telephone conversations. (The precise capacity depended on the transmission technique and the physical size of the earth station antenna.) Telesat created a ground segment of several large telecommunicationmunication earth stations to serve major cities and a thin route network of hundreds of small terminals (the forerunner of the modern VSAT). Telesat enjoys a government-mandated monopoly over domestic satellite communication in Canada.

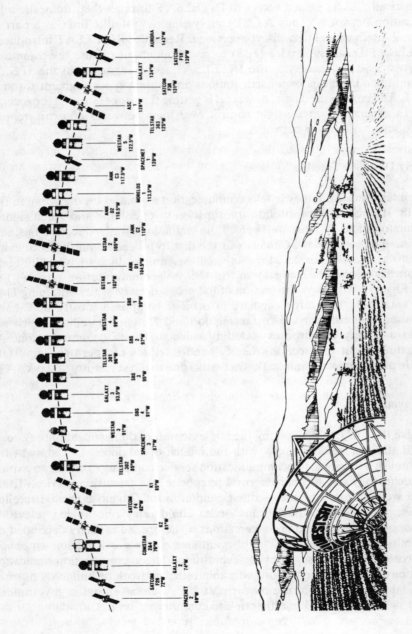

Figure 2.9 The geostationary orbital arc serving North America, showing the operating C- and Ku-band satellites. (Illustration courtesy of Westsat Communications, August 1988.)

2.3.2.2 U.S. Domestic Systems

The Federal Communications Commission followed the direction of the executive branch and opened up U.S. domestic satellite communication to competition. The evolution of competitive domestic satellite systems in the U.S. is reviewed elsewhere in detail [Elbert, 1987]. Today, there are nine different commercial companies which own and operate satellite systems in the U.S. Western Union launched the first of its Westar satellites in 1975. COMSAT and AT&T jointly established the Comstar system, with GTE sharing the use of the satellites. RCA moved aggressively into the marketplace later in the 1970s and became a mainstay in the distribution of television signals to cable television systems around the country.

Consolidation of satellite systems has begun, with some disappearing from the scene. The most successful systems are tied to a parent organization capable of offering important synergy to the operation or marketing of the satellite company. GE Americom and Hughes Communications are both successful in a business sense and are making a continued commitment to satellite communication. The parent companies each manufacture the satellites that are employed. Note that Western Union sold the Westar satellite system to Hughes in 1988. Another survivor is GTE Spacenet, which is a wholly owned subsidiary of the independent telephone holding company, GTE Corporation. The bases for the continuation of this operator seem to be the financial resources available, a strong technology base in telecommunication, and the ability to tie satellite and terrestrial networking. Finally, AT&T continues to operate domestic satellites as an adjunct to their long-distance plant. AT&T's main customers appear to be the over-the-air television networks and other broadcasters, although a strong entry into VSAT networks is anticipated.

2.3.2.3 The SBS Experience

Perhaps the most aggressive move into private telecommunication networking as a business venture was engineered by Satellite Business Systems (SBS), originally a partnership of COMSAT, IBM, and Aetna Life and Casualty. This organization was absorbed into MCI and no longer exists as an operating entity. Set up in the mid-1970s to exploit domestic satellite communication and advancing digital technology, SBS started from scratch by rapidly building a powerful organization in order to be the first U.S. company with an all-digital telecommunication network. SBS benefited from the synergies of the partners: COMSAT's expertise in satellite systems engineering, implementation, and operations; IBM's unchallenged position in data processing and data communication; and Aetna, as an innovator in insurance underwriting and a company with a desire to expand its horizons through investment in this exciting amalgamation of technologies. The three partners clearly had the economic wherewithal to bankroll SBS and the interest to stay with it through the lean start-up years.

The network architecture chosen was based on *time division multiple access* (TDMA), a versatile satellite transmission technique which efficiently uses the transponder and earth stations. (A review of TDMA architecture is provided in Chapter 4.) Digital processing was applied to the voice, data, and video traffic before it was transmitted to the satellite (uplinked). One of the advantages of TDMA is that the arrangement and routing of traffic among earth stations can be changed at any time and in response to actual demand. In addition to the earth station TDMA equipment, SBS provided PBX capabilities for customers. One of the services that SBS thought would be the foundation of its business was the provision of a piece of office equipment called the "intelligent" copier. This turned out to be the forerunner of the digital facsimile machine, which was to be the basis for the ZapMail™ offering of Federal Express (discussed later in this chapter). SBS actually funded the early development of digital facsimile, but unfortunately the product did not reach sufficient maturity to be part of the actual service offering. Instead, SBS attempted to build its business on private telephone and data networking for these large corporations.

In 1981, SBS launched its first Ku-band satellite, beginning a new stage in private telecommunication. A rendering of the satellite in orbit, showing its coverage of the United States, is given in Figure 2.10. Marketing efforts were centered on the "Fortune 100" — the 100 largest corporations in the U.S. While advocating the concept of placing a "low cost" earth station directly on the customer's premises, the architecture of the SBS network was such that each earth station represented an investment in excess of one million dollars. Figure 2.11 presents a photograph of one of the SBS customer premises earth stations. Perhaps in comparison to a standard-A INTELSAT earth station with its 30-meter antenna, the SBS terminal could be viewed as small and inexpensive. However, many customers could not absorb this level of cost, so SBS was forced to retain the majority of the earth stations and then attempted to build a business based on sharing the terminals among a number of users. Several large corporations including IBM, General Motors, General Electric, and Hercules became users of the SBS network, but revenue growth was still not satisfactory.

The opening up of the long-distance business by MCI induced SBS to offer long-distance service through their earth stations. Large telephone switches were installed at major cities. The network was then configured from the satellite system, augmented by leased circuits from other carriers. Also, a sales staff and "back office" operation was necessary to support and to build the business. A small but respectable customer base was developed, and SBS was even recognized by a consumer research group as having a good service at an attractive price. The company, however, incurred large operating losses while establishing itself as a long-distance carrier in what was developing to be a very competitive market.

Figure 2.10 The SBS Ku-band domestic satellite provides coverage of the continental United States. (Photograph courtesy of Hughes Aircraft Company.)

Referring back to the technology and architecture of the basic satellite network, SBS used conventional equipment at the interface between the public analog telephone network and the earth station TDMA equipment. Unfortunately, this was a complex task that sometimes resulted in unacceptable circuit quality. (Eventually, these problems were largely cleared up with improved echo cancellation and circuit balancing devices.) The results for the private network business were disappointing. The PBX capabilities did not work as proposed, and users were actually disconnecting themselves from the network. The name SBS became synonymous with poor satellite communication, a reputation that has hurt not only SBS, but also the satellite industry. The difficulties entailed ground equipment and network architecture, not the actual satellite links.

Because of the tremendous financial drain of supporting SBS, COMSAT and Aetna withdrew from the partnership and IBM continued as the lone investor.

Figure 2.11 A typical Ku-band SBS earth station with a 7.7 meter diameter antenna located on a customer's premises. (Photograph courtesy of Hughes Aircraft Company.)

Finally, IBM transferred SBS to MCI in exchange for that company's stock. (MCI has subsequently repurchased its stock from IBM.) MCI was attracted to SBS because of its long-distance customer base, which was fairly loyal. Then, MCI proceeded to switch those customers over to its primarily terrestrial network and dismantled the SBS earth stations. The Ku-band satellites are now being used for a variety of services, particularly commercial video and VSATs. MCI clearly has no interest in promoting satellite communication, particularly due to its competition in fiber optic with AT&T and US Sprint. The final three SBS satellites are owned by IBM, with Hughes Communications having an equity interest and marketing responsibility.

2.3.3 Modern Domestic Satellite Systems

The distressing experience of SBS is in marked contrast with the exceptional results of GE Americom and Hughes Communications. The critical difference seems to be that SBS built an entire network wherein satellites were only a part, while the two latter firms built their businesses on selling satellite capacity to users with their own earth stations. Transponders in orbit, which make up the space segment, can be used for a variety of purposes that may change over time. The ground segment, on the other hand, is less flexible, particularly in a business sense. Public switched long-distance service by satellite, while effective in international communication, is no longer an attractive alternative to the terrestrial networks. This situation was not something to which SBS could adapt very quickly, particularly after having spent hundreds of millions of dollars on ground facilities. As discussed previously, satellite communication's strength is in providing point-to-multipoint distribution of information around a country or region. Also, VSAT technology for two-way interactive networking is gaining acceptance in today's private telecommunication network environment. As long as the satellite operator stays in space, so to speak, it is positioned to move with the evolution of the telecommunication industry.

VSAT technology has matured significantly in recent years, making it entirely feasible for commercial networks employed by corporations and government agencies at the state and federal level. An example of a VSAT is shown in Figure 2.12. First-time buyers have experienced the expected start-up problems as the two-way VSAT product itself matures. For example, RF power amplifiers for the VSAT transmitter in some cases have been found unreliable and are being replaced with more rugged units. Corporate VSAT network pioneers now enjoy the cost reduction and reliability improvements that were predicted when the decision was made to undertake the project.

2.3.4 Benefits of Satellite Communication

The value and utility of satellite communication, even in the age of advancing terrestrial networks, derive from the following features, many of which are synergistic with others.

Wide area coverage. The typical coverage beam of a domestic satellite, shown in Figure 2.10, reaches every region of a nation with nearly the same signal. This provides the point-to-multipoint connectivity (e.g., broadcast) for which satellites are well recognized in the area of cable television program distribution. In the corporate context, data broadcasting is a natural extension of the technology. An excellent application of data broadcasting is the delivery of news services such as

Figure 2.12 A transmit-receive VSAT with a 1.2 meter antenna. (Photograph courtesy of Hughes Network Systems.)

Dow Jones and the Associated Press to thousands of receiving points at newspapers, television and radio stations, and brokerage offices. Uplink transmissions from any earth station capable of doing so can enter the satellite through the same beam as is used for the broadcast function.

Distance insensitivity. The cost of linking two or more points on the ground via satellite in geostationary orbit is independent of distance between the points. While the benefits of this feature are greatly realized in international communication, the more relevant context is in data communication for many branch offices. The cost of a typical terrestrial private line used in a multidrop network increases with distance. This is aggravated by the fact that a large multidrop network must be broken down into multiple lines in a star configuration to prevent losing the entire network in the event of failure at one point in the chain. The economics of satellite transmission becomes more favorable as sites are added into the network, as discussed later in this section.

Interactive communication for OLTP over satellite paths is inherently more reliable than multidrop topology because every branch office is served by its own, independent link to the satellite. The chance of a single-point failure in the hub earth station and satellite is minimized by employing redundant equipment chains which are integrated into the particular element. In general, satellite transmission has proved to have high reliability, with the availability of satisfactory signal being typically 99.95% of the time for the combined VSAT-satellite-hub path. A terrestrial private line would be expected to have an availability of 99.5%, which means that a given link in the chain is out of service an aggregate of ten times longer than for a satellite path over the same one-year period.

Simplified connectivity. Flexible communication among many points is often a requirement in a wide area network. *Multiple access* is the process whereby many earth stations transmit information in the same transponder, providing the desired wide area connectivity. (A review of multiple access methods is provided in Chapter 4.) The more flexible techniques include TDMA, Aloha, and *code division multiple access* (CDMA). Spread-spectrum modulation is a type of CDMA, relying on the ability of the VSAT receiver to pick out the properly coded information from the signals which overlay each other in time and frequency. TDMA, however, requires that only one signal be present on a particular frequency at a time. The controlling hub earth station provides a synchronizing reference signal for the VSATs which transmit individual bursts of traffic according to time slot assignments from the hub. In the Aloha configuration, VSATs transmit much shorter packets in an essentially random fashion with the hub monitoring the resulting stream in the downlink to detect overlapping packets (called "collisions"). When a collision occurs, the particular pair of VSATs is instructed to retransmit their packets with a suitable time offset to prevent another collision. As more VSATs attempt to transmit additional packets, the increasing collision-retransmission rate adds time delay and reduces throughput. Under such circumstances, another frequency channel must be added or the multiple access method can be switched to TDMA. The relative merits of TDMA, Aloha, and CDMA should be examined in the context of the particular network requirements to choose the technique which works best under expected traffic conditions.

Multiple access provides connectivity among hundreds or even thousands of dispersed locations and with a major information node such as the host computer in a hotel reservation system. A point-to-multipoint voice link on the outbound path from the hub is used by a national brokerage house to broadcast immediate news on companies or stock prices to branch offices. Called "hoot and holler," such a connection by satellite is easy to achieve with the hub simply uplinking one signal which the satellite broadcasts like a space-based radio station with a 22,000-mile-high tower. (The effect of this type of communication on the magnitude and direction of stock market swings is anybody's guess.)

Total private network. A considerable time period has passed since the divestiture of the Bell operating companies by AT&T, and many business managers yearn for the days when one very large company could provide for their communication needs. Today, a satellite network is close to being a single-vendor solution. Perhaps the best example is the cable television network of Viacom, Inc. Viacom uses its own teleport earth station in New York to uplink several television channels, one per transponder on a C-band satellite. The transponders were purchased on a condominium basis from Hughes Communications, Inc., which also continues to operate the transponders for Viacom under separate agreement. The channels are received by cable systems around the country, each having a receive-only antenna to pick up the signals and deliver them to households via coaxial cables. Viacom also owns several of these cable systems, making them a vertically integrated television program supplier to several million homes in the United States.

Bypass of the local telco. The concept of bypass is attractive for telecommunication managers who need total control of their communication resources. Placing a VSAT antenna on a customer's premises provides wide-bandwidth services without using the facilities of the local telephone company. This is economically attractive because most of the cost of service is in the switching and the final mile. Satellite links with bandwidths in the range of 56 kb/s to 1.544 Mb/s are now used extensively for digital facsimile transmission of newspaper pages of sufficient quality to allow printing plates to be made directly from downlinked signals. Newspapers such as the *Wall Street Journal* and *USA Today* could not print their issues around the country at dozens of plants if they depended on local telephone companies for covering the final mile. An additional feature is the favorable economics of using satellites for multipoint delivery of the facsimile data.

Rapid deployment. Emergency communication services make effective use of radio, the same medium used by satellites. Small transportable ground antennas and compact electronic packages permit high quality communication to be installed in a matter of hours instead of days or weeks. Rapid deployment is the primary benefit of *satellite news gathering* (SNG), a Ku-band service used by news departments of television stations and networks to originate telecasts from remote locations. SNG trucks are very popular because they can be driven to a venue and put into operation without performing frequency coordination. Within the corporate context, features of SNG lend themselves to video teleconferencing and backup communication, the latter being particularly important in the event of a fiber optic cable failure (i.e., back hoe fade) or central office fire that can cut off a corporate headquarters or major data processing center. In general, satellite links can be important to provide backup and alternate routes for important links.

Low cost per added site. In television receive-only and VSAT applications, remote sites in the initial network are relatively inexpensive to install, a fact which also applies when subsequently adding a new location. Business video teleconferencing is attractive in this regard. For example, when a site is added to a video training and teleconferencing network like that of Domino's Pizza, it need only install a low cost TVRO terminal, which is basically a consumer item available from retail outlets around the country.

The ground segment of the satellite system has provided some companies with opportunity for profitable network development, leaving the investment and technology of the satellites themselves to major operators. One of the most successful providers of satellite data networks was Equatorial Communications Company, now part of Contel/ASC. Instead of attempting to sell very expensive earth stations, Equatorial brought the end-user cost of satellite communication to below $3,000 per site. The point-to-multipoint data distribution network operates at the relatively low speed of 19.2 kb/s, which is sufficient for news wire and financial information. This concept is being imitated by other service providers, and an internal data receiver is even available for use in personal computers.

Expansion of Ku-band VSAT networks is anticipated as corporations and government agencies take advantage of the lower cost and wider bandwidths that are possible. The typical approach taken in system implementation is to purchase a quantity of 1.2-meter to 1.8-meter VSATs for installation at branch offices. Operation and maintenance of these terminals will generally be the purchaser's responsibility, but third party maintenance services are made available on a contract basis. Then, the interconnection of the VSATs with host installations would be through a centralized hub earth station. Large networks with 500 or more VSATs could generally justify a dedicate hub, which would also be operated by the user. Smaller networks could be established through what is called a *shared hub*. A satellite communication company such as Hughes Communications or GTE Spacenet operates the shared hub and offers connectivity and network management as a service to users. The third ingredient is the space segment capacity needed to complete the microwave radio links between VSATs and hub. Users may contract directly with the satellite operator, such as GE Americom, Hughes Communications, or GTE Spacenet, or users can obtain space segment as a service from the shared hub operator.

A special class of domestic satellite communication company is the teleport operator (Figure 2.13). There continues to be some confusion as to what a teleport is. Some teleports are operated as real estate projects, offering other companies office and equipment space in a facility with good terrestrial tail connections to the nearby city. Other teleports are really just earth stations for hire, available for full-time or part-time transmission to the satellites already operating in orbit. A private telecommunication network developer may find it attractive to use the facilities or services of either or both classes of teleports.

2.4 TELEX NETWORKS

One of the most important forms of information transfer is what has been referred to as *record communication*. Beyond the use of telegraphy and the Morse code, the public telex network gained wide acceptance for business use. Telex has

Figure 2.13 The Home Box Office Communications, Inc. Center is a large teleport located at Hauppage, NY. (Photograph courtesy of Tony Neste/HBO.)

been particularly popular in international communication, where time-zone differences and difficulties in verbal communication persist. In the U.S., telex is quickly being displaced by facsimile and E-mail.

Recently, the Worldcom international telex division of ITT was merged into Western Union to form the largest provider of telex services in the U.S. The hope is that this combination will allow Western Union to survive in a declining market. Up until the 1980s, the FCC barred Western Union from entering the international market.

To enhance its telex offering, Western Union introduced EasyLink, a value-added network and E-mail offering, which incorporated telex as one of the embedded services. The main benefit to telex users was that they could enter the telex network through the dial-up telephone network without having a dedicated teletypewriter access line provided by Western Union. As we move toward the twenty-first century, telex will probably remain as an international medium, much like telegram of the past, but more advanced methods like E-mail will eventually prevail.

2.5 PUBLIC DATA NETWORKS

Telex and data communication have demonstrated that there is a need for specialized services which can be accessed on demand, much like the telephone system. The public data network (introduced in Chapter 1), also referred to as a *value-added network* (VAN), is a sophisticated arrangement of packet switches and interconnecting full-time transmission lines, allowing users to send and to receive data in various formats. Two leaders in the U.S. are Tymnet, a subsidiary of McDonnell Douglas, and Telenet, a subsidiary of US Sprint. Figure 2.14 gives the basic topology of the Telenet. An important feature of a PDN is its ability to allow dissimilar terminal devices and computers to be connected to network access points, performing the conversion of the various data formats, transmission rates, and protocols by the network interface. For example, the PDN can accommodate synchronous protocols, such as IBM's SDLC, and asynchronous communication with PCs. Terminals can operate at dissimilar speeds (1200, 9600, and 56 kb/s) because the PDN network nodes can adjust rates.

Connection of users to the PDN can be arranged in two ways. A dedicated access circuit can be established between a computer or terminal and a PDN node. This is generally used for the user's host access (or in the case where the PDN is used to provide full-time connections between subscribers of a time-sharing network and the central computer of the service bureau). The second type of access is on a dial-up telephone line. Almost all occasional use is through dial-up access. Dedicated lines would be used in heavy remote access applications, such as retail point of sale.

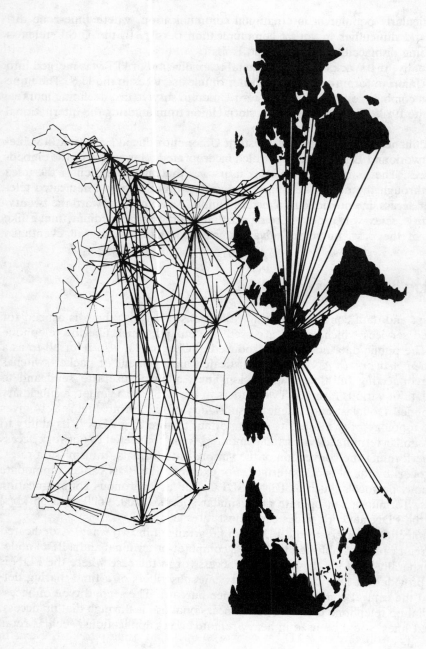

Figure 2.14 Telenet® Public Data Network, showing domestic and international connectivity. (Copyright Telenet Communications Corporation. Reprinted with permission.)

Using a PDN is almost as easy as making a long-distance telephone call. Any computer terminal with a modem can access the PDN by dialing a local telephone number in a major city. This local access line is connected to the PDN node with either a private transmission link (hardwired or by radio) or a leased line. The connection that is set up between the user terminal and the local node is continuous, but the node only processes packets of data on the network side when the user is actually transmitting or receiving information. Routing of packets within the network follows a preprogrammed plan, which is under the control of the PDN operator. From the user's perspective, the connection over the network appears to be continuous, and the PDN is said to provide a *virtual circuit* during the session. User messages, however, are divided into packets, each of which is addressed individually. The protocol also keeps track of packets that have been dispatched, and provides for automatic retransmission of packets lost in the network. (A more detailed discussion of packet network techniques is provided in Chapter 4.) The processing power of the packet nodes causes the packet network to be loaded fairly efficiently, giving an economy over directly using the long-distance telephone network on a dial-up basis. The topology of the circuits between nodes is either a mesh or a ring to provide diverse routing paths for the packets. Each node has two or more paths out onto the network and the node is able to select the path for a given packet which gives the shortest delay, or, in the event of link outage, actually permits the packet to reach its destination. Therefore, a well structured packet network is extremely reliable.

Current PDN providers are the pioneers of the technology, having invented the equipment and software which process and convey data in the form of packets. The trend is now toward a standardized PDN interface in the form of Recommendation X.25 of the CCITT (an ITU body). A user would packetize the information before handing it over to the PDN. As presented in Chapter 4, Recommendation X.25, which has the force of a standard, defines in detail the structure and addressing of packets for transmission over a PDN, both within and between countries. The user typically installs appropriate network software in a host or remote computer so that a direct X.25 connection is possible. PDN operators can provide the specialized software to be loaded directly into the accessing mainframe, minicomputer, or PC. For example, Telenet has available X.25 accessing software for every major computer brand, including IBM, DEC, HP, Tandem, Unisys, and Wang. In cases where such software is not available, an external device called a *packet assembler-disassembler* (PAD) is installed between the user device and the network. An important feature of a PDN is a closed user group, which implements a private packet network, restricting access from outsiders, over a well established PDN.

A private telecommunication network can incorporate PDN features by installing X.25 software or PADs at network nodes and using part of the existing transmission links. The nodes themselves, called *packet switches,* can be purchased from several manufacturers, including AT&T, BBN, Hughes Network Systems, Northern Telecom, and Siemens. Public data networks are also under attack by the RBHCs and AT&T, who wish to provide packet switched services directly for users over the public telephone networks. Open Network Architecture could be the "PDN killer" of the future. To counter this, the existing PDN providers are working closely with the RBHCs to entrench themselves as the more capable developers and providers of packet switched services, as mentioned previously for Bell South. Another thrust is to include higher levels of the OSI model. Telenet now provides E-mail services by using the X.400 applications layer protocol, discussed in the following section.

2.6 ELECTRONIC MAIL SERVICES

The advent of time-sharing computer systems with large amounts of on-line storage opened up the possibility for users to send and to receive messages among each other. There are many classes and versions of electronic mail service (E-mail), which effectively derive from record communication services employing teletypewriter machines (e.g., telex and TWX). Western Union first offered an E-mail service when they augmented their telex service with Infomaster, a central message switching computer system for switching and network control, which could accept messages when the teletypewriter machine at the distant end was not available. Infomaster would subsequently call up the machine automatically. It was possible to send the same message to several addressees, even to those without their own telex machines (Western Union would print and deliver the message by regular mail). Even more advanced features were developed for the U.S. government's computerized message switching system. Computers, therefore, offer powerful capabilities for switching, storing, and processing messages in textual and graphic form.

Office automation systems from major vendors, such as IBM, DEC, HP, and Wang, allow users to compose letters and documents by using word processing software. After storage, these same documents can be routed to other users, both locally and remotely over telecommunication facilities. An IBM office automation system called PROFS became rather notorious during the Congressional "Iran-Contra" hearings in 1987 because messages which supposedly had been erased on the system were eventually recovered for use as evidence.

E-mail is a very capable application for on-line computer systems, and many strategic units rely heavily on it for interoffice communication. Perhaps E-mail's main drawback is that the writer must be able to type. In contrast to information services like videotex and teletext, E-mail services do not provide information to users, but rather offer a communication system for conveying information. A complete discussion of E-mail systems and services can be found in [Caswell, 1988].

An initial disadvantage with E-mail was that the services tended to be closed to the outside world, but the situation should improve in the 1990s. While it is possible to originate a message on CompuServe® and have it delivered to a subscriber using MCI Mail, it is not possible to do the same on EasyLink. Internal systems like PROFS, HP Desk, and Wang Office can be connected to public E-mail services like MCI Mail through a dedicated access line to a gateway or on a dial-up basis over the telephone network. Sending E-mail messages internationally is possible by connection to the domestic telex network; and more recently, MCI Mail can reach foreign users in fifty-seven countries.

Progress has been made in overcoming the limitations of domestic and international E-mail through the efforts of the ITU with CCITT Recommendation X.400 on document interchange. Several computer companies, record carriers, and PTTs are experimenting with X.400, and there is the prospect that efficient worldwide E-mail operations are imminent. Such a capability was demonstrated for the first time at the TELECOM 87 international exposition in Geneva, Switzerland. A fundamental requirement of a truly universal E-mail system is the availability of directory services. In telephone networks, it is easy to use operator directory services to obtain the phone number of a party in a distant city or country — if you know the correct spelling of a name and the address. It is not this simple in E-mail because not everyone is a subscriber. The X.500 directory standard is being developed to deal with the problem through a separate message protocol which runs in parallel with the X.400 system. An originator, who would not know the address of the distant party, might first send an inquiry which X.500 could route from node to node until a match were found. A return message would indicate the proper address for the actual information message to employ. Absence of an acceptable address or addressee within the X.400 environment would be indicated by an appropriate return X.500 message, thus saving the originator the cost and trouble of transmitting the entire text of an undeliverable message.

As mentioned in the previous section, Telenet has introduced an X.400-based service called Telemail 400™. The layout of Telenet's configuration for the 1988 Enterprise Networking Event is presented in Figure 2.15, indicating how Telemail 400 and the X.400 interface provide a variety of access services along with protocol conversion. This special network demonstrates a wide variety of capabilities, including E-mail over the Telenet PDN. A closed user group with X.400 access is shown in the figure. In addition to providing internal E-mail services, this allows outsiders to send E-mail messages into the closed group by way of the X.400 gateway. An added feature of this approach is that international messaging is also possible, since the foreign X.400-based E-mail networks are reached through the Telenet PDN.

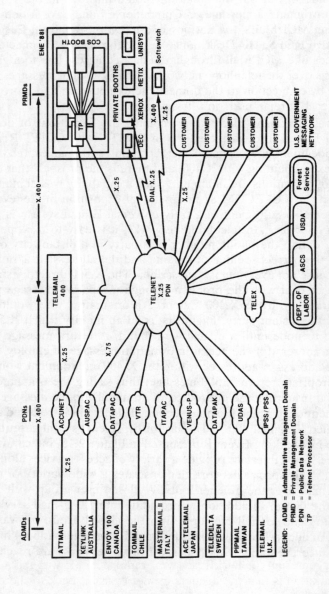

Figure 2.15 Telemail® X.400 Messaging Network as demonstrated at the Enterprise Network Event in June, 1988.

2.7 MISCELLANEOUS NETWORKS

The networks to follow are provided by various independent companies for specialized services to U.S. businesses and the public. This listing is not exhaustive, and innovative new offerings are appearing all the time.

2.7.1 Paging

The paging business occupies a small but important niche in public telecommunication service. A small receiving device called a "beeper" or "pager" can be carried by an individual, some beepers being the size of a fountain pen. When activated, the beeper gives off an audible alarm. Some models have display windows silently showing that an alarm has been set off and able to display short messages such as a telephone number to be called. To reach a person carrying a beeper and to set off its alarm, someone dials a local telephone number to reach the paging service. Special tones or voice response techniques (discussed in Chapter 3) can allow the beeper to be activated and display information without the intervention of an operator.

Figure 2.16 illustrates the basic paging system. The antenna tower used to transmit to the beeper is part of the broadcast facility of a local FM radio station. Bandwidth for the paging channel is provided on a subcarrier under an FCC licensing arrangement called a Subsidiary Carrier Authorization (SCA). This channel cannot be detected by normal home FM receiving equipment, and is therefore inaudible to normal radio listeners. The beeper, however, can detect the SCA channel anywhere within the normal broadcast coverage area of the FM station. The obvious limitation of conventional paging is the range of the FM station. Consequently, a particular pager can only be used within the local area of service.

A new class of service called *satellite paging* or *national paging* has appeared as a means of allowing the beeper to be carried to distant cities. In this case, a single 800 telephone number is provided for nationwide access. The paging signal itself is distributed to all of the cities served by a satellite point-to-multipoint data channel, which is received at the selected FM radio station and applied to an SCA channel. Conceptually, the paging information is first broadcast around the country, picked up at the local transmitter, and then rebroadcast locally. While satellite transmission is used, the beepers themselves do not directly receive the satellite signal. When departing for a distant city, a user would call in on an 800 number to inform the national paging service of the city to be visited.

Pagers represent the Dick Tracy wrist radio technology that is available today. The key limitation is that the beepers cannot transmit back to the originator, but are merely receivers and holders of messages. Putting a transmitting function into the beeper will be a major innovation in the commercial marketplace. Two-way communication by radio, however, is a reality as discussed in the following sections.

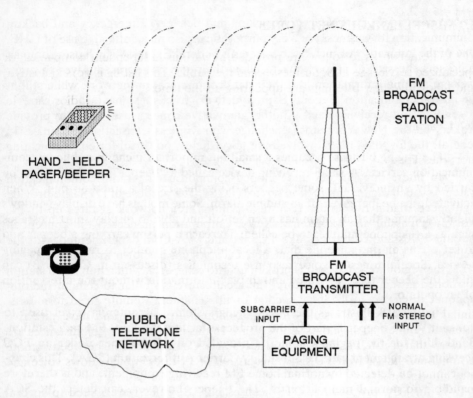

Figure 2.16 Radio paging system for use in a metropolitan area.

2.7.2 Radio Common Carriers

The general class of radio common carriers includes various installers and operators of fixed and mobile radio stations which are available for use by other parties. Microwave systems similar to those of the telcos and long-distance companies are operated by radio common carriers for use in point-to-point television signal transmission, between studio and broadcast transmitter. Telephone and data point-to-point service also can be offered. Radio common carriers are usually local or regional companies, specializing in one or more applications. Eastern Microwave, Inc., is a radio common carrier in the northeastern U.S., which has expanded its regional business by providing satellite transmission to deliver New York television station WWOR's signal to cable systems around the country.

Common carriers offer UHF mobile services, employing frequencies similar to those of public services like police and fire. Being able to communicate with vehicles and fixed points makes UHF common carrier applications attractive for service companies, particularly in trucking and delivery. Emergency and backup communication for business recovery purposes can also make effective use of UHF. One of the innovators of such radio systems is Federal Express, which uses compact radio sets the size of a notebook to allow their service representatives to enter package routing information into their data communication network while still at the customer's location. Federal Express delivery trucks also have radios capable of sending and receiving data. Another innovative concept uses radio to provide brokers on the New York stock exchange with wireless transaction terminals. In general, the low data rates involved are nevertheless powerful for communicating short messages.

An interesting but yet undeveloped common carrier radio communication system is called *digital termination service* (DTS). The purpose of DTS is to provide high speed digital radio links in a metropolitan area network at rates up to T1, possibly bypassing the local telco. Xerox Corporation pioneered the concept in the late 1970s and companies like Tymnet and M/A-Com experimented with the technology. The basic arrangement of a DTS system is shown in Figure 2.17. In DTS, the region to be covered is broken up into cells, each with a hub radio station positioned at the center. Therefore, the architecture is similar to cellular radiotelephone, discussed in the final section of this chapter. Remote fixed terminals, similar to VSATs, are installed on the customer's premises and have directive antennas pointed at the closest hub station. In this case, the intracell distances are so close that a satellite relay is not necessary. Communication between remote terminal and hub is with TDMA transmission at T1 rates, which allows voice and data integration.

In an FCC proceeding, DTS licenses in U.S. cities were issued to several companies and partnerships, much as was done for cellular radio. Except for the experiments cited above, however, no systems have yet gone into operation. The technology appears to have benefits and may still be employed, particularly as data communication needs of organizations expand beyond what is possible over the telco local loop. The technology has been developed and can be deployed when demand materializes, assuming that the frequency assignments are not withdrawn by the FCC due to lack of interest.

2.7.3 Position Location and Data Acquisition Services

A radiodetermination service which can accurately locate a fixed place or moving vehicle has been evolving in various forms. Of course, the air traffic control systems used by civilian and military aircraft have relatively effective means of determining position using ground-based operational systems. A satellite radiodetermination system using the Global Positioning Satellite System (GPSS) is being deployed for the U.S. military, and civilian terminals are appearing on the market.

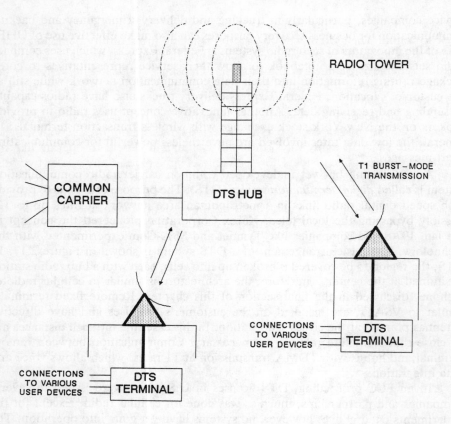

Figure 2.17 The Digital Transmission Service (DTS) is intended for short haul high speed data communication links within a metropolitan area.

The cost of using military techniques is often excessive for commercial purposes. Furthermore, return communication from the remote terminals is not provided.

In oil exploration, a very accurate position determination system has been deployed by a company called Star Fix. Using three existing C-band satellites to relay reference signals, the Star Fix system allows exploration vessels in the Gulf of Mexico, for example, to compute the location within a fraction of one meter. This commercial service is proving effective and economical for oil exploration, but will not be feasible for widespread use in, for example, the trucking industry.

A number of organizations are planning, and in one case implementing, commercial systems for locating and communicating with commercial land vehicles. Geostar has been authorized to operate a satellite position location system using a frequency band reserved for radiodetermination satellite systems (RDSS). A radio communication package aboard a commercial satellite will provide an initial one-way service capability, where vehicles transmit position and short messages back to a hub. The system will be used by trucking firms, utility companies, and others to locate objects and vehicles. In addition, a low speed two-way data transmission scheme is planned at the time that Geostar's own satellite system is implemented. Detailed information on RDSS systems and applications can be found in [Rothblatt, 1987].

Looking beyond position location, planning has begun for a *mobile satellite service* (MSS) to provide real two-way voice and data communication with vehicles and remote sites. In many ways, the MSS system will appear to operate like the cellular radiotelephone systems now available in large cities. Only one or two MSS satellites can operate in North America due to the omnidirectional coverage of the vehicle-mounted antennas. This has caused the FCC to direct the formation of a consortium for jointly implementing the system. Established in 1988, the American Mobile Satellite Consortium (AMSC) will develop and implement the MSS for the U.S., with initial operations planned for the early 1990s.

2.7.4 Cellular Mobile Telephone Service

Mobile telephone service in the U.S. prior to 1980 had been mediocre due to inadequate frequencies, noisy reception, and primitive equipment. In early systems, prior to 1960, calls were set up manually by operators. Expanded capability in the form of Improved Mobile Telephone Service (IMTS) was introduced in 1969. Capacity was still limited in IMTS because it depended on a single high tower to reach all mobile users in the metropolitan area being served. The cellular concept was studied between 1970 and 1980 by the FCC, Bell Telephone Laboratories, Motorola, and various Japanese companies. The frequency band selected is between 800 and 900 MHz (overlapping the upper part of the TV-UHF spectrum), providing 40 MHz, divided into 666 channels. This bandwidth is used several times over in a given metropolitan area, multiplying the number of channels, and thus increasing the capacity of the system. Now that the technology is mature and the FCC has made the necessary spectrum available, cellular service is establishing itself. Particularly in Los Angeles, where the freeways and automobiles dominate, cellular telephone service is popular among professionals and consumers alike. A quick survey of service rates in Los Angeles indicates that the cost of using cellular is relatively high for the average individual, but attractive for business purposes.

The conceptual topology of a cellular telephone system for a metropolitan area is shown in Figure 2.18. At the center of the hexagon-shaped area of coverage is a cell site containing a radio transmitting-receiving station capable of simultaneously establishing communication with a number of vehicles. The actual shape and size of a cell depends on local terrain and the height and transmitting power of the antenna at the center of the cell. A pair of frequencies is required for each conversation, so the capacity of the system depends on the number of pairs available

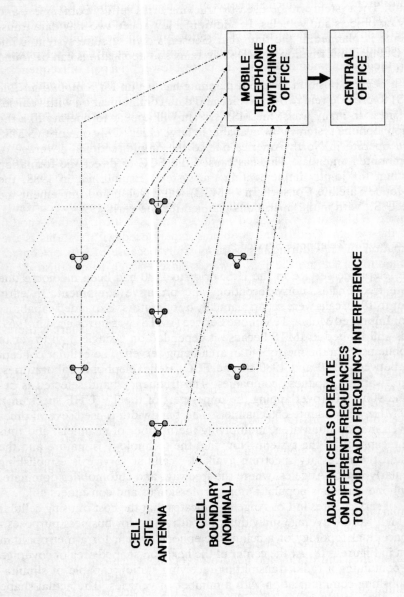

Figure 2.18 General arrangement of a cellular radio telephone system in a metropolitan area.

and the number of cells. Adjacent cells must employ different frequency channels to prevent interference. Cells which are noncontiguous and which are isolated from each other by terrain, however, can use the same channels (i.e., frequency reuse).

The operation of the cellular network is automatic and the mobile subscriber theoretically could make and receive calls as if using a normal fixed telephone. Within a cell, the car communicates with the closest tower on a pair of frequencies selected by the network control computer. There is a full-time terrestrial link from the tower to the *mobile telephone switching office* (MTSO). The MTSO houses the system that remotely controls the cell sites and a telephone switch that connects mobile subscribers to fixed telephone subscribers and each other. Another terrestrial trunk line connects the MTSO to a serving telco central office. This allows calls to be set up with local fixed telephone subscribers and those reached by long-distance.

As the vehicle moves across the boundary of a cell, a new pair of frequencies must be assigned because the vehicle will soon be out of range of the currently employed tower. This process is called *hand-off* and is critical to the smooth operation of the system. One of the problems experienced with systems is the abrupt termination of calls during hand-off, possibly because of loss of contact with the former tower before the new tower is close enough to establish clean communication. As more towers are added to a network, hand-off can be better controlled. With additional towers, cell cites will be crowded more closely together. This raises the specter of *radio frequency interference* (RFI) because of reduced spacing between towers operating on the same frequency channels. There is therefore an upper limit to the aggregate channel capacity that may be employed in a given metropolitan area.

To promote competition in the U.S., the FCC has authorized two cellular telephone operators for each metropolitan area. One of the operators would be the telco, referred to as the *wireline* carrier, and a second *nonwireline* carrier is licensed. Successful applicants for nonwireline licenses were selected by the FCC, including consortia in the mediumsized markets. After these licenses were granted, there was a wave of sales of nonwireline carriers and their licenses, resulting in consolidation to fewer, more financially solid operators to implement cellular networks. In fact, RBHCs are purchasing nonwireline franchises in distant metropolitan areas.

The primary purpose of cellular telephone is to provide good quality voice service to vehicles. The actual audio channel is analog in nature, and so data can pass at rates up to approximately 2400 b/s. Data transmission can be disrupted by propagation anomalies and during the hand-off process, which can last for a significant fraction of a second. (This aspect is considered further in Chapter 9.) Hand-held cellular telephones are available, making the service useful for a variety of other purposes. A cellular telephone can be installed within a building to allow calls to be placed, even if a disaster disrupts power and utilities. Another important feature is the "rove" facility, wherein a mobile subscriber can call from a metropolitan area not served by his or her carrier. With prearrangement, the subscriber can drive to the distant city and make cellular calls, which are eventually billed to the home account. Cellular telephone service represents a significant step forward in the kind of practical communication that anyone can employ and enjoy. Improvements will eventually come in the form of all-digital channels, which accommodate more conversations in the available bandwidth.

Chapter 3
Business Applications

The business network is one which meets the internal *administrative* or external *strategic* needs of an organization. This chapter reviews the functions that strategic units perform with telecommunication networks. The basic categories of applications include voice, data, and business video. Whether these applications are delivered by public or private networks depends on the choices and capabilities of the strategic unit that requires those applications. An interesting case is a subunit of a larger organization which meets its needs through the private telecommunication network of a larger parent organization. The parent then operates much like a public network with the subunit effectively being a "customer" for the services it requires. These issues are discussed in more detail in Chapter 8.

Throughout this chapter, the emphasis is on the "what" rather than the "how" of business communication. Using the analogy of a wrist watch, we will be speaking of what the use of a watch is (i.e., to tell time) and not how a watch works (i.e., with a battery-operated synchronous motor tied to a crystal reference oscillator). There is nothing wrong with understanding the technology of PBXs and modems, but the fundamental goal of this chapter and much of the rest of this book is to develop a framework for the reader which he or she can use to develop and to implement a telecommunication strategy. The references in the bibliography offer source material for studying the design and workings of the various classes of electronic systems used for telecommunication.

3.1 PRIVATE BUSINESS COMMUNICATION

Business communication networks represent a means to an end, not an end in themselves. Therefore, this category of network is the principal topic of discussion throughout the book. The next section defines the use of such networks for administrative purposes, while the section that follows deals with the strategic variety.

3.1.1 Satisfying Administrative Needs

In any organization, administrative communication capabilities are needed simply to do daily internal operations. For example, a parts-ordering and inventory-control system of a chain of discount stores is a fundamental element of the "back office" operation of the business. Such a system uses either a central computer or a network of minicomputers linked with data communication. Access to the system to place orders can be accomplished at warehouses and distribution centers around the country, employing remote computer terminals and the public telephone network. In the area of interpersonal communication, every medium to large sized company needs an internal telephone system composed of extension phones, intercoms, and some of the newer capabilities like conferencing and voice mail. The positive side of the administration function is that the situation is generally static, and people or machines can be efficiently used by following clearly defined procedures. The negative aspect is that even a well developed administrative system can be inflexible. When business conditions are rapidly changing (as is often the case in a competitive economy), administrative systems may not allow the organization to be responsive enough to cope with a changed environment. Fortunately, the information technologies (computing and communication) are inherently flexible, and if the administrative systems are properly constructed, they can be reprogrammed conveniently to respond to change. The human side of the equation requires a mix of creative individuals with systems and procedures for training of appropriate members of the strategic unit.

Creating the network in the first place is not an administrative function. This is because a lengthy design process, discussed in later chapters, must be undertaken before the requisite facilities can be obtained. The actual running of a private telecommunication network should be more administrative in nature. With the patchwork that exists today for assembling the elements of a private network, running telecommunication within a strategic unit is not a simple matter. There is some hope that eventually the digital backbone terrestrial networks of the telcos and long-distance companies will provide a foundation for the preponderance of administrative communication needs.

3.1.2 Satisfying Strategic Needs

Strategically oriented private networks often employ the same types of telecommunication facilities as do the administrative versions. The difference lies in the relationship between the network's functions and the organization's strategy. As an example of a strategic network, a reservation system is a powerful component of an airline or automobile rental company, allowing customers to book orders for units of service (seats on flights, or cars on a certain day at a specific rental location). Without a modern computer-based data network, such an organization can neither make sales nor manage its assets and personnel. Basic service dictates that a customer may make a booking by placing a telephone call (toll free) from any telephone in the country or possibly the world, and agents (reservation or travel) can confirm the booking while still on the telephone. Systems to do this can be purchased as a package from major vendors like IBM and Unisys, and consequently any organization with adequate financial backing can set up a reservation system capable of serving an airline or rental car agency. Recently, however, major service providers have added features to their reservation systems, which cannot easily be duplicated or purchased from outside sources. Competitors have sought help in the courts because they believe that they will never catch up to rivals with such a substantial advantage.

Commercial business units typically seek a competitive advantage, which, as discussed in [Porter, 1985] and Chapter 8, the unit can obtain if it is able to give customers more value for the price of its goods or services than competitors can. The primary benefit of a sustained competitive advantage is the ability to maintain a desired profit margin. Information technologies are attractive ways of gaining efficiency and reducing costs, potentially allowing the unit to become a cost leader in its industry. A business unit may need a private telecommunication network to implement its strategy, but, since others can copy the network design, the strategy may not yield a competitive advantage. Having the network may become a necessity as the stakes are raised by the more innovative competitors.

Competitive advantage comes into being through technical innovation, which can be protected from imitation by trade secrets, sheer investment, or a unique synergy with other units. For example, any airline can build a reservation system, but a competitive advantage may be realized if an airline purchases a rental car company for the purposes of leveraging the market power of each to aid the other. The airline and automobile reservation systems could be redesigned to allow joint bookings to be tied such that they would offer the traveler more service for his or her money. A customer may be encouraged to use the captive rental car company when booking an airline seat or *vice versa*. Units need a thorough understanding of the competitive environment in the particular industry in combination with a comparable understanding of the capabilities of private networks to do this effectively. Combining telecommunication network capabilities with organizational synergies may yield a competitive advantage which is not easily eroded by others.

3.1.3 Distinction from Public Networks

A private telecommunication network for administrative or strategic purposes is considerably different from one conceived to serve the public. (More specific characteristics of public networks were presented in Chapter 2.) Basically, a business network serves specific needs of a single strategic unit or a related group of units. The network itself is not operated to generate revenue from telecommunication services, and consequently will not be expected to show a profit. The concept of the "cost center" (as opposed to a "profit center") often applies to the private business network, meaning that the operation of the network represents a business expense. This categorization does not diminish the importance of the network for doing business because the goals of the strategic unit will probably not be realized without the network being available.

Although the network may be a cost center, users of the network are charged, directly or indirectly, for employing network capabilities (e.g., making telephone calls, transmitting facsimiles of photographs, or accessing a remote computer). The distinction between operating as a business or a cost center is blurred when charges are in excess of costs, resulting in a "paper profit" for the network. Be careful when examining such paper profit, particularly if the charges are not competitive with comparable services on public networks.

A large private network often has more capacity than needed by the implementing organization. This state offers the possibility of selling excess capacity to others, generating additional cash flow. Customers who require telecommunication network services could be attracted by low incremental prices. The network owner is presumably paying for the bulk of the expense, and is therefore motivated to get whatever revenue can be claimed, provided that significant additional expense or investment is not required to serve external users. This usually prevents reselling from being successful for the owner of the network. Sharing of excess capacity has been rewarding for some, however, and this is one of the primary network strategies reviewed in detail in Chapter 8. In general, the key to success is for the network to be versatile enough to serve a variety of needs, as they exist today and as they may exist in the future.

3.2 VOICE COMMUNICATION

A common denominator of every organization is voice communication. Since the inception of the worldwide telephone system, people have learned to use and to depend upon the telephone. Many organizations exclusively interact with customers by telephone, while others employ the telephone in the administrative background. For the local and long-distance telephone companies, business telephone usage is an extremely important source of traffic and revenue.

Aside from our familiarity with the use of the telephone, the other major advantage of the telephone network is its reliability and predictability. This author learned early in his career in telecommunication that nobody will say thank you if the telephone works well, but watch what happens if it does not work when needed!

The nature of applications using voice communication is continually changing, basically because of the sources discussed in the previous two chapters. The most basic form of voice is called *plain old telephone service* (POTS). The coined phrase "plain old" could mean that we really did not need a precise definition because everybody already knew what it was. The complexity of POTS, however, comes about from the new and exciting things that strategic units are doing and want to do through the telephone network. These known applications and many of the newer possibilities are reviewed in the following paragraphs. More detailed technical information on telephones and telephone systems can be found in [Noll, 1986].

3.2.1 Telephone Subscriber Equipment

There is a wide variety of user devices which can be connected to a conventional telephone network. In the home, we commonly find two or more regular telephones and perhaps an answering machine. Cordless telephones are also popular for people who like to roam around their home, but the office environment is much more diverse. A distinction is drawn here between telephone subscriber units and other devices, such as modems and facsimile machines, which employ the telephone network as a means of delivery.

Traditional telephone instruments convert sound into electrical signals with essentially the same frequency content. Noll gives a complete description of the operation of the instrument [Noll, 1986]. As the digital communication network is extended to the customer's premises, we can expect that the old voice instrument will be replaced with a new class of device that digitizes speech. Another important aspect of telephony is the manner in which the subscriber activates the local loop, which is the process called *signaling*. A subscriber lifts the handset from its cradle (going *off hook*); this signals the local switch (either a PBX or the local public exchange) that he or she wishes to place a call. The off-hook terminology refers to the oldest telephone instruments, where the receiver is held by a mechanical lever (the hook) and the transmitter is fixed on the base. The local switch responds with *dial tone* and the subscriber (the *calling* party) can then enter the sequence of digits to identify the *called* party. This is the familiar process for placing an outgoing call. If the called party's telephone is free and on-hook, the switch rings that telephone and the calling party hears the soft ringing tone. Conversely, the telephone system indicates that the called party's telephone is in use with the "busy" signal. During this process, the telco monitors the activity and automatically records selected information for the purpose of billing the appropriate subscriber.

Actual loud ringing of a telephone is produced by several volts of ac electricity at 20 Hz, which is sufficient to cause the bell or buzzer on the telephone instrument to sound. Removing the handset closes a circuit which indicates the *answer* to the network. The calling party hears the ringing tone stop, followed by the called party's voice, confirming that the circuit is complete. This is the protocol that we all automatically use to establish a good end-to-end connection for voice communication, which is analogous to the on-line process used in data communication networks. At the conclusion of the call, either or both parties hang up. This initiates the release cycle at the switch to ready the intervening facilities for additional calls.

The telephone subscriber relies on the telco for all management and operation of voice service. A subscriber's primary obligation to the telco is to pay the monthly bill when rendered. In many foreign countries, the subscriber is expected to contribute to the capitalization of the telephone network by paying a rather hefty installation charge (typically in the thousands of dollars). Capitalization of the local telephone networks in North America is the responsibility of the telco, which has little difficulty raising the necessary funds due to its protected monopoly status.

With the digitization of the local telephone network, subscribers are able to do things with their telephones that were hardly imagined even twenty-five years ago. Call waiting, speed dialing, forwarding, and redial are useful and familiar features available to subscribers throughout the U.S. Advances in signaling and programming of telco switches are the keys to future innovations in telephone service, particularly as the capabilities of ISDN are brought into the picture. The greatest value to subscribers, however, whether at home or in the office, is the ease of use of the telephone system to permit direct person-to-person communication such that even a child has no difficulty learning the operation of the equipment and the procedure for making calls.

3.2.2 Switching within a Building or Facility

The local telco gives subscribers fundamental access to the "outside world." For a small business with two or perhaps three people working in the same or adjacent rooms, no other form of electrical communication is required. Business locations, however, usually involve many offices, a number of floors, and even separate buildings, so direct physical contact for every communication is much too inconvenient. If different offices could have their own telephone lines with individual numbers, then people would call each other as if they were in different houses. The difficulty with such a system is that the organization would have many different numbers, and unless outsiders knew exactly with whom to speak, important calls could be lost. Furthermore, the bill for using the local telco would be amplified by the number of telco loops (one per extension).

Having a telephone which can access two or more different lines provides some of the needed flexibility. An intercom is added to permit one person to signal someone else (e.g., a secretary or attendant). Telephones with this type of capability are called *key telephones,* referring to the keys along the bottom of the instrument. A key telephone with several lines and pushbuttons allows the user to answer and to talk on one line, observe another line being rung, and then put the first line on "hold" to answer the second. A complete discussion of key systems is provided in the next section.

Obviously, a better solution to the problem of internal communication and common access to the outside world is to implement a switching capability within the office environment. A manual switchboard placed on a customer's premises was called a *private branch exchange* (PBX). This author's first real job was as a switchboard operator at a summer camp in the Catskill Mountains of New York. Camp housing and service facilities had individual extension telephones off the PBX, but the camp had only two outside lines. Even after the public telco network became automated, PBXs were still used by virtually every business in the country. Manual switching was inefficient, however, because the "stations" within the office walls required operator assistance to make outside calls and to be connected to one another.

Part of the problem of providing telephone service in buildings and campuses is the installation and maintenance of the actual wiring. As a result of divestiture, subscribers now own and must maintain the wiring in their buildings. Basic telephone wiring is done with copper cabling which must be run between each extension in every complex. Each point of connection to an instrument, called a *termination* or a *drop,* provides at least three wires for POTS. The cables run through walls and ceilings, and eventually end up in a telephone room or cabinet, where terminal blocks are available for connection to PBXs and telco loops. A systematic approach to the wiring system and careful record-keeping are important. Installation of additional wiring and terminations along with trouble shooting of problems can be time-consuming and expensive in terms of labor. Therefore, many organizations contract with outside sources, particularly the local telco, to maintain telephone wiring. Separate wiring is typically required for data communication circuits between computer terminals and the associated data processing equipment. One of the principal benefits of ISDN will be the use of one termination of a single cable to provide voice and data services. Currently, the technique of superimposing data signals over the voice bandwidth (data over voice) allows standard wiring to perform two services.

The connection of telephones over the wiring system of a building or office complex is by way of some form of circuit switching facility dedicated to the specific needs of the organization. Depending on the number of extensions and the specialized services needed, the business customer can either own the switch (e.g., key systems and PBX) or obtain it as a tariffed service from the telco (e.g., CENTREX). Modern key systems, PBXs, and CENTREX service allow office workers to do much more with their telephones, as discussed in the following paragraphs.

3.2.2.1 Key Systems

By the time the telephone became commonplace in the early part of this century, the confusion of multiple office phones and lines was something to behold. The first key systems were literally wired together by users themselves. Eventually Bell System engineers cleaned things up with the key telephone, allowing several extensions to be connected to one or more external lines. Typically, a key system services from approximately five to fifty internal stations and five to twenty outside lines. Another approach to using key systems is as a front-end to CENTREX service from the local telco. CENTREX, which provides features of a PBX, is covered in a subsequent section. The arrangement of key system elements is illustrated in Figure 3.1. The technology was rudimentary by today's standards, but electrical relays and mechanical switches made the system both rugged and reliable. The *key service unit* (KSU) performs the switching between telephone extensions and the outside world. Various features like public address (paging), music on hold, and conferencing are contained within the KSU.

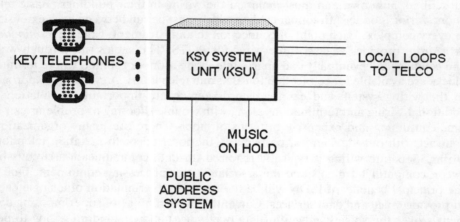

Figure 3.1 Basic key system arrangement for standard hardware (circa 1970).

The older electromechanial systems have now given way to all-electronic key systems that use transistor switching and, in many cases, microprocessor control. Photographs of a typical set of key system components are given in Figure 3.2. In current technology, the voice frequency signals between the telephone stations and the electronic KSU (or "control unit") are analog in nature. The control signals for signaling, dialing, and ringing, however, are typically digitized to take advantage of microprocessors and software within the control unit. This implies that conventional telephones cannot be connected to the system. The particular types of telephones available as stations, however, often have features not available on

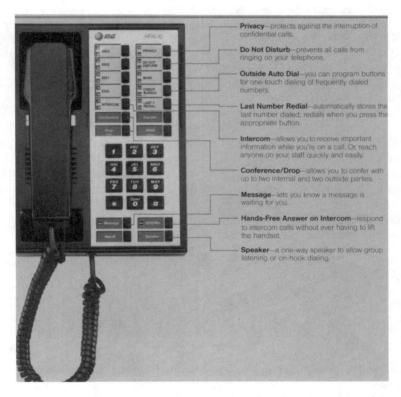

Figure 3.2(a) An example of station equipment from the AT&T Merlin Key System. (Photograph courtesy of AT&T.)

instruments purchased from a department or discount store.

A typical key telephone may have two to five line buttons. Intensive telephone activity involving a station with dozens of incoming lines, however, is common for stock and commodity brokers. A device called a "turret" is a key phone with as many as 250 line buttons. This allows the broker to accept and to hold many calls at the same time (one may even be answered).

Among the basic features of a key system is the ability to answer an outside line from any phone along with the ability to put a calling party on "hold" while doing something else (like answering a different line). Intercom facilities are also a basic need. To accommodate a variety of possible features, some manufacturers have been able to package inexpensive key systems without a central KSU. Each station therefore has the requisite "smarts" to perform the appropriate functions. The features discussed in the next paragraph, however, are only available on systems with a KSU or control unit of some kind.

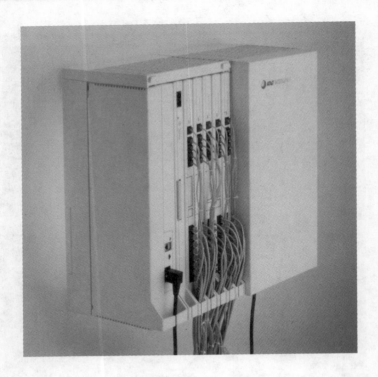

Figure 3.2(b) The Merlin II Key System Control Unit Installation. (Photograph courtesy of AT&T.)

Advanced key systems are able to perform telecommunication management functions that were inconceivable as recently as a few years ago. More sophisticated models offer *least cost routing* (LCR), *toll restriction,* and *station message detail recording* (SMDR). Each of these facilities of key systems and PBXs is explained in the following paragraphs.

Least cost routing was implemented in response to the availability of alternative long-distance services. Given a set of outgoing long-distance trunks, a computer in the KSU monitors the dialed area code and selects the particular service with the lowest expected cost for placing the particular long-distance call. Information as to the destination of the call is provided by the dialing party through the area code of the called party. Preprogrammed into the computer in the form of a routing table are the alternatives (AT&T, MCI, private line, *et cetera*) and the order in which they are to be chosen by route. For example, the routing table might have MCI as the first choice for the most distant area codes, while AT&T would be selected elsewhere. LCR was originally offered as an add-on box, but has since been diffused into key systems, PBXs, and CENTREX.

Figure 3.2(c) The Merlin Call Management System employs an AT&T Personal Computer. (Photograph courtesy of AT&T.)

Toll restriction simply means that some of the stations on the system will not have the ability to place long-distance calls. Typically, a toll call is indicated by the exchange within the same area code, or by the digit 1 for calling outside of the area code. When dialing from a restricted station, the KSU responds with a regular busy tone or a congestion tone (fast busy tone), indicating that the type of call cannot be placed. Station message detail recording is an accounting feature wherein a hard copy printer attached to the KSU can be used to run off a report of actual activity by station. This informs the telecommunication manager (or, more likely, the business manager) who is placing which calls, an important means of controlling usage and therefore cost.

The leading manufacturer of advanced key systems in the U.S. is AT&T. Their product line, shown in the photographs of Figure 3.2, is named Merlin™. Several generations and various configurations of the Merlin™ key system exist, and AT&T continues to upgrade the line in response to user requests and competition, particularly from Japan. Note that LCR and SMDR are performed by an external AT&T model 6300 personal computer.

an external AT&T model 6300 personal computer.

The advent of more powerful and cheaper microcomputers has now made possible the provision of many or all the features of a PBX within the KSU of a key system. Such a key system, which is also like a PBX, is called a *hybrid telephone system*. Ultimately, the key system might be at the low end of a full PBX product line. This would be advantageous because it should be relatively easy to upgrade a system in a well integrated product line. More information on PBXs is provided in the next section. Perhaps one important distinction between key systems and PBXs is the possibility of interfacing the customer's facility directly with the telco's switch at the T1 level. PBXs can also function as tandem switches to permit alternate routing over the private trunks. While hybrids may not offer direct T1 connection and tandem switching, they can do some of the more advanced jobs of integrated PBXs such as simultaneous voice-data communication. Based on the experience to date, organizations are apparently selecting the advanced hybrid key systems over the more traditional variety.

3.2.2.2 Private Branch Exchange (PBX)

The capabilities and capacity of standard telephones, key systems, and manual switchboards are inadequate for meeting the voice communication needs of medium or large sized organizations. Decades ago, automatic exchanges using electromechanical technology were installed as PBXs for large users, and AT&T introduced CENTREX service to respond to growing needs (as discussed in the next section). Compact electronic switches were developed by firms other than AT&T, squeezing considerable capability into an affordable package, which could be purchased by a medium to large firm. When these electronic units first appeared in the 1970s the term *private automatic branch exchange* (PABX) was coined. Today, the manual and electromechanical PBXs have all but disappeared so that there is no ambiguity when using either term in reference to the modern generation. We will continue to use PBX throughout this book. A photograph of major components of a typical PBX system, which provides in the range of 100 to 10,000 lines or more, is shown in Figure 3.3.

The capabilities of the first electronic PBXs were far beyond what was possible with CENTREX. Its principal purposes are to provide switching of calls within an organization and to permit calling to and from the outside world. Many more features and capabilities are included, allowing the PBX to play an important role in the satisfaction of an organization's strategy. The programming of the computer is very complex because the control functions must respond to the dynamics of telephone calling, answering, signaling, and a host of other functions, which are expanding in scope.

This author's introduction to the modern digital PBX came in 1976 in the Republic of Indonesia. Working on a project installing a national satellite communication system, we required an internal telephone switch within the master earth station near Jakarta, the capital and largest city in the country. Our implementation group had selected a seemingly unimportant piece of equipment from a small company called Digital Telephone. Although we were turning up a modern analog network, our little PBX was totally digital, providing the necessary analog-to-digital (A/D) conversion inside of a small cabinet. The digital switching feature was not what intrigued me so much; rather, I was awestruck with what the little machine could do. Working long hours, my team used the innovative features of that switch in the simple coordination of activities at the master station of the network. One feature was the ability to break in on a call in progress. When someone was breaking in, it was announced with a soft beep so that the talkers knew they were being interrupted. This, among other features, significantly increased our productivity and efficiency (which was important with our 12-hour days and seven-day weeks of intense work at the station).

Companies now enjoy and expect the most advanced operational features of their PBXs, not the least of which is the ability for easily upgrading the system for the next batch! The PBX has seemingly become the new corporate status symbol after the mainframe computer. Particularly noticeable in an advanced system is the telephone instrument itself. As shown in Figure 3.3, the "high-tech" look and feel of station equipment is one of the biggest selling points. Features found on stations include multiple buttons, speakerphone, display of outgoing number dialed, the number of the incoming call, messages, and even a data terminal. This last feature relates to simultaneous voice-data capability, which is now included in some PBXs and larger advanced central office switches. Conversion to ISDN interfaces is also available.

Additional programming and add-ons permit a PBX to perform roles well beyond simply allowing people within a building or campus to talk to one another. As organizations grow in size and spread out, they need to expand the internal telephone system. The modern PBX conveniently accommodates this growth by building a voice network with individual switching nodes at major locations. Certain PBXs are therefore used as tandem switches to route calls without direct trunks between final PBXs. Another important feature is a direct T1 interface for employing advanced trunk facilities such as fiber optics and satellite links. (The networking aspects of PBXs and other such devices are covered in detail in Chapter 4.) Most importantly, the PBX can be the fundamental switching and control element of business communication, providing the facilities and services that individuals within an organization use on a daily and hourly basis. The using end of the PBX is on the desk of each member of the organization, next to the bed of

Figure 3.3 The Focus 9600 PBX which is capable of simultaneous voice and data switching. (Photographs courtesy of Fujitsu America, Inc.)

the hotel guest, or connected to the voice-data terminal of the airline reservation agent. From an evolutionary standpoint, the PBX represents the most direct growth path from the humble beginnings of the telephone to the most advanced features of ISDN.

An important aspect of the PBX is its design as a system owned and operated by the user, not by the telephone company. Therefore, the programming and maintenance of the system, including the wiring, must be straightforward and within the capability of people trained by the supplier of the equipment or the organization itself. The fundamental type of programming called "adds, moves, and changes" is simply the process of adding a new station, relocating a station from one office to another, or changing a number for a station which is currently installed. To accomplish this with the telco has traditionally been a long drawn out process, but telcos and their switch manufacturers are introducing customer controlled adds, moves, and changes into CENTREX service. A PBX is provided with an operator terminal, through which connections can be conveniently specified and revised. This uses the internal computer of the PBX, running a built-in program to modify entries in memory locations. Other uses of the terminal and computer are to perform the SMDR and LCR functions in key systems. A PBX can work as an automatic call distributor for telemarketing applications, which is discussed in a subsequent section.

The SL-1, manufactured by Northern Telecom, Inc. (NTI), has been one of the most popular systems. NTI gained large market share in the U.S., even though it is a Canadian company, probably because it began heavily exporting digital switching systems before U.S. companies had fully responded to deregulation. Another company with a good position in the U.S. market is Fujitsu America, a subsidiary of the largest computer manufacturer in Japan. The Focus series of PBXs makes good use of ISDN features which Fujitsu developed for the Japanese market. Two other important manufacturers are AT&T and IBM. AT&T has a substantial market share in the U.S., and their PBX systems continue to sell well. The System 75 is designed to cover the lower end of the range, between key systems and the largest machines. Large capacity and advanced digital architecture are offered in System 85, which is also aimed at the evolving ISDN-upgrade marketplace. Some corporate customers, like American Express and General Motors, have purchased AT&T's central office machine, the No. 5 ESS. Clearly, this is only justified for major corporate installations. Looking at IBM, their introduction of the 9751 PBX builds nicely upon the foundations laid by the Rolm unit, which they acquired to improve their position in voice systems. The new PBX appears to integrate well with IBM's computer environment, simplifying functions such as network management. The possibilities and particular needs of the organization must be properly evaluated before selecting a source. (The information contained in Chapter 7 should be helpful in this regard.)

In addition to the useful and attractive features of the PBX, the buyer must consider a number of risks when deciding on this option. First and foremost, committing to owning a PBX means that the organization could become "locked into" a certain technology. Vendors attempt to convince customers that their PBXs can be easily upgraded and the investment is never really "sunk." Users fear that a simple conversion may actually require the complete replacement of a major element of the PBX, perhaps the PBX itself (called a "fork lift upgrade"). CENTREX service, discussed in the next section, has recently undergone a metamorphosis as features previously only available in digital PBXs have now become standard. When owning a PBX, the organization takes on the responsibility for keeping the telephone system working. As we mentioned previously, if the system works, nobody says anything, bad or good. If the chief executive does not get dial tone, you can imagine where the blame is directed! An uninterruptable power system (UPS) and redundancy in the telephone system, where applicable, become important to have.

3.2.2.3 CENTREX Service

CENTREX, a contraction of *central exchange,* is a private telephone service offered by the local telco. All of the switching equipment, which is located either in a central office or on the customer's premises, is owned, operated, and maintained by the service provider. Essentially all modern PBX features can now be obtained from the telco by using their switching offices and wiring. Obviously, the telco will provide these features as a business proposition, and you can expect the charges to be comparable to the cost of ownership. The principal advantage of using CENTREX is that significant capital is not committed to telephone service. Another possible benefit is that the risk of technical obsolescence is eliminated. PBX vendors will tell you, however, that their systems can be upgraded with new hardware and software to provide any new capability on the market. Modern CENTREX has an incredibly wide array of features and capabilities. The central office also provides the necessary interconnection among local trunks, private lines, and long-distance services. There is the added benefit that all maintenance is provided with the service, raising the customer's confidence level that someone will be available to resolve problems and equipment outages. A radio advertising campaign of Pacific Bell, being conducted in California, emphasizes that the equipment will be located in the central office and not on the customer's premises, making it more convenient for the telco to keep the system working properly.

CENTREX was first introduced in the 1950s when users demanded that automatic switching be made available for internal telephone service. AT&T was able to adapt the newest of their analog switches because of the introduction of stored program (computer) control. The needed feature was the ability to identify a call placed from within a particular customer's organization to another line in that same organization. This would permit abbreviated dialing, using the last four or five digits. Another important feature is *direct inward dialing* (DID), wherein each station within the unit has its own seven-digit number which can be reached directly from the outside world. *Direct outward dialing* (DOD) was also provided, allowing stations to reach outside lines directly, without going through the switchboard operator (renamed as the "attendant").

CENTREX service was essentially the private domain of AT&T until divestiture, after which it was transferred to the BOCs. The larger independent telcos also offer CENTREX. With the rapid expansion of PBX marketing and sales, CENTREX appeared to be headed into decline, and possibly would become a relic of the past. The BOCs installed modern, all-digital exchanges, however, and began to offer services comparable to those of the more advanced PBXs, and many business customers began to recognize tangible benefits. Leading manufacturers of the large central office switches, including AT&T, Northern Telecom, Siemens, and Fujitsu, are making sure that telcos can compete in the marketplace for private switching. One particular area where they have improved is in the continual upgrading of hardware and software of the switch for new CENTREX capabilities and in allowing users to control the software features directly, using remote terminals. For example, Pacific Bell now includes Centrex Management Service in the package, which extends a variety of telephone management functions to a video display at the customer's location.

The class of customer with the most to gain from CENTREX is apparently the small business user with a requirement of less than 100 internal lines. With the most modern central office exchanges, particularly those made by AT&T and Northern Telecom, the telco is able to segregate such a small group of distinct lines and still offer a range of PBX-like capabilities. Users receive full accounting of calls under *station message detail recording* (SMDR), can make changes in line assignments and locations with an on-premises PC, have the ability to program automatic route selection (akin to LCR), and can use station equipment with the "bells and whistles" of the larger PBXs. The switches also permit calls to be transferred and automatically forwarded to other stations. Several manufacturers, in fact, sell telephones and attendant stations which interface with CENTREX to provide particularly smooth control of the dialing and routing capabilities.

The CENTREX promoters (telcos and CO switch manufacturers) have considered the future by now offering data switching over telephone wires. As an early introduction of some features of ISDN, data switching could be an attractive way for small to medium sized companies to get a feel for the possibilities. A particular type of data communication network capability is called the central office LAN (CO-LAN). Employing the telco equipment in the CO and the wiring between it and the user location (i.e., data over voice), the CO-LAN simplifies the problem of interconnecting computer equipment located on multiple sites within a campus or possibly a metropolitan area. The CO-LAN is actually an entirely separate data switching capability, using either time division or packet switching within the telco's plant. Equipment of this type can, of course, be purchased and operated by the user (see Chapter 4). Perhaps the CO-LAN is but a precursor to ISDN, which allows voice and data to be routed over the same wires (i.e., 2B + D) and uses switching capability within the CO. Another useful feature available on many CENTREX offerings is voice mail, which is discussed in some detail in the next section.

Because the telco is able to invest in and operate the largest switches on the market, major corporations and government agencies also find it attractive to use CENTREX. This can make sense, even though these organizations can afford to purchase their own PBXs. As the CENTREX market continues to be defined and expanded, the future of the PBX as a generic telecommunication device may be somewhat limited. Regulatory actions of the FCC and courts are also having a decided effect on the future of CENTREX *versus* PBX, particularly if large access fees are to be levied on owners of PBXs. Like any contest, the ongoing debate over CENTREX *versus* PBX will shift back and forth. The optimum choice for any particular organization will depend on factors unique to that organization. One factor currently in CENTREX's favor is the high degree of customer satisfaction that the service has achieved in recent years.

3.2.2.4 Voice Mail

The basic concept of the telephone answering machine has been advanced by several steps with a computer-based technology called *voice mail*. Other terms which are in current usage include voice store and forward, voice messaging, and audiotex. (There might be quantifiable differences among services with these different names, depending on the equipment supplier or telco.) In addition to saving messages, voice mail supercedes the answering machine by allowing users to save, modify, and redistribute messages. For example, the head of a division can initiate a message and send it to every department head. A particular head can append his or her own comments to the message and distribute it further down the organizational ladder.

By introducing voice mail, an organization with heavy telephone usage could reduce cost and significantly increase personnel efficiency. The technology represents a partial solution to the problem of "telephone tag," the familiar process whereby you leave a message for someone who then leaves a message for you, *ad nauseam*. The voice mail equipment can be purchased as a separate package and attached to either a PBX or CENTREX service. Newer PBXs even have voice mail capability available as an add-on, and the same goes for CENTREX in some areas of the country. While voice mail is currently a relatively expensive frill, the day will come when it is a standard business telephone feature like call transfer.

Unlike analog telephone answering machines which employ conventional magnetic tape recorders, voice mail relies on processing and storing voice messages in digitized form. Consequently, voice mail operates like a computerized messaging system. Interfacing voice mail with a digital PBX is also facilitated, and integration within the PBX or central office exchange can be accomplished.

The operation of a typical voice mail system is illustrated in Figure 3.4. The called party, who is not present to receive calls, has set his or her line to forward calls to the voice mail system. An incoming call is thereby routed to the voice mail system, as if it had been set up to go to a receptionist or message center desk. The calling party is connected and hears a message from a computer voice. In new systems, the called party's voice may also be employed. The caller is asked to leave a message at the beep. Alternatively, the caller can press a key on his or her instrument to route the call elsewhere to speak to a real person within the organization. A message is held in a sequential store for access by the subscriber. The presence of messages in the queue is indicated to the subscriber by a signal light or readout on the station unit, or by a distinctive tone heard on the line the next time the instrument is used by the subscriber. To listen to stored messages, the subscriber calls the voice mail line and enters his or her private identification number. (This ensures that mail is delivered to the right party.)

Voice mail systems have added features to increase the effectiveness of this application. A message can be sent to multiple addressees who may in turn relay the same message. The sender of a message can obtain confirmation that the message was received. Messages can be edited prior to transmission, and those already received can be stored for replay or subsequent editing. The technology of speech recognition, discussed in the next section, allows the subscriber to talk to the voice mail system and, by voice print identification, be given his or her messages without keying numbers. The possibilities for voice mail are limited only by the imagination, since computer technology permits the widest possible variety of processing features.

Voice mail can be installed within an internal PBX or CENTREX system. In addition, voice mail is being offered as a national service to allow anyone to send and receive messages. This is possible because essentially everyone in the U.S. has a telephone.

3.2.2.5 Speech Recognition and Voice Response

The ultimate user-friendly feature of a computer would be its ability to understand human speech. Computer recognition of speech has moved from the laboratory into the business world, albeit on a limited basis. There are now a few systems capable of recognizing words and phrases from a very limited vocabulary. This is adequate for allowing a talker to use a rotary or pulse telephone to enter numeric data into a voice mail or voice response system.

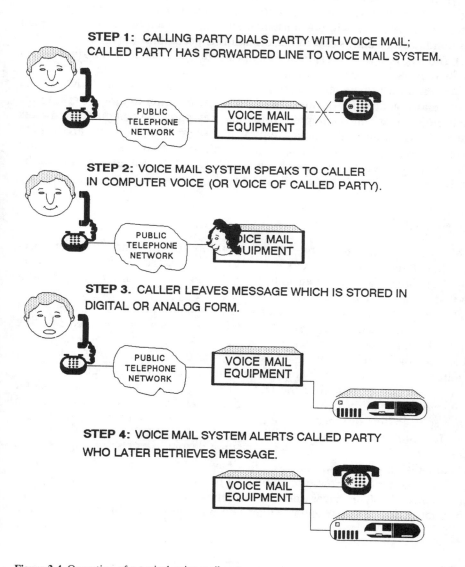

Figure 3.4 Operation of a typical voice mail system.

Voice response is a technology which combines aspects of voice mail and speech recognition. Most telephones now have touch-tone pads, obviating the need for speech recognition. If no pad exists, then speech recognition solves the problem. (The telephone network will not normally pass digits which are entered using an old-fashioned rotary dial, since the information is not within the voice frequency range.) The key pad, however, provides a powerful man-machine interface which is almost universally available. American Airlines has an interesting application for voice response to handle telephone calls for flight information (arrival times, news of delays, *et cetera*). A caller hears a computer-generated voice asking for the flight number, which is entered through the key pad. The computer then gives the caller the latest information on the particular flight. Other options are offered to the caller, and a numerical selection can be made to continue the "conversation." Alternatively, the caller can simply hang up. (After all, a computer would not be insulted!)

3.2.2.6 Autodialers

Many inexpensive telephones have the ability to store and use several frequently called numbers, so that this is fast becoming a standard feature of POTS. Autodialers, however, can be an indispensable tool in telemarketing operations which involve many outgoing calls.

A type of "hands-free" autodialer has appeared on the market employing a simple speech recognition system. These units have been incorporated into the high-priced cellular car telephones (an obvious use for them). The subscriber must first have his or her voice print established by the unit. Then, any of a dozen or so numbers can be dialed by saying the name of the person to be called.

Autodialers are applied to the trunk side of the private telephone system. This relieves the user of dialing the series of special digits needed for access and use of a specific long-distance network. One interesting configuration of the autodialer for trunk access allows the user to set up the system to operate as a virtual private network. This is not to be confused with the VPN and SDN service offerings of the long-distance carriers, discussed in Chapter 2. When a particular trunk is selected, the equipment knows the number of the distant location and directly sets up the connection. This saves the cost of a private line, but gives some of its benefits. Of course, the particular circuit will only be set up for the duration of the call.

3.2.3 Private Line Usage

Private telecommunication networks are synonymous with the use of dedicated communication links between dispersed locations of an organization. This is the domain of the *private line,* also referred to as dedicated line, private wire, and leased line. For the purposes of this book, these terms are interchangeable. A private line is distinguished from an access line to the public telephone network in that, while the telco and long-distance carrier may provide either, the private line does not use the switching capability of the public telephone network. Multiplexing, routing, and switching are typically done by the implementer of the private network, although private line carriers offer such capabilities as options. Application of private line therefore demands that the user either purchase the requisite equipment or obtain it on a lease or service basis.

Bandwidths of private lines range from 100 b/s for a teletype circuit to 3000 Hz for a conventional telephone line (capable of up to 19,200 b/s) to 56 kb/s up to T1 (1.544 Mb/s). Even greater bandwidths can be ordered, particularly if employing fiber optic or satellite communication, where 100 Mb/s or more is feasible. These wider bandwidths can be obtained, provided that the user commits for a sufficient period of time to justify special construction by the common carrier.

Private lines are often used as trunks between PBXs and CENTREX, allowing calls to be placed without entering the public network. There is obviously a trade-off between using long-distance services like MTS, WATS, and VPN, *versus* installing dedicated private lines. The procedure for performing this evaluation is fairly straightforward, and is covered in Chapter 6. For sensitive applications like data communication and high resolution facsimile, a private line may be the only means to obtain a transmission path with the requisite bandwidth and stability. The latter point has to do with the fact that a private line is established as an essentially fixed circuit from end to end, unlike a dial-up telephone call for which the circuit is set up along randomly selected channels for the duration of the call. More sophisticated modems deal with this uncertainty by automatically compensating for line irregularities through a technique called *equalization.* (Leased lines are constant and come with some form of partial equalization, depending on the particular tariff.)

Some of the common business applications of private lines are reviewed in the following paragraphs. The emphasis here is on the use of private lines to serve the voice communication needs of a strategic unit with diverse locations. Application of private lines to data communication is reviewed in a subsequent section.

3.2.3.1 Tie Lines between Locations

A tie line is a dedicated telephone circuit routed between two separate business locations. The tie line employs the regular wiring and transmission facilities of the telco, but bypasses the telco's switches in the intervening exchanges. As was shown in Figure 2.1, connections within the central office are made at the distribution frames. The tie line is connected across the distribution frames without passing through the switch. Typically, the tie line is terminated at the key system, PBX, or CENTREX to facilitate access to the line by personnel at both sites. The fundamental purpose of a tie line is to enhance communication and control, insofar as the tie line greatly simplifies calling. In the simplest arrangement, called a "hot line," illustrated in Figure 3.5(a), picking up the handset at one end causes the instrument at the other end to ring. The tie line can be used to establish an extension from a central PBX at a company headquarters, shown in Figure 3.5(b). This way, the remote site has a telephone which appears on the headquarters system as another telephone and can be dialed as if it were in the same building. Consequently, it becomes very easy to reach the people in the distant location.

(A) DIRECT "HOT LINE" BETWEEN EXTENSIONS

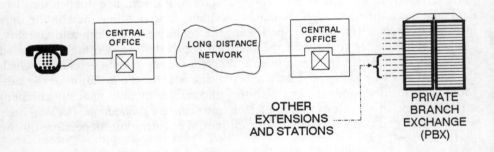

(B) REMOTE EXTENSION FROM PRIVATE BRANCH EXCHANGE

Figure 3.5 Application of telephone tie lines for communication with remote locations.

My first contact with the tie line concept came when I held a summer job with the company that my father had recently sold to a much larger firm. I noticed that on my father's desk there was a special telephone, which he explained rang directly on the switchboard of the new parent company. The parent company was located approximately 25 miles away and across the river from where we were. The telephone instrument on my father's desk was connected to a tie line which ran over New York Telephone Company's system from Long Island City in Queens to Varick Street in lower Manhattan. The connection at the parent company's manual PBX (*circa* 1963) allowed anyone at that location to call my father directly.

This tie line was an important link for exercising control of the newly purchased company.

3.2.3.2 Foreign Exchange Lines

In instances where an organization places a lot of calls into a distant city, a private line can be used to extend a normal connection to the local telco of that city. This type of private line is called a *foreign exchange line* (FX line). Conversely, an organization which needs to place many calls in a particular distant city can potentially save toll charges if it bypasses the switched long-distance service with an FX line. A variety of approaches can be used to implement FX lines. Satellite carriers like American Satellite (now Contel/ASC) and RCA American Communications (now GE American Communications) built up a substantial business in the late 1970s and early 1980s by offering private lines as FX lines. Today, the versatility of the digitized long-distance network is making it somewhat less attractive to lease permanent circuits for use as FX lines, due to enhanced bulk calling plans and virtual private networks offered by AT&T, MCI, and US Sprint.

3.2.3.3 Private Trunks

Another common and effective use of a private line is to establish a private trunk between PBXs or CENTREX facilities. Illustrated in Figure 3.6(a) is the *intermachine trunk* (IMT) used to tie together two PBXs. A private trunk can also be used to establish an FX line between a PBX and a distant central office, as shown in Figure 3.6(b). Private trunking is best applied in the largest of private telephone networks. The example given in Chapter 1 of the backbone network is one in which private trunking forms a foundation for substantial savings on both internal calling as well as calling through the network into the public system. This offers bypass of long-distance in instances where there are potential economies.

The technical properties and attractive pricing of leased T1 channels make them a logical choice for private trunking. As we mentioned previously, PBXs can interface directly with T1 trunks and can process individual DSO voice channels. In Chapter 4 we review multiplexing and digital access and cross-connect system (DACS) facilities which provide a flexibility not possible with analog voice frequency private lines. From an economic standpoint, T1 channels offer the opportunity to obtain bulk capacity at a relative discount. As we will discuss in Chapter 6, the challenge is fully utilizing the capacity through efficient loading of telephone and data traffic.

In larger private networks, PBXs can be connected by IMTs in a hierarchy much like the public network. Tandem switching, made possible by having various IMTs between PBXs and toll switches, is attractive because it creates alternate

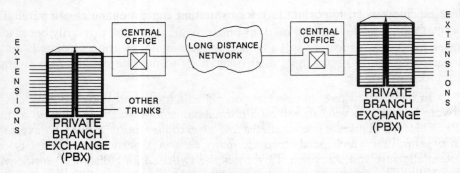

(A) INTER – MACHINE TRUNK CONNECTING PBXs

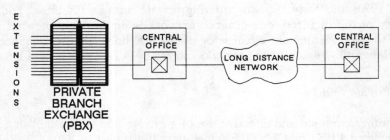

(B) FOREIGN EXCHANGE (FX) LINE

Figure 3.6 Private trunks between switch boards which are implemented with private lines from common carriers.

paths for telephone calls to take when the direct path is blocked or not functioning. The Electronic Tandem Network (ETN) is a service from AT&T for users who do not wish to own this class of switch. AT&T would provide the private lines along with the ETN services to implement a dedicated, private, switched long-distance network for major corporations and government agencies.

3.2.3.4 Special Access Provisions

A new arrangement is available from the BOCs as a result of divestiture and the development of ONA. Referred to as *special access provisions,* local loops can be provided to large corporations and other long-distance companies to provide new capabilities for employing the switches of the BOCs and long-distance carriers. The nomenclature is specified as Feature Groups A, B, C, and D. Feature Group A (FG-A) is basically a conventional access line provided by the BOC to a long-distance carrier without equal access; it uses a normal seven-digit telephone number and interfaces with a piece of electronic switching equipment such as a PBX. The FG-A type of access is being used in some private networks, primarily because of lower line costs, which are still charged by the minute. Feature Groups B and C are used by the BOCs to meet their obligations to provide equal access to long-distance carriers.

Feature Group D (FG-D) is set up to employ a single numbering plan throughout the country without consideration of local exchange designations and area codes. The technique depends on a database within the network to translate between the unique nationwide number system and the actual area code with the particular seven-digit number being called. In a private network, this allows calls to be automatically forwarded anywhere in the country where this type of access is available. In general, special access provisions with FG-A, FG-B, and FG-C are potentially attractive where available to private network operators because of reduced line costs for connections to the BOCs. This is because the tariff rates are intended for service to common carriers, representing a kind of trade discount for members of the service provider club. Deregulation, however, will cause these restrictions to be gradually removed.

3.2.4 Telemarketing

The marketing and sales approach that employs the telephone in lieu of face-to-face contact with the customer is called *telemarketing*. It is traditionally used in consumer marketing of products and services which can be described verbally. Telemarketing, however, can be a flexible and effective tool in business-to-business marketing as well. With all sales personnel contained under a single roof, this powerful system can make effective use of telecommunication facilities and personnel management techniques. The cost and complexity of establishing and maintaining a nationwide sales force is eliminated. Nonetheless, many firms are using telemarketing to work with the outside sales force to increase the "bang for the buck" of sales calls. In addition to selling, telemarketing can be used to process various kinds of customer requests, for example, for maintenance information and technical help.

Telemarketing operations are efficient due to the use of sophisticated telephone switching equipment and computers. Consequently, the cost of this form of selling has decreased over the past several years, while the costs of other forms of selling have generally increased. For example, the cost of a telephone contact may be only $2, whereas that made face to face will be $150 for a field sales call made in a single metropolitan area. The comparison for an intercity call would be considerably more dramatic due to the greater expense for travel.

To provide visual information on products or services, the seller may distribute catalogues which allow the prospective customer to make a selection prior to the telephone contact. Cable television and over-the-air broadcast channels carry "home shopping" programs that display and demonstrate products which can be ordered over the telephone with a bank credit card. Home Shopping Network of Tampa, Florida, is the current leader in this field. A photograph of one of their telemarketing operations is shown in Figure 3.7.

Telemarketing is a big business in the U.S. with a substantial fraction of all consumer sales made in this manner. While larger merchandisers can afford to have their own telemarketing departments, small companies can still take advantage of the technique by subcontracting the effort to a specialized telemarketing service provider. This type of organization maintains the operators and telephone facilities, and will arrange for the necessary outward and inward WATS lines to make and take calls.

The scope and flexibility of the public local and long-distance networks have helped make telemarketing a business success. Many of the available services and equipment employed in telemarketing are reviewed in the following paragraphs. The basic intent of these facilities is to manage the personnel who actually stand by the telephone lines used in telemarketing. In addition to telephone facilities, telemarketing employs computer equipment to assist sales agents in handling the high volume and potentially wide variety of calls. Software for personal computers makes these capabilities available for even small operations which cannot afford a centralized system. The computer screen displays a prepared script to coach the agent, and various scripts are available to match the purpose or origination point of the call. In the following discussion, the emphasis is on the telephone operations side of telemarketing.

3.2.4.1 Use of WATS

Primarily, the WATS calling program has established the basis for telemarketing. We are all familiar with employing toll-free 800 numbers in the U.S. to reach airlines, catalogue sales companies, and hotels. The fact that we can call toll free tends to encourage us to make the reservation or purchase, which is extremely important to the success of many businesses. The manner in which the telcos and long-distance carriers offer WATS was presented in a previous section. Outward WATS is used for client prospecting ("cold calling") or follow up on visits by a field sales force. Inward WATS is used in reservation systems, catalog sales, and service support. Specialized telecommunication facilities can enhance both telemarketing and WATS, as discussed in the following section.

3.2.4.2 Automatic Call Distribution

Telemarketing takes on new and expanded dimensions through the facilities

Figure 3.7 This impressive telemarketing operation is employed by a leading cable TV home shopping channel. (Photograph courtesy of Home Shopping Network.)

of an *automatic call distribution* (ACD) system. The term *automatic call direction* is synonymous and uses the same acronym. There are several variations on the concept, which are reviewed here. The basic approach of this type of facility, however, is to allow a group of operators or attendants to service several incoming and outgoing lines. Calls arrive from prospective clients, and are routed to an available agent. In a conventional PBX system, the caller usually knows ahead of time to whom they want to speak. The difference in telemarketing is that the caller is mainly interested in a product or service, and does not care who takes the call. Call direction without ACD would otherwise be handled by the switchboard operator, who would not be well versed in the caller's need. The role of the ACD is to assign the incoming call automatically to an available agent and, more importantly, to take certain steps to prevent the call from being lost — and the business from being lost, also! Call directors also assist in the management of telemarketing staff, providing statistics on how the agents and telephone lines are being utilized.

The simplest type of director is a *sequencer* (also called a *rotary switch*), which arranges incoming calls chronologically and signals the sequence to all operators. Each agent has a control console with keys indicating the incoming lines. The sequencer is, in fact, a variation of the key system. It is inexpensive, and the newer generation of key systems can be upgraded to provide this function. Selection of an incoming call, however, is in the hands of the agents, who must make decisions for themselves. Callers can easily be left hanging.

An improvement on the sequencer is the *uniform call director* (UCD). This is basically a modification of the programming for a conventional PBX. In fact, most of the current PBXs on the market can operate as a UCD. Incoming calls are directed to the next agent according to a predetermined sequence. This means that agents at the beginning of the sequence will be busy all of the time, while those at the end of the list will only receive calls when everyone else is busy. The advantage of UCD is that its facility can be provided by standard PBXs and CENTREX.

The most capable call director is the ACD itself, which is a specially designed and programmed telephone switching system. The main manufacturers of ACDs are AT&T and Rockwell Collins. The ACD can handle a multitude of functions for a large group of agents. These systems are used extensively by airlines for reservations systems, for credit verification by bank credit card companies, by the telcos themselves for customer service representatives, and for emergency public services through the 911 number operating in many U.S. cities. Telemarketing companies, which operate like service bureaus, employ ACDs in the normal conduct of their business. Unlike the UCD, the ACD will assign incoming calls in such a way as to even out the workload among agents, which tends to get the most productivity from the work force. Incoming calls can be routed directly to an available agent; if no agent is free, calls are connected to a prerecorded message or voice response system. The purpose is to prevent the caller from giving up and dropping off the line. In addition to answering incoming lines, the ACD allows agents to forward calls and to set up conference calls among each other. The ACD can be used to shift incoming calls from an operation closing down in one locale to a distant office which is still open for business. The caller only needs to know one telephone number to call. The long-distance carriers have introduced this feature into WATS so that an ACD will not be required merely for call forwarding purposes.

The ACD can play an invaluable role for agents placing outgoing telemarketing calls. ACD systems can store a list of numbers so that the agent need only activate the next one on the list to have the call go out. Automatic operation of dialing (similar to the add-on autodialers already discussed) initiates the call and sometimes responds with a computer voice to the answering party. Depending on what the called party is interested in doing, the call can then be routed to the next available agent, as if it were an incoming call. (The opportunities for this kind of system generating thousands of "junk" calls are both impressive and foreboding.)

The supervisory position of the ACD can set up and alter the assignment process, which is important to proper utilization of personnel. Analysis of the actual performance of the system and personnel is also readily available. One of the more interesting innovations is the ability of the ACD to handle a customer inquiry without the intervention of a human operator. This makes use of voice response techniques, wherein the caller is requested to enter information through his or her touch-tone pad.

As PBX systems become more sophisticated, the features of ACDs are being incorporated. This means that the same position which is taking calls can also make calls throughout the network. Telemarketing therefore can be blended with any other aspect of use of the private telecommunication network. We will be interested to see the result of such innovations as ONA on the way telemarketing is to be done in the future.

3.2.5 Audio Signal Distribution

Extension of the old public address system over a private telecommunication network is useful in a number of business situations. In times of a changing business environment, audio signal distribution allows the head office to be in close contact with the remote branch offices. Another use is in public safety services as an alert system. The Emergency Broadcast System (EBS) for alerting the public in the case of national emergency or natural disaster is a good model for this type of network. Another category of audio distribution is used in advertising as a delivery mechanism of commercial messages into special locations. The example given below is of a simulated radio network for background music and commercials into grocery stores.

3.2.5.1 Hoot and Holler

An audio point-to-multipoint circuit used in voice announcement purposes has been called a "hoot and holler," "shout-down," or "squawk-box" channel. Services of this nature are not generally available on a tariff basis from the telcos or long-distance carriers. The channel can be pieced together, however, by using either terrestrial or satellite transmission. In fact, this is the kind of application where a satellite has a special advantage, because its time delay has no consequence and the service is of a point-to-multipoint nature.

The hoot and holler channel is not intended for reception by the general public. Consequently, audio quality is not as important as intelligibility (e.g., the ability to be understood). The channel is used extensively in the securities business, as illustrated in Figure 3.8, to connect branch offices into the home office. Bulletins are announced over the circuit by specialists such as economists and stock analysts. Brokers at the remote branches monitor the channel through a local public address system. The next step up from hoot and holler is *private video broadcasting*, which is a form of video teleconferencing. This application is considerably more costly

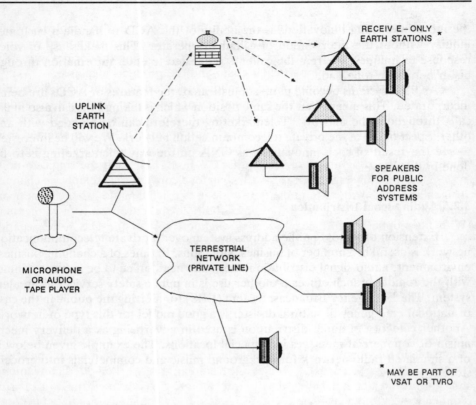

UPLINK
EARTH
STATION

RECEIV E – ONLY
EARTH STATIONS *

SPEAKERS
FOR PUBLIC
ADDRESS
SYSTEMS

TERRESTRIAL
NETWORK
(PRIVATE LINE)

MICROPHONE
OR AUDIO
TAPE PLAYER

* MAY BE PART OF
VSAT OR TVRO

Figure 3.8 Hoot and holler broadcast audio channel used for making announcements to remote offices.

to implement and to operate than audio, but offers the added dimension of the television picture. The hoot and holler channel, however, can easily be combined with a teletext broadcast so that the visual dimension is gained (for graphic images) at a substantially lower cost.

3.2.5.2 Private Audio Broadcasting

The audio broadcast channel is currently used extensively by radio networks to distribute national programming and advertising to local commercial radio stations. This satellite communication technology has been adopted for *private audio broadcasting,* which, unlike hoot and holler, is intended for listening by the public. This type of audio broadcasting is referred to as "private" because the channel is delivered to stores within a specific chain or to some other closed system engaged in a certain activity. Programming on the network is aimed at this particular market and can be paid either through subscription or by advertisers. Examples of subscription channels include the Musak background music service and cable television music channels that are not available over the air. Advertiser-supported channels can be heard in some supermarket chains in the U.S., using the format of a radio station, tied directly to the particular chain and the products available.

An important distinction of private audio broadcasting is that the audio quality must be vastly superior to that of hoot and holler. The latter would use a standard voice channel over the long-distance network (an audio bandwidth of 3 kHz and a ratio of signal to noise of typically 40 dB) as compared to what is referred to as a "program" channel used by radio networks and in audio broadcasting. The bandwidth of a program channel is either 7.5 or 15 kHz, and the signal-to-noise ratio is substantially higher (60 to 70 dB) than for a conventional telephone channel. This means that, without satellite transmission, a private audio broadcast network employing the terrestrial network can be both costly and difficult to implement. With fiber optic networks extending throughout most of the U.S., the situation may improve, and terrestrial digital audio program channels will become attractive for this application.

3.2.6 Voice Cipher

In a telephone circuit, when the voice signal leaves the customer's premises, it often follows an indirect path to the final destination (and *vice versa*). The wire and cable in the outside plant are accessible to anyone with the interest and ability to tap in and overhear the conversation. Radio links offer the added convenience for the intruder of not having to dig into the ground nor even make a physical connection. In the case of satellite transmission, the voice signal is radiated throughout a region of a hemisphere with the obvious consequence that interception is easily assumed. The information may be protected from unauthorized interception by altering its form. *Encryption* is the means for preventing the reception and interpretation of sensitive material in digital form. For analog signals, *scrambling* is used, wherein an interfering waveform is added to the signal, or the signal itself is rearranged in some random pattern.

We use the term *voice cipher* to refer to techniques for encryption geared to human speech. Of course, if the speech is digitized at the source, it can be protected by conventional digital encryption equipment using, for example, the Digital Encryption Standard (DES) developed by the U.S. National Bureau of Standards. The DES algorithm is unclassified, and is available in commercial integrated circuit form. At present, DES can only be used in the U.S. and its allied countries. Other such algorithms have been invented, and are marketed by foreign companies.

In using analog telephone networks, voice cipher can still by provided by two different approaches. An approach which is simple and inexpensive, but only offers a low level of protection, is called *privacy*. In privacy, the threat is assumed to be from a listener without sophisticated methods and resources. These include casual listeners, who operate switchboards or work for the telco or long-distance carrier. Such listeners would not be expected to record the message, nor would they use computers to reconstruct the conversation after the fact. One approach to privacy is to take the speech waveform, cut it up into analog segments, and then rearrange the segments in a seemingly random sequence, as illustrated in Figure 3.9. The transmission is audible, but sounds completely garbled to the casual listener. At the receiving end, the descrambler knows the correct sequence and simply rearranges the segments. The speaker's voice is mostly recognizable, but somewhat distorted, after descrambling. Also, a time delay is introduced because the transmitter and receiver must collect a series of segments before rearrangement is possible. Applications include business communication on domestic and international telephone circuits and use in mobile radio public services (police, fire, *et cetera*).

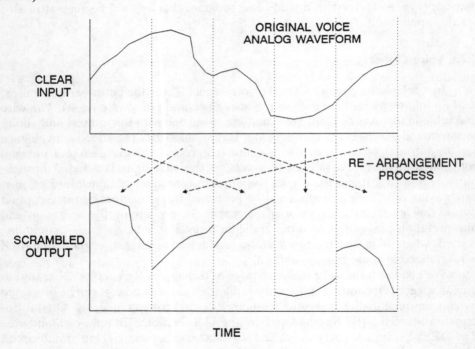

Figure 3.9 An example of the technique of voice scrambling.

True encryption of voice over a conventional 3 kHz analog voice circuit is possible by using the technique of the *vocoder*. This class of device was conceived decades ago for use in defense and national security communication. The analog signal from the microphone is first digitized, and the resulting data are heavily compressed to reduce the bit rate to as low as 2400 b/s. Underlying the vocoder is the detailed structure of human speech (frequency content and time variation), which although passable within 3 kHz, actually uses considerably less bandwidth and time. The basic principle is selectively filtering within the spectrum and volume range to extract the real information content. A linear predictive algorithm is also employed, resulting in some delay of speech. As in the voice privacy example, the delay is because the speech samples must be stored at one or both ends of the circuit. Early units yielded an intelligible voice at the distant end, but the voice was mechanical and the speaker could not be identified. The technology has been improved greatly, and vocoders which pass recognizable voices are available on the commercial market for less than $5000. When the speech is digitized and encrypted, the resulting data can be sent through a modem over normal telephone circuits or multiplexed with other data.

3.3 DATA COMMUNICATION

The most rapidly growing facet of business communication is the processing and transmission of computer-based information. Between 1986 and 1987, the rate of data communication growth was 16% as compared to 13% for the overall telecommunication industry. Today, expenditures for data communication services and equipment (excluding the computers and peripherals) represent approximately 20% of all telecommunication expenditures. As mentioned in Chapter 1, strategic units employ computers, remote terminals, and data communication in the routine performance of ever more facets of business. Some of the fastest growing and most successful corporations (IBM, DEC, and HP, for example) manufacture data communication hardware and they program the computers and communication processors that are vital to the effectiveness of the data communication aspect of private telecommunication networks.

The focus of the following discussion is on the use of data communication in terms of the flow of information among computers and terminals. How the computer is specifically used is not really relevant when examining things from this perspective. In Chapter 8, however, we will examine how strategic units employ private telecommunication networks in the pursuit of their strategy. Bringing together the information herein with that in Chapter 8 should begin to complete the picture.

There are currently five generic technologies for connecting terminals, PCs, and other devices within a narrowly defined network such as may exist on one site. These are the *concentrator,* the *LAN,* the *data PBX,* the *integrated voice-data PBX,* and the *matrix switch.* (More information on these alternatives can be found in Chapter 4.) In this section, we focus on the LAN approach because of the wide acceptance gained in the past few years. Perhaps this is because of the wide bandwidth obtainable on a LAN (depending on the transmission speed, 80 Mb/s being at the upper end of the range for commercial hardware), or because of the declining costs per connection (ultimately, PCs will have built-in LAN ports). The range of functions of the LAN could be provided by other technologies such as data switches, voice-data PBXs, and concentrators. Also, there is the "top-down" approach which is obtained from large computer networks connected to host computers (e.g., the IBM approach). The fundamental attribute of the LAN is that it is a "bottoms-up" approach, developing from the small-scale use of the personal computer and evolving into a rather elaborate PC network in the extended case.

With most organizations relying heavily on personal computers, the demand for local area networking has increased substantially. The first stage of LAN implementation is usually the connection of a few PCs with a local hard disk *file server* containing a shared database and software for word processing, spread sheet analysis, and possibly E-mail. A laser printer on the same LAN allows users to generate high-quality hard copy output. This type of application system makes efficient use of software and opens up uses of the computer medium within a department-sized group, as both an economy move and a way of efficiently using the local work force. Extending this concept further, a LAN can connect sophisticated microcomputers, called *work stations,* to perform high-speed scientific computations and simulations. Access to a "supermini" can be provided on the same LAN. In general, the size of the LAN is limited both in terms of the number of devices that can be connected and the distances that can be traversed. It is easy to imagine that there would be a strong desire to add remote terminals and devices to a LAN. Several products generally referred to as *bridges* have appeared on the market to provide the linkage between two or more LANs. Connecting a LAN into a different data communication network environment of a minicomputer or mainframe computer, however, is much more difficult. The particular type of access is called a *gateway.*

The well integrated data communication environments like IBM's SNA and DEC's DECnet/DNA are extremely important in business applications because of the power of the host computers, and the rapidity with which information is processed and distributed among a large quantity of terminals. Problems of serving small users with a large network and connecting dissimilar networks are addressed by the largest computer manufacturers with their layered architectures and by networking specialists who offer proprietary bridges, gateways, and communication software. These aspects are also covered in detail below.

3.3.1 Generic Applications

The following applications are generic utilities for performing the types of tasks that strategic units find of value. Data communication architectures and network topologies will often be optimized for the network's range of applications. While a difficult challenge to realize in practice, combining different applications on one network provides an economy of scale in terms of computers and telecommunication facilities. The next section reviews the most prominent data communication environments to see how the applications may be supported in the real world.

3.3.1.1 Time-Sharing

Time-sharing computer systems have been with us for a few decades already, allowing several users to share the computing capacity of a single machine. The technology which permits time-sharing is called *multitasking* wherein the computer's operating system interrupts the performance of one task after a specific time interval, shifting to others in sequence. The mainframe computer can therefore be shared by a number of users, each of whom is running completely independent tasks on the system. These users may be located in the same room or building, but are more likely to be in different parts of the city or even the country. When moving off-site, efficient data communication becomes a requirement. Time-sharing was actually first offered as a utility type of service, with companies like Tymshare, GE Information Systems, and Computer Sciences Corporation having established themselves as leaders. On one of their systems, it is not only possible to share the computer, but also to use prewritten programs which are available in an on-line software library. Such applications are used extensively in the engineering and scientific areas, where access to sophisticated analytical programs and even supercomputers is a necessity.

Time-sharing is available on most larger computers, even if they are not part of a public utility. The Time Sharing Option (TSO) of the IBM mainframes is very popular among users. The same computer can support several different tasks, seemingly at the same time. Of course, the more tasks and processing per task, the slower the computer appears to be working. In actuality, the computer is not slowing down, but rather the individual tasks are not getting as much of the central processor's attention because of the increased workload.

Data communication is the means for allowing time-sharing users to access the central computer from a distance. Another term used to define this approach is teleprocessing. They would use the computer to perform some task, such as running a financial analysis program or printing statistical reports. There would not normally be a relationship among different users, except possibly that they would be part of the same organization. Some companies sell excess time (e.g., allow the use of a lightly loaded multiprocessing computer) to outside organizations. You can imagine that the early hours of the morning would be particularly good candidates for this type of business, since it retains the full power of the mainframe for the work day.

Looking briefly at time-sharing network topology, which is covered in more detail in Chapter 4, the data communication is arranged in the form of a star. At the center or hub of the star is the time-sharing computer. The rays emanating from the hub are data lines to the remote sites. Each line terminates at a port on the central computer. As will be discussed later, a single leased circuit can traverse each termination in what is termed a multidrop line. Operation of the network is basically the same as if the users were physically located in the same building. Perhaps a unique aspect of time-sharing is that the remotes are actually serving people who wish to employ the central computer as a resource.

This definition of time-sharing is very basic and essentially represents an arrangement which was perhaps popular in the 1970s. Today, virtually every central computer, whether a minicomputer or mainframe, operates in a multitasking mode, since this economizes on processors, file storage, and software. Systems geared toward *on-line transaction processing* (OLTP) would also have capacity available for time-sharing. In general, unless one is talking about time-sharing as a public utility, it is taken as a given for all centralized computing.

3.3.1.2 Batch Mode

Before time-sharing was invented, and before personal computers were even thought possible, there was *batch mode,* wherein users submitted computing tasks as packages and awaited their outputs to be delivered. The term is also applied to any manufacturing process where the product is made as a group or "batch," rather than on an assembly line. In computing, a user prepares his or her job off-line (without using the computer in the preparation), either on punch cards or as a listing stored in a file on magnetic disk or tape. The job is submitted to the mainframe for processing as a single task. All of the jobs awaiting execution by the computer are arranged in an input queue, usually on a hard disk which is readily accessible by the computer. A particular job is executed in turn by the computer, and the results are delivered where directed by the user. For example, the user may direct that the output be printed locally at the computer center. Data communication facilities will permit the user to input the job from one location and have the printing done at that or a different location, which is also distant from the computer. If done properly, batch processing can be used to perform a variety of tasks and to reduce transportation costs as well. A popular form of batch mode operation of data communication facilities is called *remote job entry* (RJE). This is a capability offered on the previous generations of IBM mainframe computers, and was popular during the late 1960s and throughout the 1970s. Today, RJE is still used for heavy computing and output delivery, where the actual job might be performed during nonworking hours.

3.3.1.3 Database

Access to remote databases is perhaps one of the most revolutionary applications of computers and data communication, introducing new opportunities in information management, library services, research, and finance. The database itself is a large collection of textual and numerical information which has been entered through a computer into an on-line storage medium. Typically, the database is stored on large hard disk files and is directly accessible by a minicomputer or mainframe. New information is continually entered for updates and to expand the database. The key is to have useful information available on the database to encourage users to access it. Users can select information by using powerful search algorithms, saving time as compared to conventional indexes. The "care and feeding" of a database is clearly an involved business. In many cases, this updating is done continually, such as in financial databases that contain stock prices. Of greatest interest are relational databases which provide the greatest flexibility. A type of database which does not change very often is an on-line encyclopedia, and in fact such files are available for purchase by users on optical disk.

The smallest computers, including PCs, can now contain extensive databases by using large hard disk files and optical disk readers. In either case, the key to utility is to provide some mechanism to update the database. A data broadcast link, possibly through teletext or videotex, could transfer data updates to the PC database without involving an operator. At any time that a user wishes to interrogate the database, the latest information would still be available. The bulk of the data, however, would exist in a more permanent form. Periodically, the entire "permanent" database could be replaced with a new one. This technique greatly reduces the data communication requirements because most of the information will reside locally and only updates will be sent — in this case, via an inexpensive broadcast.

3.3.1.4 Inquiry-Response

The *inquiry-response* classification of data communication application is a subset of OLTP. A good example of inquiry-response is an airline reservation system, which was discussed in Chapter 1. Hundreds or perhaps thousands of computer terminals are connected to a host computer through a diverse data communication network. Typically, this network is merely a facet of the public telephone system, employing either leased lines or switched long-distance service. Alternatively, an existing public data network could be employed. An operator at a remote terminal may "inquire" of the host what flights are operating between two particular cities. This short message produces a "response" from the host in the form of a listing of possible flights and their times. The operator can select one of the flights from the menu and then request a booking. Upon receiving the request for the booking, the host checks (in a relational database) for availability of the type of seat; if it is available, the host makes a firm booking. It responds accordingly back to the operator so that the client can be told that everything is "set." From the other general perspective, the entire "transaction" between remote terminal and host is carried out in an automated fashion, with several such conversations being allowed, seemingly at the same time.

IBM and DEC are well recognized as innovators and they have large installed bases of transaction-based data communication systems. The architectures of the computer network are reviewed in a subsequent section of this chapter. Tandem Computers Inc. of Cupertino, California, has gained a devoted user base in inquiry-response systems. Tandem's strong points are the speed with which the high quantities of transactions are served and the reliability of the computer hardware, which makes use of parallel processing techniques and on-line redundancy.

Inquiry-response and database access are quite similar in network structure, even though the applications may be different so far as the users are concerned. Perhaps the amount of data in the response is considerably larger in the database approach. Nevertheless, there would be many users connecting themselves to the host and database during the same period of time. Two important points apply. First, the system must deliver its responses in a relatively short time interval so that the the user is not overly inconvenienced. This time period, between when the inquiry is made and the response comes back, is called simply the *response time*. If the system provides a service to paying customers, a long response time will be unattractive in the competitive information service marketplace. The second important facet is reliability. Both types of applications involve many users wishing to maintain business (or, in the case of a service like videotex, a continuous flow of revenue for the network operator). The data lines, communication processors, and most importantly the host computer, must be capable of maintaining service in the event of telecommunication outages and equipment failures.

3.3.1.5 Electronic Mail

The sending and receiving of messages, letters, and documents in textual form without the use of writing or printing on paper is generally referred to as *electronic mail* (E-mail). As discussed in Chapter 2, an E-mail message is addressed to a particular individual or set of individuals, and transmitted over a data communication network. The destination is referred to as a "mail box," an obvious use of the analogous end point in regular paper mail. This definition presumes that the message information is in the form of character data using a coding format such as the American Standard Code for Information Interchange (ASCII). In the very detailed reference on E-mail by Stephen Caswell [Caswell, 1988], however, facsimile and voice mail are also included. Since these noncharacter techniques are decidedly different and covered elsewhere in this chapter, we will restrict the definition of E-mail to those methods which employ character data.

Efforts by domestic and international standards organizations could someday result in a universal E-mail system, providing the advantages of the postal system and the facility of the personal computer. At present, E-mail is still somewhat specialized in its applications. Some are confined within private organizations, some can be accessed by individuals with personal computers, and some are available to the general public as adjuncts to other services like telegram and overnight delivery. The ways in which E-mail is applied in the context of the strategic unit are reviewed in the following paragraphs.

Messaging

E-mail is fundamentally used to pass messages among different users. A message is a complete set of information which is intended to go from one address to another address over the network. An example of a message might be a telegram being sent over the public network or a memorandum from one employee to another within a single organization. Note the distinction here between public and private messaging.

The basic manner in which the messaging service is used is shown in Figure 3.10. At the originating end, a user composes a message on a PC or terminal by using simple text editing or more elaborate word processing software. When the message is completed, the user establishes a connection to the E-mail host, which then instructs the user to provide an addressee. If the messaging system is restricted to a previously authorized set of users (i.e., "closed"), this may be sufficient to determine the destination; otherwise, the user must also indicate the destination address as well. When the user tells the system to "send" the message, the entire ensemble of information is applied to the network for routing to the destination host or mailbox.

The network, shown as a cloud in Figure 3.10, can take two basic forms. In the simplest arrangement, the network is, in fact, a host computer to which all users can be connected. The message is simply stored on magnetic disk as if placed in a pigeon hole or mailbox, awaiting its accessing by the addressee. When the addressee checks to see what messages are waiting, he or she activates a PC or terminal connected to the same host. The message can be read from the magnetic disk, reviewed on the screen, and either stored or deleted from the system. The other possible arrangement is that of a real data communication network with various nodes through which the message will be switched before it reaches its final destination. In particular, the packet network is the most common implementation of the universal type of messaging system (which is discussed in more

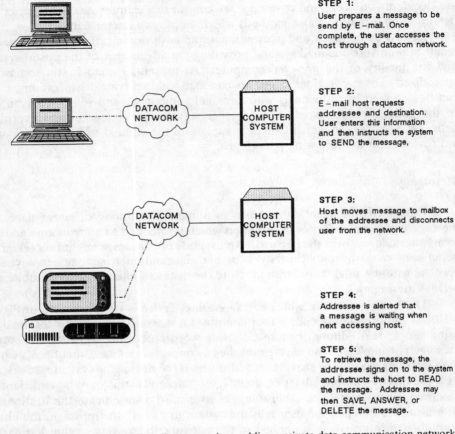

STEP 1:
User prepares a message to be send by E-mail. Once complete, the user accesses the host through a datacom network.

STEP 2:
E-mail host requests addressee and destination. User enters this information and then instructs the system to SEND the message,

STEP 3:
Host moves message to mailbox of the addressee and disconnects user from the network.

STEP 4:
Addressee is alerted that a message is waiting when next accessing host.

STEP 5:
To retrieve the message, the addressee signs on to the system and instructs the host to READ the message. Addressee may then SAVE, ANSWER, or DELETE the message.

Figure 3.10 The handling of E-mail messages in a public or private data communication network.

detail in Chapter 4). A local user at the receiving end would then connect to the storage device to read the message and to dispose of it in some appropriate manner. Depending on the network and the nature of the message, the receiving node transmits an acknowledgment back to the sender. In public messaging such as MCI Mail, a hard copy of the message can be printed at the distant end by the network operator and then delivered to the addressee either by local mail or (for an extra fee) by messenger.

The X.400 application layer protocol now implemented in the OSI model (discussed later in this chapter) offers means for allowing private E-mail networks to communicate with their public counterparts throughout the world. In creating a private network using X.400, an organization can still maintain control of access by allowing only one gateway into the public domain. An example of a public network offering of X.400 is the TeleMail 400 offering of Telenet, discussed in Chapter 2. Individuals in small organizations can use the E-mail service by accessing TeleMail-provided mailboxes over Telenet's X.25 public data network. From that point, E-mail can be routed domestically or internationally. Private E-mail systems using X.400 can also send mail to TeleMail mailboxes of users who are not part of the private network.

Many organizations have effectively employed E-mail messaging systems to replace written internal correspondence. In particular, telecommunication and computer companies with multiple widespread locations have pioneered the technology, almost out of necessity. More than one of these companies have turned their internal messaging systems into public utilities, and now operate them as a service for profit. Chapter 8 reviews this concept for leveraging a network through a strategy which we call *reselling*.

Probably the original form of this type of application is from the use of manual teletypewriters in telex, where the biggest innovation was automatic answer-back and receive operation. The actual message, however, could only be recorded on paper as the receiving machine typed it out. The computerized version is clearly superior because a variety of other features are available. These include the ability to save messages, to send messages to multiple addressees, to modify messages before and after transmission, and to perform statistics on the message flow. Messaging is a built-in feature of office automation systems so that many organizations can employ the technology, even if it is not currently in use. Apparently, there is a barrier to using messaging in some cases due to inadequate quantities of user terminals, insufficient computer capacity, or simply the discomfort of exchanging written correspondence over computer terminals. These barriers can be surmounted as more and cheaper terminals are deployed and people become more comfortable with them. At present, intermediaries like secretaries, terminal operators, and telecommunication service providers can employ messaging in effective ways.

Document Transfer

A *document* in the context of document transfer is not a flexible piece of writing created with a word processor. Rather, it would have a standard format like a tax return, with only specific items filled in differently for a given transmission over the network. In a commercial application, one document could be the special form used by an automobile dealer to order a particular make and model of car. A system like this was developed for the Buick division of General Motors. The salesperson sits in front of a computer terminal alongside the customer. On the screen appear alternate selections, and the users are coached by the local computer through the choices. After all selections are made, the completed document is sent on the network to the ordering center in Detroit where the order is processed. Since the document is standard, only the data specific to the customer needs to be sent, greatly reducing transmission bandwidth and storage requirements. This is more like a database access system, where the form is really a computer screen layout and the particular entries become nothing more than records.

One of the fastest growing document transfer applications is in an industrial area called "just in time manufacturing." Through the use of data processing and data communication systems, the production of a complex item such as an automobile is compressed so that inventory of parts and materials is minimized. This reduces the manufacturer's costs for maintaining those inventories, specifically in financing the purchase and for physical storage and handling. Instead, orders are placed with suppliers within a preset time prior to when the material is actually needed. Of course, the supplier must be very efficient; otherwise, the supplier would need to store the material. The difference is that the expense of storage would be on the supplier's side of the ledger. To facilitate quick ordering and delivery, the manufacturer and supplier could employ a common ordering system using standard document formats. Transfer of such documents would be done electronically by document transfer applications on a data communication network. Ideally, the document would be initiated almost automatically by a computer-based manufacturing system, which predetermines the types and amounts of various materials needed for a run of some final product.

Standardization of document transfer formats and, more importantly, protocols, comes under the title of *electronic data interchange* (EDI). Major data processing and networking vendors like GE Information Services, McDonnell Douglas Information Systems, IBM, and AT&T are pursuing their individual strategies to maximize market share. Due to the importance of just-in-time manufacturing, the potential business volumes in the years ahead are large. Manufacturers the size of General Motors can create their own EDI networks by installing appropriate software on existing data networks and developing the requisite applications. Niche players like Sterling Software service specific industries like health and pharmaceutical companies.

E-Mail over Videotex and Teletext

E-mail can be offered as an adjunct to videotex and teletext, the two public information offerings described in Chapter 2. Recall that videotex is a two-way interactive service employing the public telephone network, while teletext is a data broadcast medium, which usually accompanies a local television station's signal. Any of the popular videotex services assign mailboxes to subscribers so that they may "leave" messages for one another. The messages are routed between switching nodes in the access network and are stored on a host computer where they may be read and handled in any of a number of ways. A number of the videotex services provide a means for the message to be printed in hard copy form and subsequently mailed to an addressee who is not a subscriber to the service. This is particularly useful where the addressee may be a subscriber, but one who does not "sign onto" the videotex service frequently enough to keep current.

To apply E-mail over a teletext system implies that messages are broadcast to all locations. This works well with a private network for delivering notices to branch office locations, customers, and other affiliates. For example, E-mail via teletext could distribute automobile repair manual updates and new product information to dealerships. Interspersed with broadcast notices might be E-mail messages for particular addressees. If privacy is required, the transmission over the broadcast link will be coded and only the addressed receiver can recover the contents. Response from the remote site to confirm delivery or to request other information would be made over the telephone network on a dial-up basis or a PDN.

E-mail and videotex apparently have much in common for public network application. Ultimately, the two modes may blend together. Access to E-mail will be facilitated by making packet networks universal through innovations like X.400 and ONA.

3.3.2 Data Communication Environments

The starting point for nearly every discussion of data communication is the particular computer network environment that the user is employing. Perhaps someday there will be one universal environment and all computer systems will understand how to "talk" to one another with a common "language." Here and today, however, there are several key environments which must be reviewed. In any major organization, you will usually find that one of the following environments is the mainstay, but, quite typically, there is a second or even a third, which was adopted by one subgroup or is a vestige of the past that continues to hang on. This latter point is appropriate because a working system (producing revenue) will not be shut down until the successor system is installed and working as well or better than its predecessor.

Environments are either *host-based* or *peer-to-peer,* as illustrated in Figure 3.11. The host-based environment is most effective where remote users wish to access a common collection of resources and applications. Prior to the advent of minicomputers and microcomputers, this was the only option because of the expense of installing and operating a mainframe computer at each site. An example would be a centralized banking system on the mainframe at headquarters to maintain bank accounts, with remote terminals located in various branches around the

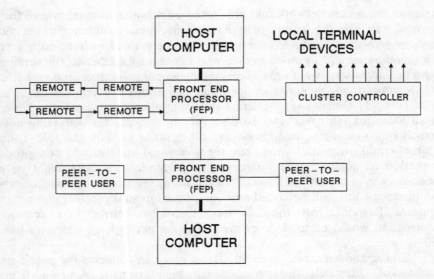

(A) HOST – BASED ENVIRONMENT WITH PEET – TO – PEER SERVICE CAPABILITY

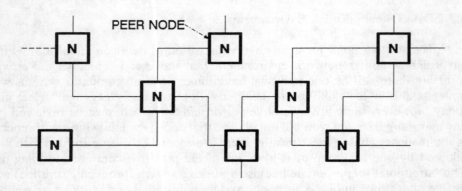

(B) PEER – TO – PEER ENVIRONMENT

Figure 3.11 The two basic computer environments: host-based and peer-to-peer.

state. Everything is done on the mainframe, since remotes are dumb — they display information directly as it is received and send information only when the agent enters it on the keyboard.

In peer-to-peer environments, there would be any number of nearly independent operations around a region. The data communication environment would allow the dispersed computers to send and to receive information, and all terminals connected to a local computer could access any other computer in the network. Minicomputers and microcomputers are becoming so powerful and inexpensive that this networking approach has become particularly attractive for handling data communication needs. In effect, messaging is blended into computing, a very powerful concept for a dispersed network. Introducing the peer-to-peer approach into our banking example, each branch would have a minicomputer to handle internal needs such as account maintenance, billing, and word processing. This minicomputer would communicate with one at another branch over a network to handle a transaction when a customer of one branch wants to do some business at a different branch. A centralized database could still be maintained at a host or on a larger minicomputer at the main branch. Due to the flexibility and modularity of the peer-to-peer approach, the host-based environment is gradually giving way, except in special situations.

The most common way to refer to a particular architecture is by the levels (i.e., a layered architecture) that data pass through when moving between user and computer or between one computer and another. In Figure 3.12, we see a hypothetical end-to-end connection involving several elements in the data link of a host-based environment. The host and remote are at their respective ends, while the actual information is routed over a data network. The blocks indicate generic functions of data communication in a layered architecture. We will come back to the concept of the layers when the seven-layer OSI stack is defined. Within the data communication cloud is the collection of network nodes (shown as circles), which can either be nodes of a PDN or minicomputers in a private peer-to-peer environment. Entry to and exit from the network is through a processing device or protocol converter, which, in this example, is a *packet assembler-disassembler* (PAD). The end user's terminal connects on the right with a communication controller and the host mainframe on the left through a *front-end processor* (FEP).

At the lowest level of the data communication architecture is what we usually refer to as the *physical layer,* which involves the actual (physical) connection at the communicating device. There is a physical layer connection of each block element in Figure 3.12. In personal computers, the typical physical layer interface is the RS-232 standard, employed for rates between 300 b/s and 19.2 kb/s. Another common standard for very high speed data is the DS-1, which sends at 1.544 Mb/s. The actual data network is composed of nodes and interconnecting data lines (4800, 9600, or 56 kb/s). Between nodes in the network cloud, communication is managed by using the second layer, the *data link* layer. The purpose is to provide synchronization, error detection and correction, and supervision (i.e., letting the other end know that this end is ready to receive data). To navigate the data from gateway to gateway, the network layer follows rules for routing information from end to end. Recommendation X.25 identifies these layers of nearly end-to-end

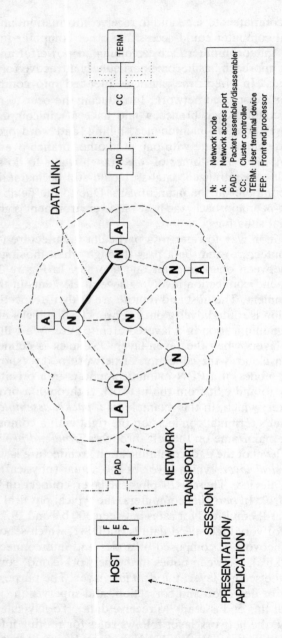

Figure 3.12 Generalized architecture of a layered packet-oriented data communication network.

N: Network node
A: Network access port
PAD: Packet assembler/disassembler
CC: Cluster controller
TERM: User terminal device
FEP: Front end processor

DATA LINK

NETWORK

TRANSPORT

SESSION

PRESENTATION/
APPLICATION

communication. Higher layers are required, however, to allow the host to serve the remote.

Adopted in 1975 by the CCITT, Recommendation X.25 for data network interface has now gained fairly wide acceptance. As defined in the CCITT documentation [CCITT, 1985], X.25 specifies in great detail the interface between a terminal device, such as a computer or communication controller, and a data communication device which is attached to the actual network. The basic operation of an X.25 network is reviewed in the next paragraph and discussed in detail in Chapter 4.

Typically, a message to be sent via X.25 is large enough that it must be broken down into packets of essentially uniform size. There are two fundamental network layer routing modes for packets in X.25. In the *datagram* mode, each packet of the message is addressed to the same destination, but routed individually over the network. A given node accepts datagram packets one at a time, sending each along the most appropriate path at the time of action. Congestion on the best path causes the node to employ an alternative, according to the routing table stored in its memory. The message is reassembled at the destination by collecting the packets as they arrive, storing them in a random access memory, and delivering them arranged in the proper order. Because packets in the same message are numbered, requests for retransmission of packets which contain errors or are missing can be sent back through the network to the originating node. In the *virtual circuit* mode, which is more common, a setup packet is sent through the network to establish a particular routing which is remembered by the nodes along the path. Packets which make up the message follow this predetermined path until all packets are accounted at the destination. During the time that the virtual circuit is active, other packets can time-share the various segments of the path because the message is not sent as a continuous stream of data.

The end-to-end connection is usually provided by smart communication controllers or computers which contain X.25 compatible software. In Figure 3.12, the external device which makes the connection to the particular network is a PAD; it operates at the network layer of the architecture and provides an interface between a variety of end user equipment and the packet network. For example, a user installs and operates a PAD at each major access point to an X.25 packet switched network. With appropriate software in the host computer, the messages can be applied to the PAD, which, in turn, divides the data into packets and appends appropriate headers for routing. Also, as we mentioned previously, the function of the PAD is often provided within the computers themselves, obviating the need for this external device.

As discussed in Chapter 1, the FEP is part of the user's host support system, which actually controls the remote terminals and other computer type devices in the data communication network. The FEP typically operates at the transport layer of the architecture to set up a reliable end-to-end connection. There are still higher layers, however, which implement the actual data communication applications running on the computers of the network. These concepts are reviewed in more detail in the following discussion of the seven-layer international standard architecture called OSI.

3.3.2.1 Open System Interconnection (OSI)

A standard layered architecture now under development and early use is the OSI model illustrated in Figure 3.13. In fact, we will continue to use this model as a frame of reference for understanding other such architectures. OSI is an acronym for *Open System Interconnection,* and was approved by a worldwide committee comprising many companies and telecommunication organizations called the International Standards Organization (ISO). The OSI model is also recognized by the CCITT, which is one of the two technical standards committees of the International Telecommunication Union (ITU). Conceptually, the layers of the OSI model follow the pattern of Figure 3.12. In the following discussion of the model, we assume that the data are created at the host, processed, and moved through the network to the terminal at the remote. The peer-to-peer type of connection would also follow a similar pathway and connections between peer-to-peer data processing equipment could be arranged similarly.

Beginning at the top of the model, the actual data to be employed by the computer application is made available for transmission from the host. This would consist of a block of bytes (a byte is eight bits, which in OSI terminology is called an octet). The number of bytes depends on the particular application and would only be recognizable to the computer and program creating and using it. A good example of such a block is an electronic mail message. *Presentation* layer service is probably done in the host, so the first action is to add bits called a *header* to the front of the message. This header might identify the destination device such as a specific computer terminal at the remote. At this point, the block would be referred to as a packet with an address for a *logical* device such as a computer memory location used as an E-mail mailbox. To reach the distant logical device desired, however, the network must translate this block into a routing over the network, since the logical address is only known to a few elements of the overall system. The *session* layer header, added by the FEP, identifies a major node or location in the network, which is the address on the network of the remote location where the particular terminal is located.

Below the session layer is found the actual functional structure of the data communication network, with packet switched nodes, modems (required for analog channels), and transmission lines. The *transport* and *network* layers are involved with packet routing and verification of data integrity, respectively. Finally, the

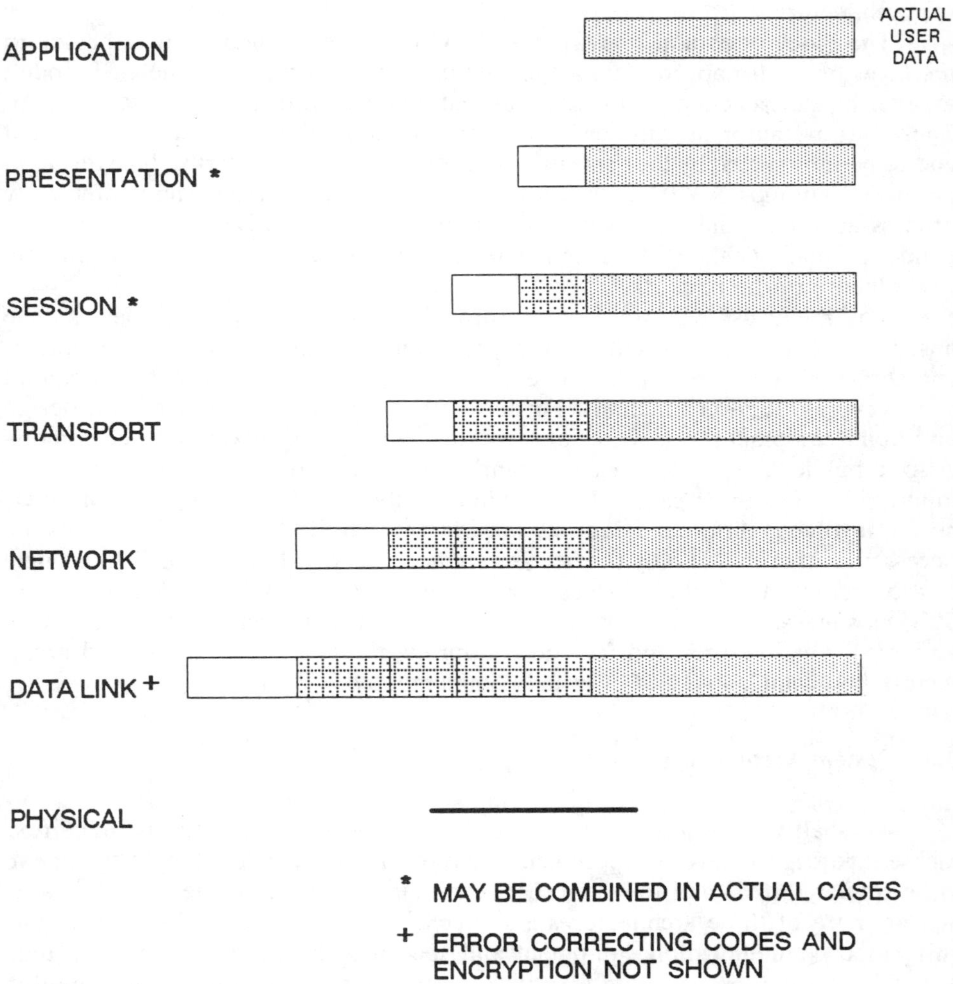

APPLICATION

ACTUAL
USER
DATA

PRESENTATION *

SESSION *

TRANSPORT

NETWORK

DATA LINK +

PHYSICAL

* MAY BE COMBINED IN ACTUAL CASES

\+ ERROR CORRECTING CODES AND
ENCRYPTION NOT SHOWN

Figure 3.13 The layered protocol stack of the Open Systems Interconnection (OSI) model.

data link layer operates on each specific path between data communication equipment, providing error control and monitoring of performance of the particular link. The accepted data link protocol is called *high-level datalink control* (HDLC), which is discussed in Chapter 4. The physical layer does not alter the bit pattern of the data link layer, but simply defines the type of connector, the wiring to pins, and the actual electrical characteristics at each pin. In relation to the OSI model, Recommendation X.25 provides for the physical, data link, and network layers of the architecture.

The block of data at one layer of the OSI model in Figure 3.13 is passed to the layer below for appropriate action. At the next lower layer of the OSI model, another header is added to provide information needed for that layer to perform. These bits are automatically removed by the corresponding element at the distant end to be interpreted by the "smarts" of the device. These headers reduce throughput by 50% or more. Without the headers, however, the network cannot efficiently process and route the information on its own. The alternative is to use an independent signaling channel to disseminate routing and synchronization information to the nodes of the network.

You might ask why we must go through all of these layers and levels. The answer is that this process is the only way that standard data communication devices and communication procedures can evolve. Otherwise, a business network cannot serve several different functions. Furthermore, a computer vendor cannot design and build equipment, nor write software that can be used by different customers. Also, a public value-added network cannot serve the breath of the business community. Finally, there cannot be any kind of efficient data communication operations involving different countries around the world. None of this would be necessary if there were only the need to connect one or a few terminals to a single host computer, and if the control software had to work for but a single application. Still, as suggested above, data communication networks must deal with a wide variety of possible uses and facilities, requiring that some sort of layered architecture be promulgated and eventually implemented.

3.3.3 System Architecture

We shall review some of the other key data communication architectures, each employing the layered approach. The difficulty is that while they may appear to be similar to OSI and to each other, communicating across the boundaries of two or more of these architectures is difficult if not impossible without very sophisticated communication equipment and special software. Therefore, the tendency for a given organization is to pick a particular architecture (and computer company) and stick with it for the long term.

We will review the prominent environments in the U.S. data communication industry. Only IBM and Digital Equipment Corporation (DEC) are presented as proprietary systems, followed by the generic environments developed by government and international standards organizations. Several other proprietary environments are in active use, such as those of Codex (Motorola), Data General, General Datacom, Hewlett Packard, Tandem, UniSys, Wang, and Xerox. The reader should refer to the technical literature that may be obtained from the vendor or can be found in reference material such as listed in the bibliography.

There are two reasons for discussing the IBM and DEC environments in detail. First, these two companies are effectively the industry leaders. Second, their environments represent the two basic alternate approaches for tying a data communication network together. IBM evolved a host-based architecture while DEC is associated with a peer-to-peer architecture. Note that IBM and DEC are adopting some of the attributes of each other's architecture to capture the benefits to be derived and, not insignificantly, because their customers are demanding that the two architectures be able to work together for particular applications.

3.3.3.1 IBM (SNA)

IBM supplies central computer systems in three size ranges: mainframes (the 4381 and 3090 series), mid-sized (System 38, the 9370 series, and the recently announced AS/400 series), and minicomputers (primarily the System 36). A standardized architecture is employed to allow information to pass among these variously sized computer systems and with remote terminals. This is the Systems Network Architecture (SNA), originally developed for application with the 370 series of mainframes in the mid-1970s. Today, SNA has the largest base of computers, telecommunication lines, and terminals, mainly because of the dominance of IBM in the computer marketplace. More detail on SNA can be found in [Cypser, 1978]. With the greater reliance on personal computers by organizations (e.g., IBM's customers), a face-lift of SNA has been announced and is being developed under the name Systems Applications Architecture (SAA). The theory behind SAA is that any computer program written on a PC within the SAA environment can be subsequently run without modification on a minicomputer, mid-sized, or mainframe employing the SAA operating system. This is a very radical concept, but one which users such as this author have long thought to have been possible. The structure of SAA is still undergoing development, and hence will not be considered further in this book.

Fortunately for IBM, its designers created a versatile architecture in SNA, effectively allowing network elements to communicate among one another. The nomenclature of SNA distinguishes between two classes of elements or nodes in a network: a *physical unit* (PU) is an actual piece of IBM or compatible equipment, while a *logical unit* (LU) is a function within an element with software (or firmware) which can be addressed by another element. Consequently, a PU is something that can be seen and needs electrical power to operate, while an LU has a specific (unique) address, like a telephone number, and performs a data processing or data routing function. A PU could contain one or more LUs.

The computers of other vendors and the other architectures cited elsewhere in this chapter cannot easily interface with SNA. The exception is through "gateways" between SNA and the other systems, provided by IBM or another supplier.

DEC, in particular, has introduced SNA gateway products for use in their customers' networks. The underlying data protocol that helps make SNA so powerful is Serial Data Link Control (SDLC). The results of SDLC have led to its adoption as a subset of the international standard data link protocol, HDLC.

The most basic arrangement of an SNA network is illustrated in Figure 3.14(a), which has been simplified to emphasize the critical elements. Central to the operation of the network is the host mainframe computer. Each of the LUs in the network has a unique address, which allows devices to send and to receive messages encoded under the control of the common software systems of SNA. Connected to the host is a communication controller or front-end processor. Much of the programming and protocols of the network are managed by the FEP, but the actual information and application programs reside within the mainframe. For example, the network supervisory software in the FEP is called the Network Control Program (NCP), while support for applications is managed within the host by Virtual Terminal Access Method (VTAM). The FEP can be connected to another FEP at a second host computer to allow the two machines to support the network and to back up each other. Communication between the mainframes is on an equal status or peer-to-peer basis. The bulk of the communication and information processing, however, is done to support a multitude of remote sites and terminals, as illustrated below the hosts in the figure.

The device at the remote site which interfaces between the line to the FEP and the actual terminals is called a *cluster controller*. Because of the host-based topology of the network, terminals and cluster controllers at one remote cannot communicate directly with those at another remote site. All communication is by way of a host computer. This is a key aspect of the traditional SNA approach. Coordination of data between two elements (an FEP and a controller, for example) employs the SDLC protocol.

In response to competition and customer demand for distributed processing, IBM has introduced features and software which allow peer-to-peer communication without passing through a host. Such an arrangement is illustrated in Figure 3.14(b). At the remote site might be located a mid-sized or minicomputer with terminals attached to it. This computer then could have a direct connection to another minicomputer, without passing through the host. We can imagine that the protocols, addresses, and other supporting overhead must be quite substantial in comparison to the relative simplicity of having all communication passing through a large mainframe. Equipped with the appropriate system software, a peer is referred to as LU 6.2, and the protocol for interconnection is called Advanced Peer-to-Peer Communications (APPC). The family of devices and software products relating to LU 6.2 and APPC are the fastest growing aspects of IBM's SNA business.

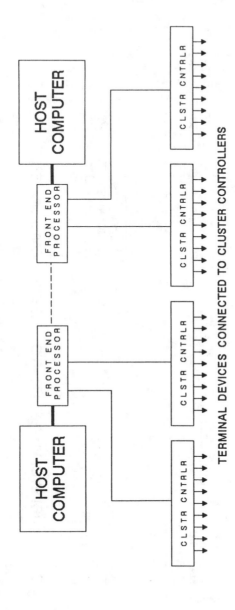

Figure 3.14 Simplified IBM data communication network environments employing Systems Network Architecture (SNA).

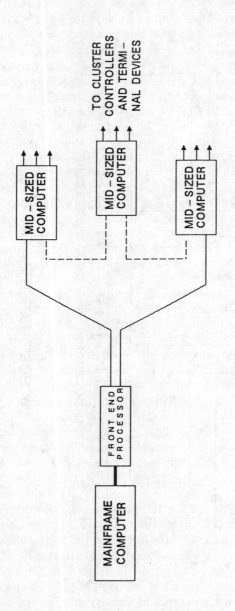

(B) EVOLVING PEER-TO-PEER IBM ENVIRONMENT

Figure 3.14(b)

A comparison of SNA layers in relation to that of the OSI model is provided in Figure 3.15. The physical and data link layers are functionally comparable, and SDLC is on each point-to-point connection between two communicating elements. The higher layers of SNA are decidedly different from OSI due to the host-orientation of the former. Note how OSI includes the network layer to handle routing of packets from node to node, and the transport layer to establish an end-to-end connection for sending or receiving and accounting for messages (consisting of one or more packets). In comparison, SNA uses the path control layer via the host to establish a connection in the form of a virtual circuit to a remote and to route packets between nodes. In SNA terminology, a *session* is an established connection via a virtual circuit from host to user, allowing an application to be run on the host computer. Higher layers are more specific to the information processing functions of the network. SNA data flow and the OSI counterpart session are used to maintain the connection during the application. The superficial similarities of SNA and OSI give way to significant differences in detail under careful examination. (The more careful the examination, the more differences are exposed.) Connection between systems running the different architectures would be at the highest level, requiring the use of elaborate software running in each system.

A recently introduced aspect of SNA receiving much attention is NetView, a facility for implementing an automated network management system. The system employs the resources of a mainframe in the network, which is consistent with SNA's host-based architecture, and is described in detail in Chapter 5. In addition to making NetView an environment for monitoring and controlling IBM products and their interconnecting links, the offering can also manage hardware elements provided by other vendors. The control link between the NetView host computer and the non-IBM elements is through an IBM-compatible personal computer running a software product called NetView-PC. Interfacing with a personal computer running NetView-PC would be by a separate low-speed data communication link. This implies that the non-IBM equipment must meet tight interface specifications and have software features in their *network management* (NM) system, which are compatible with NetView-PC. This author witnessed a demonstration by General Datacom of a network using their own proprietary NM system connected into NetView via an IBM PC. While not every conceivable control and monitor feature was passed over the link to the the IBM host, it was nevertheless a positive step.

Connections among PCs in an IBM-supported data communication network are provided through their LAN offering called the IBM Token Ring. Each PC is equipped with an adapter board, and special cable is used for the interconnection. While Token Ring uses an industry standard interface, its particular structure is geared toward IBM networks, and ultimately SNA. A Token Ring LAN can be connected to an FEP or computer, and thereby gain access to the SNA environment. This is very attractive to an organization with a heavy commitment to IBM, providing an efficient means to allow PCs to access mainframes and through APPC

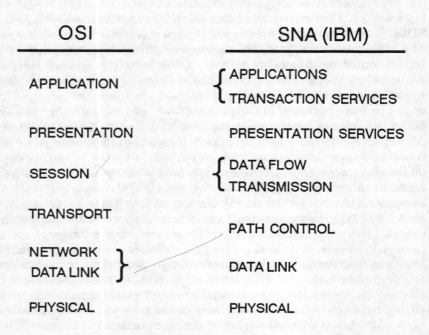

Figure 3.15 Correspondence between protocol layers of OSI and System Network Architecture (SNA).

to other PCs and minicomputers on the larger network. As an early adopter of IBM's networking architecture, American Express Travel Related Services has received an award for creating a highly effective worldwide computing environment for its wide diversity of service businesses.

3.3.3.2 DEC (DNA, DECnet)

Digital Equipment Corporation (DEC) has earned the reputation as the foremost manufacturer of minicomputers. The DEC minicomputers, particularly the VAX series, are very powerful for their size, and the way in which DEC computers can communicate among each other is also well recognized and respected. A small, compact, and relatively inexpensive version of the fastest minicomputer, called the MicroVAX, has been available for a few years.

The first to adopt DEC computers were engineers, scientists, and universities wishing to have real-time computing on a budget. Obviously, you could perform just about any task on an IBM system. DEC, however, was able to produce hardware and software for this class of applications which were lower in initial cost and could be reprogrammed rather easily as the system needed to be expanded. For example, DEC minicomputers were adopted for use in monitoring and controlling communication satellites of the type employed in domestic satellite networks. Because of continued enhancements in performance and reductions in cost, DEC systems are finding their way into a wide variety of commercial applications. As testimony to this, DEC is perhaps the fastest growing computer company of its size in the world.

The basic communication facility is called simply DECnet, which was introduced in 1975. The basic communication approach is the transmission and routing of independent packets, similar to the X.25 protocol. In contrast, the SNA routing structure uses a virtual circuit, where packets follow an identical path after the first packet passes through. The overall data communication environment is referred to as the Digital Network Architecture (DNA). The fundamental difference between DNA and SNA is that DNA is designed around peer-to-peer communication while SNA is host-based. Local area networking by using the Ethernet standard has also been adopted. DEC is now offering a means to connect VAX computers into an OSI network, through a gateway capability. Likewise, there is a gateway now available to connect into an SNA network, as mentioned previously. For example, a combined SNA/DNA network was implemented by Corning Glass to allow users throughout the network to access applications on any of their computers. The DECnet arrangement between remote locations routes data at relatively low rates, taking advantage of the peer-to-peer architecture. On the other hand, the SNA hosts run heavy computing tasks, sending results over higher speed lines to printers and other user devices. The Ethernet connection and gateway provide the interface between the two subnets.

Illustrated in Figure 3.16 is a typical DNA network, showing how minicomputers are connected directly to one another. There is effectively no master computer; information is routed from minicomputer to minicomputer (e.g., node to node) by using routing information attached to each of the packets of data. Terminals, other devices, and gateways to other networks are connected to the closest node. We may easily comprehend why this peer-to-peer structure was adopted by DEC, a company which basically makes minicomputers and not mainframes. A strong feature of DNA is that new DEC minicomputers can be added as nodes to expand the network and its computing power.

While DNA is very different in detail from SNA, and does not tie directly into it, it is nevertheless a layered architecture. This is shown in Figure 3.17, where DNA is related to the OSI reference model. The data link layer of DNA employs a higher level protocol called Digital Data Communications Message Protocol (DDCMP). At the routing layer, messages are routed individually in a single packet datagram. (This mode, while provided for in X.25 is not actually used therein.) There is a network hierarchy of two levels, arranged in stars with their hubs

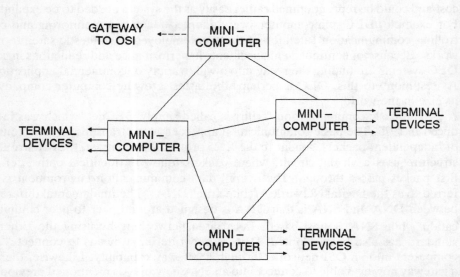

Figure 3.16 A simplified DEC data communication network employing Digital Network Architecture (DNA).

interconnected by additional data lines. While the messages are handled individually at the routing level, a virtual circuit can be established at the next higher layer of DNA.

One of the attractive things about the DNA layering is that the terms used are rather descriptive of the function of the particular layer. Also, there is nearly a direct correspondence between the OSI model and DNA. Keep in mind, however, as with SNA, that DNA does not tie directly into OSI. Data format to be routed from DECnet into an OSI network must be changed within a gateway.

3.3.3.3 Application Layer Protocols

Efforts by major industrial organizations, governmental bodies, and international standards groups in the data communication field are producing fruit in the form of common application structures. The X.400 application layer protocol, as previously described, is a case in point. Any of these high-level protocols ride atop a layered architecture, notably OSI. The definition of a higher level application layer protocol is a set of procedures and message formats that is developed around a specific group of tasks. In the case of X.400, the application is electronic mail. Two other well recognized higher level protocol sets are described in the following sections. These are the Manufacturing Automation Protocol–Technical Office Protocol (MAP/TOP) and the Transmission Control Protocol–Internetwork Protocol

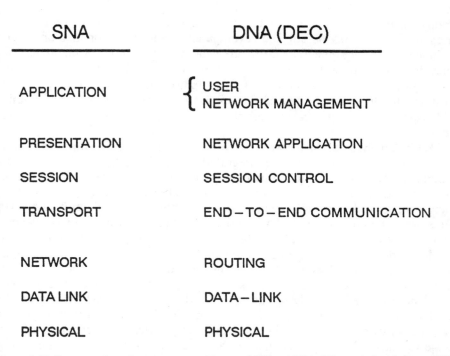

SNA	DNA (DEC)
APPLICATION	{ USER NETWORK MANAGEMENT
PRESENTATION	NETWORK APPLICATION
SESSION	SESSION CONTROL
TRANSPORT	END – TO – END COMMUNICATION
NETWORK	ROUTING
DATA LINK	DATA – LINK
PHYSICAL	PHYSICAL

Figure 3.17 Correspondence between protocol layers of OSI and Digital Network Architecture (DNA).

(TCP/IP). The former is intended to meet the widest variety of needs in the industrial and engineering environment, while the latter is geared toward allowing totally incompatible data communication environments to communicate effectively and to function ("interoperate") with one another. These examples are representative of the problem and solution to which data communication is applied, and they are all undergoing considerable development and initial use in a variety of applications.

MAP/TOP

In an internal development program initiated in 1983 by General Motors Corporation, MAP was conceived as a means of allowing *computer integrated manufacturing* (CIM) systems to function effectively. The job of the MAP environment was to handle information flow for slow-speed, low-volume uses on one end of the spectrum, all the way up to the highest speed links characteristic of large computer systems. Message formats are standardized, and the document transfer applications cited in a previous section will be used extensively. To facilitate the range of speeds and degree of interaction required, the network backbone was envisaged as being LAN-based. The development program was opened up by GM to include network equipment and software vendors as well as other manufacturing corporations who could generate an economy of scale. More recently, GM has begun to pass the sponsorship for MAP development to industry groups. Clearly, even if the topology and utility of the application network were high, nothing would happen on a large scale unless the unit cost of network connections was reasonably low.

To implement MAP on a cost-effective basis, companies like Intel, Motorola, and Concord Data Systems produce integrated circuit chip sets, which can be built into units and even within automation equipment, allowing connection to a coaxial cable LAN. Third party vendors integrate products to customize MAP systems for particular manufacturing environments, not unlike how home and building security systems are installed in the U.S. Having the critical pieces of technology available in chip form along with a standard architecture (e.g., MAP) fuels the application.

MAP's development cycle has paralleled that of OSI, and the leadership has attempted to keep MAP within OSI's seven-layer structure. MAP, however, is an outgrowth of the perceived need of a group of powerful users led by GM. With financial backing and a ready market, the MAP group can push standards and technology development in a specific direction. The kinds of uses envisaged for MAP were practical in nature. For example, the protocol can recognize specific message formats used to send controls to an industrial robot or automatic milling machine. Forms used for inventory management and parts ordering (i.e., document transfer) can be considered in the protocol. All of these particular uses are geared toward manufacturing in an automated factory.

The other higher level protocol set, Technical Office Protocol, was proposed by Boeing Computer Services, and is gradually being merged into MAP. Behind TOP is a drive to integrate wideband and narrowband services in an engineering work station environment. Being able to tie a finished engineering design to the manufacturing process is an exciting prospect. A concept can be taken from the computer screen, to manufacturing plans, to final product, without a single piece of paper being involved. The TOP network approach allows data and video services to share the common LAN coaxial cable.

Development of MAP is ongoing and a common standard still evades the industry. The current generation is called MAP 2.1, for which considerable work has been done. Many installations were made using MAP 2.1, but its principal drawback (if indeed this is one) is that the lower layers of the OSI model are not rigorously followed. MAP 3.0 is coming up as a major revision, which carefully follows OSI. The advantage of MAP 3.0 is that integration with future networking systems and products is more likely ensured, since 2.1 may ultimately expire. Organizations now considering a direction in the MAP environment have a difficult choice to make: choose 2.1 for currently available hardware at reasonable cost, or choose 3.0 for future compatibility and the prospect of waiting for workable products.

The first demonstration of MAP 3.0 by a number of vendors and application developers took place in June 1988, at the enterprise networking event in Baltimore, MD. A network linked dissimilar computer systems running such applications as engineering design, materials resource planning, invoicing, manufacturing, and final assembly. Users who attended this landmark gathering were generally impressed with the rapid progress of implementation of this OSI-based generation of MAP/TOP.

TCP/IP

Whereas MAP/TOP is the result of industrial demand and OSI stems from international standards-making efforts, the TCP/IP suite of protocols was developed under U.S. government sponsorship. TCP/IP is actually a full set of protocols covering all of the necessary layers for effective data communication. The intention of TCP/IP was to establish uniform data communication applications over entirely different computer systems and communication network architectures. This is the same objective as OSI, but TCP/IP predates OSI and is a precursor to it. The Defense Advanced Research Projects Agency (DARPA) of the U.S. Department of Defense has funded and encouraged development of a nationwide data communication network called *ARPAnet* (for Advanced Research Projects Agency Network), which now involves most major U.S. universities and government research establishments. Over this network, which is packet switched, users initiate data communication sessions and exchange information with the application-layer protocols of TCP/IP.

TCP/IP proved itself as an effective capability for using various computers to perform tasks which otherwise would not be possible. There is a belief that TCP/IP will be supplanted by OSI when it is used on a widespread basis. TCP/IP, however, is well ingrained in government-sponsored networks, and for commercial users, it is an effective linkage for applications that must function now. Uncelebrated though it may be, TCP/IP is currently integrated into systems employing the Unix operating system developed at AT&T Bell Laboratories. Communication between machines running Unix is with TCP/IP, and therefore multitudes of computers and data communication links already employ TCP/IP on some basis. Extension to more elaborate networks and applications, involving equipment of multiple vendors, would appear to be relatively easy to initiate. There are approximately 150 vendors of TCP/IP products, and the protocols are used extensively in local area networks and wide area networks as well. Bear in mind that TCP/IP preceded OSI, having been designed in 1973 in conjunction with ARPAnet. OSI took the layering concept from TCP/IP, which itself was not made final until 1980.

The basic composition of the TCP/IP suite of protocols is depicted in Figure 3.18. Application-layer activities at the top are designated *upper layer protocols* (ULPs). One standard application is *telnet,* a terminal-to-host communication software package which allows PCs to access central computer systems over the TCP/IP structure. *File transfer protocol* (FTP) expands upon telnet by facilitating the efficient and error-free transmission of bulk data files. The next layer down is TCP at the transport layer for access into the data communication network. This follows the concept of the OSI network layer, but is not compatible with it at this time. TCP can set up a virtual circuit over the network, connecting the ULP application at one end to its counterpart at the other. The final layer in Figure 3.18 is the network layer, in the form of the *internet protocol* (IP). Owing to developments on ARPAnet, IP transmits data in independent packets (i.e., the datagram mode). This approach is the most flexible for dealing with the wide variety of transmission modes and systems experienced in domestic and international networks. Since IP routes packets individually, the role of the TCP layer above is to collect packets, arrange them in correct order, and deliver them to the user. Procedures for recovery from errors and disruptions are also included in TCP.

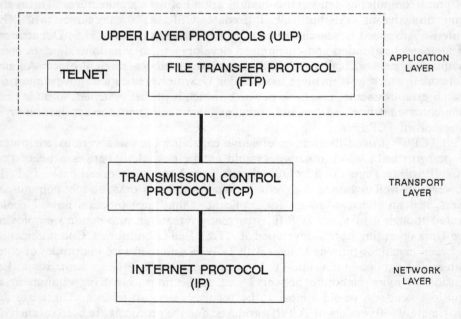

Figure 3.18 The TCP/IP "Suite" of protocols developed under ARPAnet.

The real power of TCP/IP lies in its simplicity and proven performance after many years of operation. From a performance standpoint, TCP/IP can employ the widest possible variety of transmission systems, including standard telephone channels, high frequency radio, satellite, and undersea cable. The datagram mode tends to make the system very powerful in terms of its ability to recover from poor circuit quality, variable time delay, signal fading, and link breakdowns. (This is, in fact, the environment that tended to exist when the system was developed and first proved out.) Software companies are producing commercial products which employ the TCP/IP suite to connect various devices into networks so that more elaborate applications can readily be developed. The general feeling is that since TCP/IP is here and it works, using it for real tasks makes a great deal of sense. A philosophy of "if it ain't broke, don't fix it" would seem to exist among advocates.

GOSIP

In response to OSI inroads with government agencies in general and the General Services Administration (GSA) in particular, the U.S. government is adopting a standard application layer set called Government Open Systems Interconnection Profile (GOSIP). The National Bureau of Standards (NBS), acting as technical consultant, has endeavored to gather requirements from government users and is preparing a protocol suite based on the OSI layers. Note that NBS is also operating an OSI test bed laboratory, where commercial devices and network facilities can be exercised under standard conditions. The intention at some point is to impose GOSIP on major government procurements of data processing and data communication systems. Quite logically, GOSIP is intended to supplant TCP/IP because of adherence to the newer internationally recognized protocols. The benefit to the government would be from the availability of a newer generation of more advanced network products. Also, there could be commonality with facets of the MAP/TOP development activities within the industrial sector. As an interim measure, the U.S. Navy is procuring gateway software and products to permit connections between TCP/IP and GOSIP.

3.4 BUSINESS VIDEO COMMUNICATION

Video communication is expanding from the commercial arena into the business world. In particular, the whole area of private broadcasting and video teleconferencing has been given a lot of attention in the U.S. and European trade press. To date, the actual use of these capabilities has been rather limited. Video transmission equipment is still very expensive. (You cannot now reach for your video telephone, as was suggested as far back as 1964 during the New York World's Fair.) Furthermore, in most office settings, you cannot walk down the hallway and enter a video teleconferencing room. Yet, many organizations make effective use of video in their daily business. The telecommunication facilities used to carry business video can be either satellite or terrestrial based. Recently, the RBHCs have been promoting video as an aspect of ISDN. Three particular areas of business video communication are discussed in the following paragraphs: *private video broadcasting, interactive video training,* and *teleconferencing.*

3.4.1 Private Video Broadcasting

Organizations, public and private, employ video transmission to broadcast information to many locations and multiple audiences. These broadcasts can be divided into those intended for reception by the public and those only intended for viewing by an internal, private audience. Private broadcasting represents a form of marketing communication, destined primarily for the news media and television networks. Using commercial satellites, the broadcast can be picked up by virtually any television station in the country and then rebroadcast over the local airwaves. A satellite receiving dish is needed to pick up the broadcast. The entity sending the broadcast should also inform interested parties of the time, satellite, and frequency channel. This medium has been used for press conferences, allowing the presenter to control some of the important aspects of distribution. Obviously, the delivery of the broadcast is at the discretion of the final distributor so that viewing by the general public cannot be guaranteed. Production of such broadcasts proceeds like the type of advertising commercial or press conference being emulated from normal over-the-air television.

The private broadcasting category is definitely controllable as to both content and delivery. An organization using this medium will often own the video production facility and uplink earth station. At the headquarters of the organization are some type of television studio and tape facility so that "programming" can be originated. Production crews, who are either full-time employees or hired by the event, can gather material on location on tape using portable cameras. Examples of the type of material involved include interviews with corporate executives, displays and demonstrations of new products, and training sessions in new procedures or government regulations. The cost of producing a private broadcast ranges from minimal, with an established studio and internal staff, to hundreds of thousands or even millions of dollars for an elaborate annual report.

A private internal broadcasting network is often used to disseminate corporate-wide information. Weekly or monthly "pep rallies" are common in dispersed organizations. One such network supports the merchandising efforts of a large retailer. Another is used by a national brokerage to unify the organization. To maintain privacy of transmission, the video and audio portions can be scrambled by using one of several techniques, including VideoCipher I or II and B-MAC. More information on scrambling systems and video broadcasting can be found in [Elbert, 1987].

3.4.2 Interactive Video Training

This business video medium is directed toward the training needs of a large organization, including corporations and government agencies. Interactive training differs from classes on public television in that students can speak with the instructor over a dedicated telephone connection. Two-way video is currently very expensive in terms of equipment and transmission lines, and is not required in any case. (A discussion of two-way interactive video teleconferencing for meetings is provided in the next subsection). One of the first applications was in off-campus college courses offered for credit. Large technology-based companies like Texas Instruments and Hughes Aircraft Company maintain internal classrooms, which receive terrestrial microwave transmission from a local university. The authorized frequency band is called the Instructional Television Fixed Service (ITFS), which is defined as a terrestrial point-to-multipoint service. During conduct of the class, the instructor stands in front of a camera and takes questions over a telco-supplied dedicated audio line to each remote classroom. The obvious reason for using video broadcasting is to reduce travel time of students and to extend classes to employees who cannot otherwise attend. The University of Virginia and Virginia Polytechnic Institute, as well as other institutions, have taken this a step further by employing satellite distribution to dispersed sites.

The Interactive Satellite Education Network (ISEN) was put together by IBM to implement a nationwide capability for corporate training and teleconferencing (see Figure 3.19). Instructors at four different studios around the country can originate classes independently and simultaneously. At twenty remote receiving sites, some collocated with the studios, are found classrooms with space for sixteen students each, making for a comfortable class experience. In total, four classes involving eighty classrooms with up to 1280 students can participate. Using eight television channels simultaneously would be prohibitively expensive were it not for the fact that a single C-band transponder is shared by eight video transmissions, each digitally compressed to the T1 level. An audio and data response return link is established from each receiving site back over the satellite.

The interactive satellite education network allows the instructor to be seen and heard by the student in classrooms networked for the particular class. In addition, the instructor has a second video channel, which simultaneously displays a computer-generated graphic, 35-mm slide, or color transparency. A facility for showing a film or video tape over the second channel is also provided. To assist the instructor in evaluating student comprehension, a key pad on each student's desk permits the entry of a real-time response to a multiple-choice question over the air. Students may indicate to the instructor that they have questions by pressing another button, and only if the instructor activates the microphones can students be heard in the selected classroom. In effect, the instructor has considerable control

Figure 3.19(a) The instructor's studio which originates two video channels for the IBM Interactive Satellite Education Network. (Photograph courtesy of Hughes Communications, Inc.)

of the class and students may interact with the instructor as if they were present at the same location. Experience with the network has been excellent, and student surveys consistently show a preference for this approach over conventional classroom lectures.

An obvious feature of a private video education network is that it can also serve as a broadcasting medium for announcements and promotions, as discussed in the previous subsection. All that you need do is invite the outside participants to the regular classrooms where the broadcast can be viewed. Having the voice return link would permit the media or outside customers to ask questions of the people on camera. (Where else can you talk back to your television?)

3.4.3 Teleconferencing (Meeting)

The concept of the *picturephone* has been applied on a limited basis by a number of pioneering organizations. Instead of a person-to-person link, however, teleconferencing provides a full-time video link between two groups of people

Figure 3.19(b) The ISEN classroom at an IBM facility showing dual video displays and individual student control units. (Photograph courtesy of Hughes Communication, Inc.)

conducting a meeting. (In fact, AT&T recognizes the attractiveness of this approach by referring to their offering as Picturephone™ Meeting Service.) The basic premise is that face-to-face meetings across the country are too expensive because of the value of people's time and the cost of air travel. Holding group meetings between a pair of sites apparently tends to improve communication and productivity. The point is made that, without teleconferencing, many potentially important participants will not be present due to the limitations of intercity travel.

A typical arrangement of a video teleconferencing facility is designed to conduct a joint meeting at two locations (see Figure 3.20). The facility at each end is specially arranged for this purpose, with participants sitting on the side of a table that is opposite to the cameras and monitors. Teleconferencing meets a particular need which cannot be satisfied by any other means (short of extensive travel). A good example is a regular weekly review meeting for a critical project. At this meeting, the attendees review the results of the previous week and agree upon actions for the coming week. Interaction can be critical in a rapidly moving project, where days or even weeks of effort can be for naught if group discussion were not

Figure 3.20 An example of two-way interactive video teleconferencing for meetings. (Photograph courtesy of Hughes Communications, Inc.)

possible. The video element is important for giving the feel of a face-to-face meeting, communicating visual clues from facial expression and body language. Images from hard copy and computer graphics can be displayed for all participants to see or even to modify. In a successful video teleconferencing meeting this author attended, a participant on the other end made clear, in real time, through his body language that he was not pleased with the price being offered for a particular telecommunication service.

The two-way interactive teleconference is usually transmitted by digital means. This is done so that bandwidth can be reduced, as discussed for the interactive education network cited in the previous subsection. Furthermore, the digitized signal can be protected from unauthorized reception through encryption. The quality of compressed digital video is generally thought to be adequate for the purpose of conducting meetings. This is in contrast to the broadcast quality of video signals in private and interactive video broadcasting. Today, several commercial organizations and the U.S. government rely upon this technology. Due to the high cost of *coder-decoder* (codec) terminal equipment and that of transmission (currently at the T1 level), however, no nationwide public access network yet exists. This severely limits the utility of video teleconferencing. Trials of personal teleconferencing akin to picturephone are appearing. As transmission bandwidths are reduced through improvements in compression and the cost of codecs and rooms decreases, greater use and acceptance will most probably follow.

Chapter 4
Digital Integration of Networks

The trend from analog to digital communication has nearly taken its full course, and today most communication applications are digital in nature. Digital presents the opportunity for efficiently packing the various signals into common streams of data, a result of the integration of various services on a digital wide area network. In addition to being efficient, digital integration typically employs low-cost, bulk long-distance transmission capacity at the T1 or DS-3 level. These aspects of the technology make it the cornerstone for coming generations of private telecommunication networks, as was emphasized in Chapter 1.

Digital integration is possible because digital data from a variety of sources can be combined in binary form. Each type of information — voice, low speed data, facsimile, video, *et cetera* — has particularities, however, which the network must accommodate. At one end of the spectrum, we have conventional telephone communication in the form of POTS. The local exchange and long-distance networks accommodate telephone calls containing voice traffic in an efficient and reliable manner. The hierarchical structure of the switching systems is geared toward setting up a circuit for each call (e.g., circuit switching). Once established, the circuit can carry voice frequency traffic, such as speech, low-speed data from compatible modems, and facsimile (which also uses modems). Digitization in this context simply means that the voice frequency band is sampled at a rate of 8000 times per second, and each sample is converted by PCM into seven or eight bits of digital data. This process of encoding is elaborated in the next section.

Digital communication goes many steps further when user information is digitized at the source and applied to the network at rates much higher than a voice frequency channel can carry. Here is where the need and opportunity for digital integration arises. Alternatively, the user may have only a small quantity of data to transmit to one or more distant locations, not justifying a dedicated connection. For relatively infrequent transmissions, the dial-up telephone network can be both acceptable and lowest in cost. The next level of service and cost is to arrange for a dedicated access line to a public data network to connect to a variety of occasional users or distant hosts.

Due to differences in the bandwidth, timing, duration and routing of information among these modes, digital integration is no simple matter. In fact, no standard architecture does everything for everyone. That is not to say that concepts like ISDN will not someday produce the desired general approach, and efforts in that regard should be respected and pursued. At present, however, the telecommunication practitioner or manager must approach digital integration with the variety of tools and services currently available on the marketplace.

We begin with a review of the technologies of signal processing and digital transmission. This is an overview of relevant, proven approaches on the commercial market. We assume that the reader has a basic concept of voice and data communication. If not, we suggest that a basic text (such as [Noll, 1988]) be examined. Later in this chapter, we will elaborate the classes of equipment and systems that implement digital integration, including terrestrial and satellite transmission technologies.

4.1 DIGITAL TECHNOLOGIES

To understand the technologies used in integrated digital communication networks, we first need to examine the processing and switching elements in a hypothetical end-to-end link. Each significant element (illustrated in Figure 4.1) processes, connects, routes, and carriers information through the network. The particular chain shown in the figure covers the transmitting end of the link from the information source (such as a telephone instrument, facsimile machine, or ASCII terminal) to the transmission channel over cable, microwave radio, or satellite. A discussion of the capabilities of these three long haul transmission media is presented at the end of this chapter.

There is typically symmetry between transmitting and receiving ends so that the order is reversed on the other end of the transmission channel. This was not shown in Figure 4.1 to simplify the picture. For example, on the sending end, the last block of the chain is the modulator, while the first block on the receiving chain is the demodulator. We have encoding and decoding, compression and decompression, multiplexing and demultiplexing, and so on. Each of the block functions is encountered either directly or indirectly, and sometimes the order of elements can be altered, depending on the architecture employed by the particular equipment manufacturer or network operator. The function of each element is discussed in subsequent sections of this chapter.

The information generally should be reproduced at the distant end with satisfactory accuracy or fidelity, within an acceptable time delay, and without loss of content. A big advantage of digital transmission is that the information is regenerated along the way, which eliminates noise accumulation associated with analog systems. With regard to accuracy, every communication system introduces distortion to some extent. Therefore, the network is engineered to certain standards of reproduction. This is most appropriate where the network takes analog information from the input source. The most stringent quality standards apply to television signals intended for broadcasting. Likewise, high fidelity audio for FM stereo broadcasting also must meet demanding standards. Voice communication is tolerant of all kinds of distortion and noise, and hence a good digital network will deliver excellent quality. A possible exception is when the PCM channel is further compressed, resulting in a significant degree of quantization error (explained further on in this section) or clipping. This is particularly a concern when the telephone channel must pass in-band data at 9600 bits per second.

Requirements on time delay vary between signal types. From a physical standpoint, every communication path will introduce time delay. The speed of light cannot be exceeded, and a long-distance connection typically has an intrinsic delay of 5 to 10 ms per 1000 miles (depending on the transmission medium). In the case of a link over a geostationary communication satellite at an altitude of approximately 22,300 miles, the delay is nearly independent of ground distance and approximately equal to 240 ms. To these propagation delays must be added processing delay within digital multiplexing and switching, as well as that of any computer storage used in the case of data. If the data to be carried cannot tolerate time delay, either the data must be transmitted at a very high rate or distances must be held to an absolute minimum. Often, however, the problem of time delay can be overcome with digital compensation equipment within the communication channel. For example, some data communication protocols require immediate response from the receiver for every character sent. Long delay over a satellite link would slow the transmission speed to a snail's pace. A delay compensation unit (DCU) on the sending end can, however, be used to respond back to each character from the data source, while sending continuous blocks of data over the satellite link. Similarly, voice communication is affected by time delay over terrestrial and satellite circuits. The problem of echo stems from voice signals being reflected back from the distant end and being heard by the talker, albeit at an attenuated volume. The longer the time delay, the more objectionable is the echo. Data communication using modems over a voice frequency channel is also degraded by echo. To remove echo, both ends of the circuit are equipped with echo cancellers, which adaptively remove echo before it can be returned to the sending end of the circuit. A discussion of this consideration can be found in [Elbert, 1987].

The problem of reliably routing information over a network can be divided into two parts. In circuit switching, the connection is set up at the beginning of the call and maintained continuously for its duration. The critical aspect is call setup, which is dependent on a signaling system that instructs the intervening switches to make appropriate connections. Once established, the circuit is essentially physically stable. There are phenomena which can affect circuit performance during the call. In terrestrial cable circuits (fiber optic or copper), there is little that can change. Microwave radio paths over terrestrial and satellite systems experience signal fading due to a variety of random radio propagation phenomena. During a fade, signal quality will diminish and occasionally drop below the standard for acceptability. When links are properly engineered for telecommunication services, the occurrence of this will be less than 0.1% of the time.

The following paragraphs describe the function of each key element in the processing and switching chain illustrated in Figure 4.1. Because the emphasis of this chapter is on functional capabilities (end results), the discussion will be necessarily brief. Each topic can be studied in detail using references identified in the bibliography. For example, we review different voice compression techniques (e.g., PCM, ADPCM, and RELP), but do not cover the details of the algorithms, nor the electronic circuits used to generate them. What is important in the context of this book is the relative merits of each.

4.1.1 Signal Processing of Voice, Data, and Video

The three functions of source encoding (to convert from analog to digital), compression (for bandwidth reduction), and encryption (to provide security) are part of the topic of signal processing. In some of the more modern equipment, the first two functions are combined together. Encryption, however, may or may not be provided, depending on the need for transmission security. Current approaches for each are reviewed in the following paragraphs.

4.1.1.1 Source Encoding

Obviously, information to be sent through an integrated digital network must be in digital form. Source encoding from analog to digital (A/D) is often performed by a device called a PCM codec, the combination of an encoder with a matching decoder for the receiving path. The A/D conversion process was introduced at the beginning of this chapter, and is explained in detail elsewhere [Schwartz, 1980]. When codecs were introduced prior to the advent of the integrated circuit, they were bulky, costly, and consumed a large amount of electrical power. Today, a voice codec is embodied in a single chip and can easily be made part of a telephone instrument. Most telephone systems place the codec in the multiplexer, PBX, or exchange switch. Standard PCM transmits one voice channel at 8000 samples per second, with eight bits per sample, resulting in a transmission rate of 64 kb/s out of the encoder. Seven of the bits identify one of 128 discrete amplitude levels in a scale arranged in nonlinear steps to enhance voice quality. The remaining bit is available for synchronization and signaling. This structure is in common use and explains why only 56 kb/s are available out of 64 kb/s for clear data communication

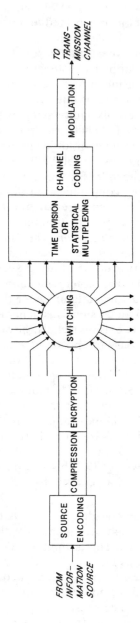

Figure 4.1 Digital signal processing and switching elements in an information source-to-transmission channel chain.

over a PCM channel. The extended super frame (ESF) arrangement being introduced by AT&T into their digital backbone is expected to open up the DS0 channel for the full 64 kb/s of data throughput.

Standard PCM produces excellent voice quality for the frequency range of 300 to 3400 Hz used on the telephone loop. The listener will hear no noise on the circuit provided that errors on the transmission link are controlled or corrected. A threshold error condition occurs at a rate of one error in 10,000 bits received (10^{-4}), wherein there are approximately six faulty samples per second reaching the receiving end. The effect is an audible "click" or "pop" of randomly varying strength. This is not particularly noticeable or objectionable to the telephone subscriber; in fact, the normal error rate is 100 or 1000 lower, resulting in clicks and pops occurring every 15 or 150 seconds. Further link degradation produces rapid increase of the click rate, making communication difficult, particularly for in-band data. The main type of distortion under error-free transmission is introduced by the fact that the voice frequency information input is forced to fit within 128 amplitude steps. This produces *quantizing error,* wherein the signal recovered by the decoder is slightly askew because the encoded signal was not a perfect description of the input. Quantizing error does not affect speech particularly, but rather in-band data transmission using data communication modems running at speeds above 9600 b/s. This aspect is discussed later under the topic of modems.

Encoding of voice frequency and audio signals is generally quite satisfactory using PCM. The process of encoding television signals, however, is considerably more complex. Using PCM, it is necessary to encode at a sampling rate which is at least twice the highest frequency of the input signal. For a video signal which lies between 20 Hz and 4.2 MHz, this theorem of communication demands that the sampling rate be 2×4.2 MHz $= 8.4$ MHz. The number of quantizing levels for video has been investigated, considering that the color components occupy considerably less bandwidth than the black-and-white (luminance) components of the picture. A transmission rate of 135 Mb/s is generally recognized as needed for broadcast quality video, acceptable to the commercial television networks. Compression of television signals down to 45 Mb/s has only a minor effect on picture quality, but an encoding standard acceptable to the commercial networks has not been adopted as of 1988.

Other forms of information can be encoded with the A/D process embodied in PCM. Graphic images from photographs are scanned with photosensitive detectors capable of reproducing a gray scale. This is input to a PCM encoder. Groups 3 and 4 facsimile machines operate in this fashion. The Group 3 facsimile machine is designed to operate over an analog telephone circuit using data modems on both ends. These machines operate at speeds up to 9600 b/s, allowing the transmission of a page in approximately 20 seconds. In contrast, the Group 4 facsimile machine interfaces directly with a digital line such as from DDS or a PDN. Quality of reproduction is almost as good as that obtained from a photocopying machine, and speed of transmission is under one second with a 56 kb/s data line. Extremely high resolution is obtained with specialized laser scanning and reproduction systems used to transmit photographs and original pages of newspapers and magazines. These require transmission rates in the range of 56 kb/s to 1.544 Mb/s, depending on the available bandwidth and desired speed of document delivery. Such high rates must be obtained on a digital backbone employing either terrestrial fiber, microwave, or satellite communication. The local loop must also be capable of these rates. An example of newspaper page transmitted via satellite using this technique is shown in Figure 4.2.

4.1.1.2 Compression

Reduction of transmission bandwidth is the function of compression, which involves taking PCM samples and processing them to remove redundancy and constant elements of the original signal. One consequence of using compression is that the end-to-end circuit must incorporate compression-decompression on both ends.

If the number of samples per second is maintained, the number of bits per sample is reduced. The other approach is to reduce the number of samples per second. A given compression technique would do either or both reduction processes. To work, the algorithm makes assumptions about the nature of the information input. The more that is known about the input, the more compression can be accomplished. This is a double-edged sword, because, as the algorithm is fine tuned to the input signal type, the channel will be less tolerant of other signal formats which may occasionally appear at the input. The classic example of the dilemma is the transmission of voice band data through a telephone channel. Standard PCM will properly pass 9600 b/s. As compression is introduced, the redesigned channel still matches speech characteristics. The modem data signal, however, differs appreciably from human speech with a deleterious effect on possible throughput. Therefore, any compression system should be tested thoroughly for the full range of signal types that the network is expected to carry.

The voice compression algorithm that is gaining acceptance in private telephone networks and even some public networks is *adaptive differential PCM* (ADPCM). An international specification was adopted by the CCITT, and various manufacturers now produce equipment which generates ADPCM during the A/D process. Alternatively, a device called a *transcoder* takes two standard PCM streams of 24 channels each (i.e., each at the T1 rate), and compresses that by using binary mathematics to one T1 stream. Each voice channel now employs 32 kb/s instead of 64 kb/s.

Figure 4.2 Remote printing of the newspaper, *USA Today*, is made possible by satellite transmission: (a) the uplink in Roslyn, VA to plants around the country; (b) a front page on the laser facsimile machine.

A compression algorithm using voice switching called *time assigned speech interpolation* (TASI) was developed by Bell Labs in the 1950s for use over transatlantic telephone cables. The basic approach is illustrated in Figure 4.3. In the example, six input telephone circuits are switched onto the long-haul link containing only three duplex channels when there is actual speech activity on the input line. The receiving end (on the right) has a switch bank that is directed by the sending end for properly matching the switched pattern. For example, the talker on line 1 is active and the switch has been set to connect this input to circuit A. Initiation of activity on line 2 is detected by the speech detector, and that line will be connected to an available circuit. This not unlike an electronic game of musical chairs. It tends to work, however, because on the average talkers are active only 40% of the time. The size of the bundle actually must be at least 20 circuits, in which case the compression of two-to-one is achieved.

Experiments with a digital version of TASI, called *digital speech interpolation* (DSI), were conducted over international satellite systems along with the development of TDMA, discussed later in this chapter. Instead of switching "spurts" of speech, DSI routes speech on a sample by sample basis. There is no need for analog speech detection, and consequently switching is essentially undetectable. In effect, the speech samples from the codec output are examined by the DSI sending unit for the presence of information and then routed to the destination in much the same way as a packet switched network. The fast packet switching technique to be discussed in a subsequent section easily provides the DSI function because of the manner in which it routes voice communication from end to end.

There are other forms of compression which can be applied to particular information classes. In facsimile, it is convenient to scan a document and to transmit data only when there is image information present. White areas are compressed because the sending unit can indicate to the receiver that a specific portion is blank. This speeds up transmission by a significant factor so long as the document has enough blank space. In data communication, compression is accomplished in much the same manner. The statistical multiplexing scheme to be discussed later is a means of compression much like DSI.

4.1.1.3 Encryption

The third type of signal processing to be considered is encryption for security. This topic was introduced in Chapter 3 for voice communication. Codes and ciphers are a recognized means of protecting information from interception by unauthorized parties. In modern digital communication, encryption is easily accomplished on the information. The process of encryption is purely mathematical and can be performed at the transmitting end of the channel. The user may encrypt his or her data within the computer system, and then route it safely over any network. Multiple levels of encryption could be done, which would be the case if encrypted

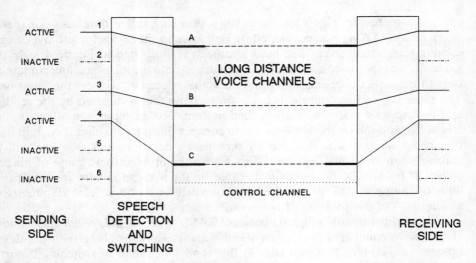

Figure 4.3 Speech interpolation is used in long haul transmission to compress the number of channels required.

user data enter an encrypter at the next level, for example at the transmitting end of a larger public or private network.

Encryption involves rearranging the bits of the information data stream according to a *pseudorandom* process. The rule for this random rearrangement is established within the hardware or software that performs the operation. Furthermore, the process is synchronized to a particular starting point, which can be altered to enhance security. The initiation process is tied to a digital string called the *key*, which is analogous to the key to a safe containing the information. The key is entered into the cipher equipment prior to initiating an encryption session. If you know the process but do not have the key, decryption is mathematically possible but practically impossible; likewise, the key without the process is equally worthless. Security therefore involves the protection of three things: the information in the first place, the encryption process, and the key. A new technology eliminates the key through what is called the *public key* process, which is built into the encryption device.

The National Bureau of Standards has produced the Digital Encryption Standard (DES) for use in commercial applications. As mentioned in Chapter 3, the algorithm is unclassified, and chips to implement it can be purchased commercially. At the present time, export of DES devices from the U.S. is subject to government restrictions. Other algorithms are used in the area of financial transactions.

When encrypted, the data cannot usually be recovered by unauthorized parties ("compromised"). Because ciphers are mathematical processes, however, it is possible to decipher the information with a great expenditure of time and energy.

For most business applications, there is not much risk of unauthorized access through such a compromise. Nonetheless, there is very sensitive and valuable information to be had when large financial transactions or government security is involved. Therefore, it behooves the user to understand the risks involved before making a decision on an encryption system.

4.1.2 Switching and Multiplexing

The topics of switching and multiplexing are perhaps the most complex and critical to the design of a digital communication network, particularly where multiple traffic modes are provided. Figure 4.1 indicates that switching is applied ahead of multiplexing in the channel. This order can be reversed, however, depending on the composition and intended routing of the traffic through the network. Some of the newer and more interesting network access systems even combine the two functions to enhance flexibility and network traffic loading efficiency. In the following paragraphs, we will review a variety of techniques used extensively. Many such techniques are combined within the same network, or even the same piece of equipment. We begin with a brief discussion of the initial access point into the network. The assumption is made that the traffic has already been digitized, compressed, and encrypted, as appropriate.

4.1.2.1 Network Access

Two basic concepts for accessing a digital telecommunication network are presented in Figure 4.4. A standard access arrangement used in traditional networks provides a dedicated port for each incoming access line, shown in Figure 4.4(a). The device which takes the N access lines and connects them to M trunks is called either a concentrator or multiplexer. An important point to keep in mind is that this arrangement gives each user's device a stand-alone connection to the network. We will show in our subsequent discussion that the concentrator or multiplexer can take on several forms, and it may or may not provide a switching function as well.

Shown in Figure 4.4(b) is a network access approach into a broadcast medium. One very common and important broadcast medium is a communication satellite transponder providing access to many earth stations within the footprint of the satellite antenna. Another example is the local area network, wherein communication devices are connected to a common coaxial cable loop or bus. In both cases, any port can transmit into the broadcast medium (airwaves or common cable) and can receive the transmissions of all other stations as well. This simplifies direct communication among a large user community. In a satellite network, the community can be located throughout a country, continent, or hemisphere. In a LAN,

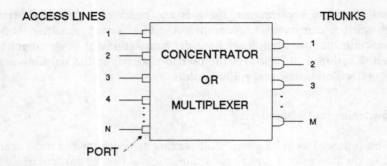

ACCESS LINES

1
2
3
4
N

CONCENTRATOR

OR

MULTIPLEXER

PORT

TRUNKS

1
2
3
M

(A) ACCESS THROUGH A DEDICATED PORT

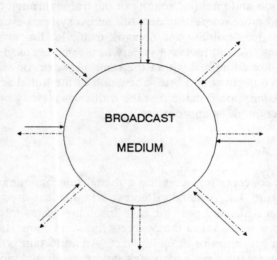

BROADCAST

MEDIUM

(B) ACCESS THROUGH A COMMON CHANNEL

Figure 4.4 Network access concepts.

the community would generally be within the same building or campus, and to extend beyond a local area usually requires some type of point-to-point link to interconnect LANs. A critical aspect of the common broadcast channel approach is that stations may inadvertently transmit on top of one another. The two approaches used to prevent such interference are either to synchronize and time assign the transmissions, or to provide a method (called *contention resolution*) for having stations retransmit in the event of a collision.

The focus of the following discussion is on the *dedicated port* approach for long-haul terrestrial networks and on the *common broadcast channel* approach for satellite networks. Applying the common broadcast channel to LANs is normal for connecting computers and peripherals within a building. In the current context, such LANs can be connected to a long-haul network with a suitable access port of the type indicated in Figure 4.4(a). Alternatively, the LAN can send packets bound for external locations over a satellite broadcast data channel. Products which accomplish this are available from Vitalink Communications Corporation of Fremont, CA.

4.1.2.2 Circuit Switching

A circuit switched network is used to set up the traditional type of continuous connection between end points. Historically, circuit switching was used primarily for voice communication. With the advent of end-to-end digital communication, direct connections are now possible between digital devices operating at 56 or 64 kb/s, which is considerably in excess of what can be passed through the normal voice frequency range of today's telephone circuit. The basic architecture for circuit switching has been covered in Chapter 2 in the context of the public telephone network, and is presented in Figure 4.5 in simplified form. The end-to-end circuit is set up on demand through a hierarchy of switching nodes. At the bottom of Figure 4.5 is shown the simplified construct of a switch, with N input ports and M output ports. If N were greater than M, there could be instances when users who would be accessing from that side might not be connected to any of the M trunks. This would occur when the M trunks are 100% loaded. Such a switch is said to be of the blocking type, which is typical for local telephone exchanges and PBXs. The other nonblocking type of switch has N equal to M. This would be found in some tandem switching equipment, wherein trunks could always be served by the switch.

The dynamics of circuit switching have to do with call set-up, because, when a circuit is established, there is no contention or interruption of service. Setting up of calls is performed with a signaling system. In analog networks, signaling is primarily done with audible tones arranged in twelve distinct pairs, which can pass through the same channel that carries the speech. *Dual-tone multiple frequency* (DTMF) is the term used for tone signaling within the voice frequency bandwidth of the telephone channel. The tones are first introduced by the caller through the touch-tone pad, and they may or may not be retransmitted by the central office. With in-band signaling, the tones are picked up at the local office, and used to route the call through the network. This routing goes successively from switch to switch, each of which must detect the area code from this tone format and then select the appropriate outbound trunk that is headed in the correct direction. Finally, the last switch connection is made at the distant end office and the local

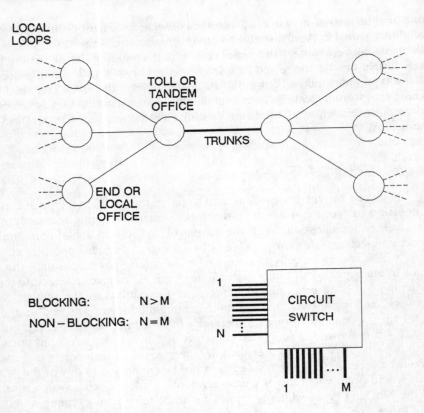

Figure 4.5 Basic topology of a circuit-switched network.

subscriber is rung. As discussed later, links with common channel signaling do not pass the DTMF signal, but rather extract the dialing digits at the originating central office.

A routing table is stored in each switch, containing the outbound trunk choices in decreasing order of priority for each distant area code that can be reached via that particular switch. These are like routes and alternate routes at forks in the highway. From the routing table, the processor within the switch selects the best fork to take to reach the destination, and the same process is repeated until the path is completed. The concept is elegant in its simplicity because the routing information stored in the switch is relatively static, changing only when a new area code or node is added somewhere in the network. Routing tables are used in virtually every type of multinode network, whether for voice or data.

In the all-digital circuit switched network, the signaling is not transmitted in analog form over the traffic-bearing circuit. Rather, all signaling and network control information is sent through a separate data communication circuit in a technique called *common channel signaling* (CCS). The current approach being taken for CCS is to construct a very reliable packet switched network, which is carried over an essentially independent facility to enhance reliability further, since a failure of the signaling network will bring down the traffic bearing network. AT&T first introduced common channel interoffice signaling in 1976, a system now referred to as common channel signaling system No. 6 (CCSS-6). Work done by the CCITT in developing an international standard for signaling (which relates to connecting together the telephone networks of different countries) has led to the introduction of CCSS-7, also called SS-7. This system is becoming more important domestically and internationally, and it is one of the cornerstones of ISDN. Use of SS-7 is offered with the newer telephone switches, and it is expected to become the standard for common channel signaling in the developed world in the near future.

The OSI reference model was used to construct SS-7, applying packet switched architecture with the datagram mode. Because the datagram mode is extremely effective, even with relatively unreliable transmission facilities, the probability of getting the call set-up information through the network to each intermediary node is very high. Call set-up time is a critical aspect of circuit switched operation. In older networks with in-band signaling, call set-up times were measured in multiples of seconds, and even tens of seconds. Now, with SS-6 and SS-7, call set-up time has been greatly reduced. Future improvements will come from increasing the transmission speed on the common signaling channel because this determines the time delay between when the request is first made at the origination point and when each node has received the request. SS-7 is also attractive because the same data channel will be used for network management services, as discussed in Chapter 5.

In telephone service, call set-up time is a relatively small fraction of the total duration of the conversation. Therefore, a set-up time of one second is no problem. With an all-digital channel, use of the circuit switched network will be very attractive for all sorts of data calls. At 56 kb/s, a single second is sufficient to send the information contained on approximately two pages of this book. Imagine being able to make calls of as short as one second or even less. A call set-up time of one second, however, may be unacceptable for data communication applications involving hundreds or even thousands of repetitive calls of extremely short duration. This is therefore another motivator for using SS-7 and providing the highest possible speed for its transmission.

4.1.2.3 Packet Switching

Packet switching is the most recognized information routing technique for data communication. As mentioned in Chapter 3, packet switching networks employ network interface standards which are built on a multilayer architecture. The TCP/IP suite of protocols covered all layers and was the first generally recognized standard to do so. In the case of X.25, there are only three layers: the physical, the link, and the network. This architecture properly divides the critical functions associated with routing information in packets so that reliable information transfer is possible. Layers above X.25 are being promulgated through working groups of domestic and international committees in specific areas. Examples are discussed in Chapter 3 and include standards for E-mail under X.400, the MAP/TOP activities for manufacturing automation in the U.S., and the GOSIP standard for U.S. government data communication procurements. The upper layers of the TCP/IP suite of protocols are being used above the X.25 network architecture.

Figure 4.6 presents the relationship between the layer and the associated network element interface for the packet network. In the upper left quadrant of the figure is situated the host of an X.25 packet switched network with the principal layers of the architecture indicated. Application, presentation, and session services reside in this host to serve data communication users who are connected to it over the network. The network itself is managed with either X.25 or the Internet protocol of TCP/IP. On the actual link to the closest access *packet switched node* (PSN), the HDLC link-level protocol provides reliable synchronous communication on a point-to-point basis. The physical link to the node is indicated with a solid double arrow, which is the access line to the packet switched node of the X.25 network.

Two PSNs are indicated at the bottom of Figure 4.6 within which the physical, link, and X.25 network layers reside. The protocols and transmission systems which the PSNs employ are not of particular interest to the user because packets are automatically routed to the destination in the most appropriate manner. This is an X.25 network, and the PSNs thus provide a virtual circuit for messages which are routed over the network. The box labeled *internal to PDN* contains the customized or proprietary network logic of the particular packet network. As discussed elsewhere, packet networks are designed for optimum performance, and may in fact not employ a standard protocol set like X.25 between the PSNs of the network. To the outside (shown along the top of the PSNs), however, the standard X.25 interface exists to allow users to send messages to one another and to access one or more hosts containing data processing applications which are of interest. The figure assumes that the user is connected to the PSN through a remote processor (upper right). This element would contain the appropriate interfacing software for the X.25 environment along with higher layers of OSI or TCP/IP, as appropriate. As an alternative to a communication processor, the user may access the network through a PAD if the required X.25 software cannot be provided within the local facility.

In summary, the network layer establishes a virtual circuit via the packet switching nodes. Between the packet switching nodes are found individual point-

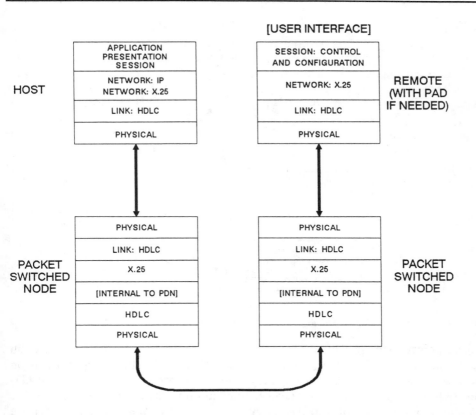

Figure 4.6 Relationship between the data communication network elements and layers.

to-point paths using a link-level protocol such as HDLC. The actual port-to-port connection of cables and radio gear is by way of the physical level and its hardware definitions at the connector boxes.

A simplified packet switched network is illustrated in Figure 4.7. The message in the form of a block of data for transfer from one computer device to another over a distance could amount to a few bytes, as would be the case in a meter monitoring system. For the block to be measured in the thousands of bytes, however, is not unusual, such as when an entire document is transferred from a large database to a remote terminal. Functions of the host (shown at the lower right of the figure) include breaking the message down into packets, placing the address of the distant access location on the leading side of the packet, and applying the packet to the network entryway. Shown at the upper left are user devices or terminals which communicate by using an asynchronous or IBM bisynchronous (BSC) protocol. These inefficient protocols are converted to X.25 with a PAD.

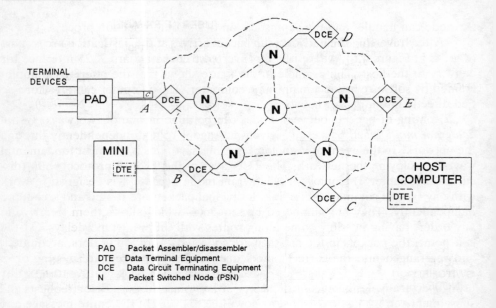

Figure 4.7 Packet switching terminology.

Access points to the packet switched network in Figure 4.6 are labeled A through E; the packet entering at A is to be routed to C, as indicated by the address on the leading segment of the packet. Any external device (e.g., host, mini, PAD) can route messages to any other device. In addition to the address, a count of the amount of information in the packet and a *cyclic redundancy check* (CRC) are placed in front of the packet. The purpose of the CRC, which is recomputed for each point-to-point link, is to allow the next node to verify that the data transfer between nodes was without error.

Entering at A, the PSN reads the destination address from the header of the packet. The node then determines the best outbound path to reach the destination. There are typically a number of alternate outbound paths to reach the destination, but one may be optimum because it involves fewer intervening nodes or it costs less to employ. To make this selection, the intelligence within the node uses a routing table and optimization criteria contained in the node's memory bank. As in circuit switching, this table gives outbound paths for every possible entry-exit port of the network. Having made an optimum selection, the node sends the packet on that route to the next node. The route selection process is repeated at each successive node. Another important function of the switching node is to hold packets in a queue if the selected outbound path is occupied. The queue is maintained in a buffer memory store in the node. Critical performance parameters for a node include the size of the buffer store, the number of packets per second that the node can handle, and the sophistication of its route selection process.

After traversing the network, the packet arrives at the destination access port (e.g., C in Figure 4.6), where it is reconverted to the standard X.25 interface for delivery to the host. The example in the figure shows that the interface is implemented by software inside a mainframe computer. The computer can therefore be tied directly into a packet network.

Routing of packets between nodes can be done in one of two ways. In the *datagram mode,* each packet in a given message is routed independently through the network. In the event of a single packet message, this is the most fundamental way of employing the network. The TCP layer of the TCP/IP protocol suite (discussed in Chapter 3) employs the datagram mode because it is designed to work with any type of communication link. Individual packets are routed and accounted independently. They are numbered by sequence, which allows them to arrive in any order, having possibly come along routes with different time delays. At the exit point, the packets in the message are collected, stored in buffers, accounted, and rearranged into the correct time sequence. If any packets are missing or in error, the port sends a request for packet retransmission back to the entry port by using specifically designed control packets. Many such retransmission requests and subsequent retransmissions will slow down the delivery of the entire message and can overload the network. Nevertheless, the technique makes for extremely reliable transmission on an end-to-end basis.

The other packet transmission mode is called the *virtual circuit mode.* In this approach, a call set-up packet is first sent through the packet switched network to establish the best available routing. Subsequent packets of the end-to-end connection, which contain the actual user information, follow precisely the same routing. A virtual circuit identification is assigned to this particular message, which is maintained by each node along the route. Subsequent packets contain the virtual circuit identification, which informs the intervening nodes of the outbound route to be selected. The virtual circuit mode is used in SNA and for most applications in OSI.

The element functions are usually created by software or firmware running in computers of various sizes. As was mentioned previously, the PAD function is often placed within the host computer or peer node in a data communication environment. An interesting example of a PAD is the compact modem with X.25 PAD built in, shown in the photograph in Figure 4.8. A user can easily access an X.25 network such as Telenet® through a simple dial-up connection. Moving into the network to examine a PSN, we encounter a piece of equipment approximately the size of an oven or small refrigerator. The photograph in Figure 4.9 shows a typical PSN for use with the X.25 standard interface. The transmission speeds and network protocols employed between PDNs are designed for maximum throughput and reliability, and may not conform to the details of X.25. This is because the X.25 interface, while clear in its definition of packet structure and routing, does

Telenetics 2400bps Modem with X.25 PAD Provides low-cost, direct addition of remote offices as dial-up terminals to X.25 network.

Figure 4.8 Low-cost 2400 b/s modem with built in X.25 PAD. (Photograph courtesy of Telenetics.)

not provide the greatest efficiency in the actual switching and routing of user data. Packet switch manufacturers such as Hughes Network Systems have developed their own proprietary architectures to accomplish these aims. As a consequence, PSNs of different manufacture often do not interoperate on the network side of the interface. Another approach, offered by Codex, attaches an X.25 interface to a switching *statistical multiplex* (STAT MUX) network, allowing the user to operate in an X.25 environment, while maintaining flexibility over a network of leased lines. A detailed discussed of the standard STAT MUX and an introduction to the more advanced switching STAT MUX are given later in this chapter.

4.1.2.4 Time Division Multiplexing

Multiplexing in telecommunication networking is the basic process by which several independent communication channels are combined into one larger (higher speed) channel. The purpose is to reduce the number transmission links needed to carry the channels over some distance between nodes or access points. In this sense, it is the true embodiment of network access shown in Figure 4.4(a). The arrangement of the elements of the source-to-transmission channel depicted in Figure 4.1 shows the multiplexer to be located after the switching function. This is the typical approach taken in telephone networks. There is a trend in larger private networks, however, to place the multiplexer on the customer's premises and bring it under the control of the strategic unit. The switching function can then

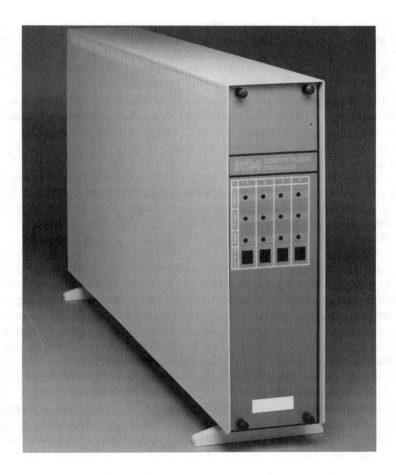

Figure 4.9 A typical X.25 packet switch node. (Photograph courtesy of Hughes Network Systems.)

be performed on the combined stream of digitized channels, which is possible because modern digital switches are implemented with a T1 interface. The switch is able to select channel assignments from the combined stream. This particular aspect is expanded when we review the operation of the digital access and cross-connect and the high-level multiplexer.

The first multiplex approach, called *frequency division multiplex* (FDM), was introduced prior to World War II. FDM can take from four to six hundred voice channels, which are translated individually in frequency. They would be stacked in a baseband starting at 60 kHz and ending at an upper frequency approximately equal to the number of channels multiplied by 4 kHz. FDM is an analog multiplexing technique, but the channels themselves can contain digital data from a voice band modem. (That aspect is covered later in this chapter.) A wideband service can be provided by combining the baseband bandwidth of a number of 4 kHz channels. This might be used for high fidelity audio programs in radio broadcasting and for T1 transmission within the FDM baseband.

While FDM has been used extensively and is still found on many existing telephone networks, its digital counterpart is of most interest in private telecommunication networks. *Time division multiplex* (TDM) is completely analogous to FDM in that time slot assignments are used instead of frequency channels. A common practice in data communication is to use the term "bandwidth" with reference to time slot capacity in TDM. Technically, baseband bandwidth in kHz is proportional to the data rate in kb/s, where the proportionality constant depends on the encoding format. While data rate can be converted to bandwidth by using this mathematical conversion, such usage of the term is really a throwback to the days when FDM bandwidth was the measure of capacity utilized.

The TDM information stream, also called a baseband, is formed by taking the individual data channels, each operating at a basic rate of typically 64 kb/s, and adding them to produce a combined channel data rate, which is approximately equal to the number of channels multiplied by 64 kb/s. The increase in data rate is obtained by compressing the individual bits in time by a factor equal to one over the number of channels. For telephone communication, FDM is more efficient in terms of its baseband bandwidth occupancy for the same number of channels. In the case of data traffic, however, the TDM format is more efficient in terms of the number of user data channels that can fit in the baseband and more reliable in terms of accuracy of reproduction.

An example of a simple four-channel TDM multiplexer is shown in Figure 4.10. At the left, each digital information channel enters the equipment through a compression buffer. The purpose of the compression buffer is to store a block of input data and read it out again at a higher data rate equal to the combined rate to be delivered at the output of the multiplexer. Reading out of the buffer is commanded and timed by the sequential switch, which keys the buffers one at a time, according to a periodic sequence preprogrammed into the unit. The combined output data stream with the channel information now arranged in the desired sequence is shown at the right.

Operation of the other side, which is the demultiplexer, is the exact opposite. The high-speed sequence of data channels enters another switch, which sequentially directs the channels to their assigned expansion buffers. Each expansion buffer holds a block of data and then slows the channel speed to the original continuous rate as it delivers the data to the associated access line. Throughout the process, the data remain in their original digital form and are only changed in speed for transmission over the long-haul path between multiplexer and demultiplexer.

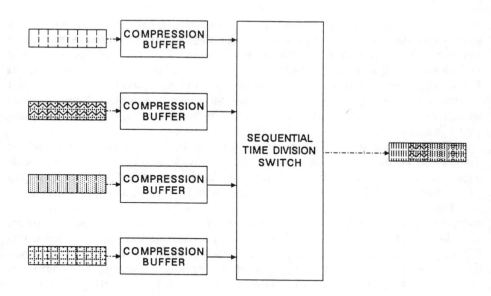

Figure 4.10 Basic operation of a time division multiplexer.

The previous example shows a simplified arrangement for multiplexing four uniform data channels. In practice, the TDM multiplexer operates on a higher quantity of channels for more efficient use of the outbound link. The standard T1 baseband (discussed in the next paragraph), contains 24 channels multiplexed together. A variety of access line data rates can operate. For example, some of the access lines can operate at 64 kb/s, the standard for PCM and the ISDN basic rate bearer (B) channel. Others may be submultiplexed from lower speed DDS type of circuits by using rates of 2400, 4800, 9600, and 19,200 b/s [Glasgal, 1983]. Digital access lines at 56 kb/s, 256 kb/s, and 1.544 Mb/s are also fairly common in special data communication applications. The multiplexer can combine all of these different rates into a single stream, provided that the access port for the particular rate is properly configured for the input and output speeds. Note that rates other than 56 or 64 kb/s are for dedicated private lines and cannot be connected directly to a switched network.

The standard T1 format is used extensively in North America. Recall that there are 24 PCM channels, each operating at 64 kb/s. This baseband is also called a "digroup." A complete frame of channel samples contains 24 × 8 bits, or 192 bits. To this is added a 193-bit (called the F bit) to synchronize the frame, producing an output data rate of 1,544,000 bits per second, the standard T1 rate. The 193rd bit is capable of playing another role in T1 networks: that of a signaling and testing data channel in the *extended super frame* (ESF) format.

Since many private telecommunication networks must connect to points outside North America, we should consider the digital hierarchy used elsewhere in the world. The international standard used in countries outside of North America and Japan multiplexes thirty 64 kb/s voice channels plus two signaling channels for an aggregate rate of 2.048 Mb/s. Referred to in the United States as the E-1 channel (for European digroup), this baseband also can contain lower channel speeds than 64 kb/s using submultiplexing or higher speeds by direct multiplexing [Schwartz, 1987].

Commonly found in T1 networks is the class of TDM multiplexer called the *D4 channel bank*. The device is configured to provide 24 telephone channels (one digroup) using PCM encoding, with 128 quantization levels and μ-law companding (a nonlinear algorithm for enhancing voice quality). The PCM codec is manufactured as a plug in printed circuit card, there being space for 24 such cards in a compact chassis. Instead of a voice channel card, it is possible to place a card for a 56-kb/s duplex data channel. This card uses seven of the eight available bits within the basic channel. The eighth bit, which is the least significant bit, serves double duty for supervisory signaling used in telephone networks. Newer channel banks employing the ESF or the ISDN *primary rate interface* (PRI) will leave the full eight-bit word for communication traffic at 64 kb/s. Other cards for lower data rates in private network applications can be inserted to subdivide the 56 kb/s for use on dedicated circuits such as private lines. Changes to the card configuration of the standard D4 channel bank must be done by maintenance personnel, either by altering wire straps on the card or by inserting a different card.

Programmable versions of the D4 bank have recently appeared on the market. For example, individual input ports to the MegaMux, manufactured by GDC, can be reconfigured by an electronic instruction. Remote control through the *monitor and control* (M&C) system can be employed for this purpose. Channel assignments can be made and changed within the T1 frame, which is important when rearranging an entire network. This level of flexibility aids static reconfiguration (i.e., not automatically in response to traffic demand) of multiplexers located throughout the network. Dynamic reconfiguration, where channels are assigned to access lines in response to traffic demand, is now possible with a new class of intelligent multiplexer. (We will discuss these devices later in the chapter.)

A critical aspect is the timing of the bit streams of the TDM channels as received from and delivered to the access lines. Over the digital trunks, timing is usually not a problem because the network represents a fairly uniform environment. Access lines, however, are typically connected to a variety of different user devices running at internally generated speeds. Also, there is the complexity of connecting some of these lines to other networks emanating from different devices and even different countries. When timing is not consistent (which, unfortunately, is the general case), the buffers in the multiplexer will be forced to slip a bit or even an entire frame of data. This causes the information in the slip to be irreversibly lost.

There are several techniques used to overcome timing differences and incompatibilities. First, the timing of the incoming data can be used to synchronize a clock running inside of the interfacing device. To run the network on the other side of the interface at the same clock rate may be appropriate. In most cases, however, there are too many different inputs for this method to be practical. More likely, the network timing is used to synchronized the clock of the input device, which is the approach taken when accessing the public network. If this is impractical, a device called an *overflow buffer* can be located within the access port. This buffer holds incoming bits, which are then read out at a timing rate which is perfectly synchronized with that network. If the incoming clock rate is less than that of the network, blank bits are "stuffed" into the input stream; these excess bits are detected and removed at the distant end. Conversely, an incoming bit rate slightly higher than the network's will result in overflow of the buffer. When this occurs, the buffer is reset to zero (i.e., all held data are thrown away), and the process of communication again proceeds until the next overflow. The greater the size of the overflow buffer, the greater can be the difference in clock speeds. Eventually, a higher input clock speed will cause buffer overflow. Infinite buffer capacity, which precludes overflow, unfortunately causes infinitely growing time delay for the data to pass through.

Private network facilities can obtain their clock reference from the public network. In large private networks, an independent cesium standard can be employed as a reference. Because of the high accuracy of the cesium standard, the incidence of frame slips will be minimized. On the other hand, the public network of AT&T, for example, will experience frame slips on a routine and random basis. This forces data communication network operators to allow for such disruptions in the architecture of the network. (For an example, see the discussion on fast packet switching later in this chapter.)

4.1.2.5 Statistical Multiplexing

Born out of the demand for greater digital data capacity on the standard telephone circuit, the *statistical multiplexer* (STAT MUX) has revolutionized the industry. The original approach was to combine several low-speed data lines onto a single outbound trunk running to one particular destination with identical STAT MUXs on both ends. The more advanced switching STAT MUX is covered later in this chapter. The basic arrangement of STAT MUXs on a telephone circuit is shown in Figure 4.11. In this example, there are four independent data channels connected between two locations over a common leased line. Because an analog circuit is assumed, there must be a modem on each end for interface of the trunk connection of the STAT MUX. The statistical aspect of the device is its ability to take several input lines operating at a combined burst rate, which is higher than the rate provided by the modem and leased line. For example, input lines 1, 2, and 3 might have nominal speeds of 2400 b/s, and line 4 would be 4800 b/s. The modem and leased line will be capable of 9600 b/s, which is less than the sum of the rates of the four inputs.

To squeeze more data into what appears to be an inadequate line, the STAT MUX only passes data from input to line when the particular input has active data applied to it. Typically, the user on a given line is not transmitting, such as when the terminal device is idle. This would happen when a distant host was responding to the terminal with a bulk file transfer. At the bottom of Figure 4.11, three slices in time are shown. In the first frame of data, lines 1, 2, and 3 are active and connected to the circuit. Line 4 had not been active, so it was not connected. Then, during the second frame, line 3 was not active, and in frame 3 line 1 was not active. Actual STAT MUXs handle a dozen or more low-speed devices so that there is a better statistical mix of input line speeds and activities. In general, most data communication "conversations" do not involve simultaneous transmission in both directions. Of course, if the data communication application has more simultaneous activity on the input than the modem can handle, the STAT MUX can do nothing but indicate to the terminal device that the line is not ready for data.

Having STAT MUXs on both ends is necessary because of the signaling that is conducted between them to control transmission back and forth. The STAT MUX on the left must let the one on the right know which lines are active and the time slots that are carrying them. This implies a TDM time frame format, where input line assignments are changed, depending on data activity. Another technique that is used in statistical multiplexing actually employs packet transmission, analogously to the LAN. The port of the destination is indicated as an address attached to the packet of data by the associated input port.

A more sophisticated STAT MUX has recently appeared, which allows these devices to be used in a star or mesh network. The switching STAT MUX is a programmable device to establish data channels between user devices as if they were attached to a circuit-switched or packet-switched data communication network. As an example, the 6000 series of STAT MUXs manufactured by Codex could be used as nodes to build a wide area network. Rather than packetizing the message and addressing the individual packets, the switching STAT MUX retains the TDM format to transmit data with high throughput efficiency between nodes. The common signaling channel provides the means to route the channels of data and to make changes in the network configuration, even on a call-by-call basis.

STAT MUXs are still used extensively on leased lines. Their real importance, however, was as a precursor for more advanced digital networking techniques. Packet switching, high-level multiplexing, and other approaches still only in trial stages rely on statistical multiplexing to squeeze more traffic into a transmission link or switching node than is possible on a fixed assignment basis.

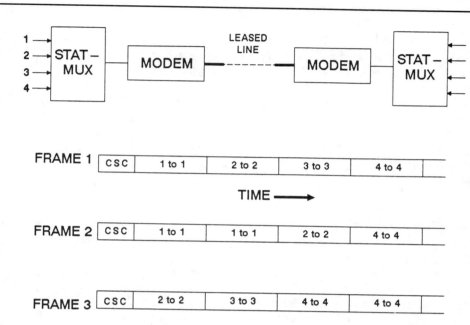

Figure 4.11 A statistical multiplexer (STAT MUX) used for dynamic combination of several low speed data lines onto a single leased telephone circuit.

4.1.2.6 Digital Access and Cross-Connect System

A class of switching device now in wide use in T1 networks is the *digital access and cross-connect system* (DACS), illustrated in Figure 4.12. It is used to interconnect T1 channels with one another by rearranging DS-0 time slot assignments. The concentrator presented in Figure 4.4 is very close in function to the DACS. Unlike the circuit switch, the routing provided by the DACS is held constant as calls are set up and taken down. As shown in the figure, the DACS's capability is expressed in terms of the number of input T1 lines *versus* the number of output lines, the particular example being a four-by-four device. Below the block diagram is provided the relationship between the input time slots and those appearing at the outputs. For example, input access line 1 has time slots A through F; these slots have been moved around ("interchanged") using digital memory and switching within the DACS, and routed to various time slot positions among the four output lines, numbered 5 through 8. All input time slots have to be accounted and made to appear in the outputs. The particular pattern of output time slot assignment can be programmed into the DACS, and any position is permitted.

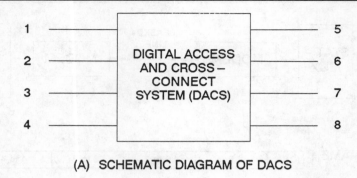

(A) SCHEMATIC DIAGRAM OF DACS

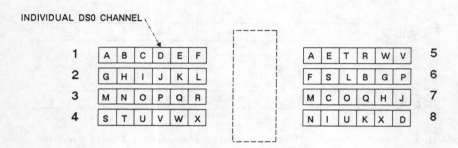

(B) RELATIONSHIP AMONG TIME SLOT ASSIGNMENTS BETWEEN PORTS
WHICH CAN BE CHANGED BY PROGRAMMING

Figure 4.12 Application of Digital Access and Cross-connect System (DACS) for use in T1 time slot interchange.

The way that the time slots are rearranged is comparable to the action inside of a digital switch such as a PBX. A complete T1 frame of information (containing twenty-four DS-0 channels with eight bits each) is stored in a matrix memory in the DACS. These stored time slots are then read from memory onto the output lines in the order which produces the desired time slot assignments. The pattern of input to output is programmed into the controller that instructs the readout process. Using a local or remote video display terminal, this assignment can be changed.

DACS are employed at various points in a typical T1 network. The local and long-distance carriers use DACS throughout their networks to interconnect individual DS-0 channels between T1 lines. These units are monitored and controlled from the network management centers, where the carriers also control their switches and transmission links. The flexibility of being able to alter the time slot assignment, and therefore the routing of channels, is important to the carriers because they can reroute traffic and thus better manage the capacity of the network.

Strategic units that operate private telecommunication networks can also make effective use of DACS. The basic application is as a means to divide a T1 access line for reaching multiple destinations. Without a DACS, there would need to be a full T1 line for each destination in the network. Fewer T1 lines are required from the telco to reach the DACS at point of presence of the long-distance carrier. By arrangement with the long-distance carrier, the individual DS-0 channels within the T1 could be routed to different destinations, using the DACS to perform the subdivision and redirection into the long-distance carrier's network. This is illustrated in Figure 4.13, which presents a customer access to the digital private line facilities of a long-distance carrier. In this example, the customer provides his own switching and the carrier provides T1 transmission capacity between five different cities. The PBX at a user location in City A is connected over a T1 access line to a DACS in the office of the long-distance carrier. Depending on the offering of the carrier, the DACS may be owned by either the carrier or the customer. In either case, the customer can be provided with control of the DACS through a capability called *customer controlled reconfiguration* (CCR). The individual DS-0 channels of the T1 access line can be connected through the DACS to T1 lines going to any of the four distant cities. Through CCR, the customer can alter the routing of the traffic. The particular arrangement in use at a particular time must be consistent with the routing of traffic by the PBX; otherwise, calls will be switched to incorrect destinations.

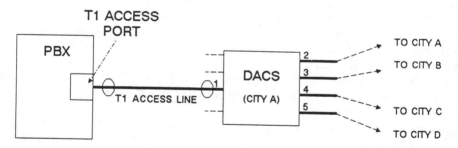

Figure 4.13 Application of telco-owned DACS for efficient access to a digital long distance network.

4.1.2.7 High-Level Multiplexing

The devices previously described (PBX, channel bank, STAT MUX, and DACS) provided the backdrop for the creation of a totally new class of device, which we call the *high-level multiplexer*. Other names for the device include intelligent multiplexer, T1 multiplexer, and T1 resource manager. This author was particularly impressed when he first came into contact with the technology in 1985. It seemed to represent the true embodiment of integrated digital services, being able to combine voice, data, and compressed video on a flexible and efficient basis. The first such unit on the market was the Integrated Digital Network Exchange (IDNX) made by Network Equipment Technologies, Inc. (NET), of Redwood City, California. Some other leading manufacturers include Digital Communications Associates, Fujitsu, GDC, and Timeplex. In fact, this product has become so popular that every leading manufacturer of multiplexers and switches has been forced either to manufacture their own product or to distribute someone else's. The role of the high-level multiplexer is to manage the T1 transmission resources of a private telecommunication network. It does so by performing multiplexing functions as well as those of the DACS and STAT MUX, all in one package. Other features include encoding for PCM and ADPCM, multiplexing of low-rate data channels, dynamic bandwidth allocation, and digital speech interpolation (DSI). Actual circuit switching at the DS-0 channel level in response to calling is performed in an external circuit switch, this being the PBX. The purpose is to give the private network developer the set of tools needed to create any capability desired over a T1 backbone.

Figure 4.14 presents the basic arrangement of the high-level multiplexer, along with an example of its use in a T1 backbone network. On the local access side, user devices are connected to port cards within the unit. This is identical to the approach taken in the D4 channel bank. Examples of port access include a T1 channel for a PBX, a high-speed data line operating at 256 kb/s, several low-speed data lines, and some analog tie lines coming from some other switch. Entry onto the T1 backbone network is shown with trunks on the right-hand side of the high-level multiplexer. The information applied to the ports is processed within the unit and connected to the prescribed output trunk. Therefore, this traffic can reach any destination, as illustrated in the simplified diagram of the backbone T1 network. The trunks themselves can be provided by fiber optic cable, satellite links, or terrestrial microwave systems.

Implicit in this type of network is the use of a high-level multiplexer wherever T1 trunks terminate. This closed arrangement is needed because the multiplexers employ a specialized *common signaling channel* (CSC) through which monitoring and control information is passed. For example, when a new access line is added, other nodes in the network must be informed as to its routing and time slot assignment. Instructions for doing this are entered through a video terminal connected to one of the nodes of the network. The CSC then automatically informs the other nodes, which make appropriate adjustments in their control programming. The CSC provides the pathway to instruct the nodes of the assignment of individual time slots within the T1 frame. Because some of the bandwidth of the trunk is used, the net data throughput is somewhat reduced. For example, the

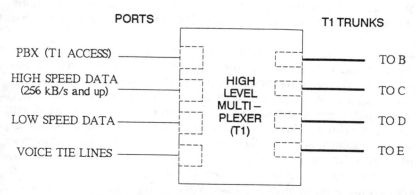

(A) FUNCTION ARRANGEMENT OF NODAL EQUIPMENT

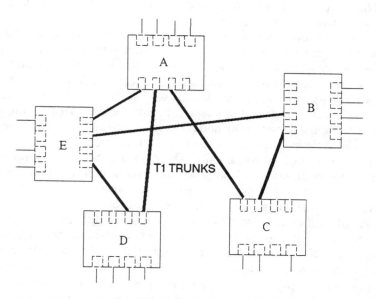

(B) EXAMPLE OF A MESH NETWORK

Figure 4.14 The high-level multiplexer used in backbone T1 networks.

maximum port data rate that can be assigned to a single T1 trunk is 1.344 Mb/s. Because only seven of the eight PCM bits are needed for telephone service, a total of twenty-four telephone channels can be carried over a T1 trunk along with the necessary CSC. Capacity can be increased to as many as ninety-six telephone channels per T1 through the combined use of ADPCM and DSI, which are available as options.

Multiplexer design has been evolving, and shortcomings found in the first units are being eliminated. In particular, forcing every trunk to be a full T1 can impose considerable expense on the private network operator. Trunk cards which operate at fractions of a T1 now can be had. Another difficulty was that only fully fledged high-level multiplexers could be used to connect trunks. These units are quite flexible and powerful, but they cost up to $100,000 per site. Network Equipment Technologies has introduced a scaled down IDNX, which is more appropriate for branch office installations due to the reduced capacity and cost. In addition, GDC's high-level multiplexer can be connected at a node to a less sophisticated T1 multiplexer to reduce equipment expense. When using such "unintelligent" multiplexers, the dynamic routing features and some of the reconfiguration ability are lost. This may still be attractive, since the entire network can be managed from a single network operations center.

On a technical level, the backbone using the T1 multiplexer can deal effectively with the timing and synchronization problem mentioned earlier. This is important to the design and operation of an intelligent network because of the wide variety of possible port devices and the high capacities per node. Furthermore, each T1 node has its own set of port devices so that the network is exposed to a wide variety of clock speeds and opportunities for clock slippage problems. The basic approach is to make one of the nodes the reference and have the other nodes derive their clock timing from it. A secondary reference node would be programmed so that it could automatically take over the function in the event of primary reference node failure. Timing for the reference node would be derived from one of the port access lines, typically the one connected to a public network or host computer (depending on the network application). Software control will allow fine adjustment of clock speed over time, as some drift can be expected. In the *plesiochronous* mode, the reference is only loosely tied to the external line to allow some shift in clock timing, but not so much that average clock rates are very different over an extended period.

The high-level multiplexer offers a range of dynamic routing features, which can be attractive in networks with large amounts of time-varying traffic. Dynamic bandwidth allocation is performed in a node to accept traffic from ports according to demand for connectivity over the network. This procedure is similar to statistical multiplexing. There is a wide variety among vendor products as to how this particular feature works, so careful examination of the equipment specifications behooves the prospective buyer. Statistical allocation of capacity is potentially

209

beneficial for saving on trunk capacity between nodes; you quickly reach a point of diminishing returns, however, as more dynamic features are added. If traffic characteristics were predictable, an approach that provided most of the benefit of dynamic bandwidth allocation would have a few different "canned" network configurations which could be loaded into the nodes at specific times of day when rearrangement might do the most good.

Along with dynamic bandwidth allocation, the nodes can reroute time slots from failed trunks to operating trunks. The nodes can therefore provide tandem connections and diverse routing. In some networks, this feature may well sell the high-level multiplexer over the more conventional product. Another feature being incorporated is the ability to use the multiplexer as a DACS. At present, the high-level multiplexer is a product in flux. Current offerings have attractive features, many of which have been described. We can anticipate that more such features will be incorporated presently. Eventually, some or all functions of the high-level multiplexer will migrate to the PBX and DACS.

4.1.2.8 Fast Packet Switching

Fast packet switching is a telecommunication technology which combines features of statistical multiplexing and packet switching. Unlike normal packet switching networks, which are designed for data traffic, existing fast packet devices operate directly at T1 rates, and are capable of passing voice and data in real time. The products currently on the market emulate the functions of the high-level multiplexer, such as the IDNX previously described. While we use this as a framework to explain how fast packet works, there is considerable interest in expanding the technology to transmission rates in the hundreds of megabits (or gigabits) per second. The potential for applying fast packet switching to intelligent public networks of the future is considered in Chapter 9.

Current fast packet technology provides transmission of all information through the network at 1.544 Mb/s in packet form, which is considerably different from the fixed time frame format used in TDM. Fast packet nodes connect packets to different T1 trunks by means of the address included with each packet. Because of the packetization, bandwidth need not be assigned ahead of time to particular channels. The T1 transmission speed, framing, and synchronization are retained to be consistent with the public networks.

We use as the prime example of this technology the Integrated Packet Exchange (IPX), manufactured by Stratacom, Inc., of Cambell, California. A photograph of a typical IPX rack is presented in Figure 4.15. This particular company entered the T1 node marketplace a few years after NET, having had the benefit of observing the success of the IDNX. Both systems employ synchronous T1 links operating at 1.544 Mb/s. The IPX, however, allows each frame of 193 bits to be

Figure 4.15 The Integrated Packet Exchange (IPX) provides fast packet switching over a T1 network. (Photograph courtesy of Stratacom, Inc.)

handled independently which provides flexibility in routing of information. A comparison of the frame formats used in TDM and by the IPX are presented in Figure 4.16(a). At the top is shown the standard PCM frame of the D4 channel bank, with the framing bit (F) and 24 DS-0 channels of eight bits each. Two fast packet frames are indicated because the IPX handles voice information slightly differently from data. Basically, voice traffic is somewhat less critical in that a lost packet (one which does not arrive at the destination) will not break channel operation. To the listener, a lost voice sample may have no audible effect; in the worst case, it could cause a brief pop. Data communication, however, may not be able to tolerate a lost frame, and consequently the system treats data frames with considerably higher priority than voice frames.

Looking at the fast packet frame, we see the F bit and a sixteen-bit destination address. Just in terms of binary count, sixteen bits would give a total of 65,536 possible destinations in the packet network. Following the address is an eight-bit word to indicate priority and a *cyclic redundancy check* (CRC). The former enables priority information to be transported through the network ahead of other information. For example, data cannot be dropped and would have higher priority than voice. The CRC, as discussed previously, is an error detection code which enables error rate monitoring, an important feature in network management. As shown in Figure 4.16, the data packet has an additional eight-bit time stamp to allow the node to control transmit time in low-speed channels operating at 64 kb/s and less. There are twenty bytes of low-speed data per frame, while the count is twenty-one for speeds above 64 kb/s.

The fast packet approach takes information from ports connected to nodes and applies it to the T1 trunks only when information is actually presented. This is, in fact, the basic reason for using packet switching. Voice traffic is sent only when there is an active sample coming from the source. Recall the digital speech interpolation concept presented earlier in this chapter. The fast packet switching has an inherent DSI capability for voice traffic. Because conversation is active only 40% of the time in a given direction, after a circuit is established between telephone switches, the fast packet system will more than double the traffic bearing capacity of the T1 trunks. By using ADPCM, the capacity can be doubled again. Stratacom advertises that the IPX can provide ninety-six telephone channels over a single T1.

Fast packet switching can also compress data channels through the statistical multiplexing process inherent in the technology. Data connections can be configured for continuous transmission or for transmission with *repetitive pattern suppression* (RPS). The RPS mode eliminates packets that contain continuously repeated information. (Repeated information is, by definition, *not* information, since its content is already known after the first element is sent.)

The architecture of fast packet nodes in a T1 network is basically the same as for high-level multiplexers such as the IDNX. Nodes can be configured with redundant standby equipment so that failed units can be replaced through the operation of the network management system. Alternate routing would be provided with nodes connected in the arrangement that was shown in Figure 4.14. Fast packet would appear to have an advantage in the area of network synchronization. Recall from our previous discussion that the difference in timing between ports and trunks will cause a frame to be lost, on occasion, when overflow buffers are reset. In TDM, a frame contains information for twenty-four or more individual channels. Therefore, loss of a TDM frame affects several channels, and thus has a measurable effect on the network. However, because a fast packet frame contains

Standard D4 Frame Format:

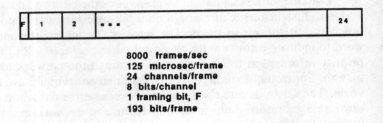

8000 frames/sec
125 microsec/frame
24 channels/frame
8 bits/channel
1 framing bit, F
193 bits/frame

FastPacket Frame Format:

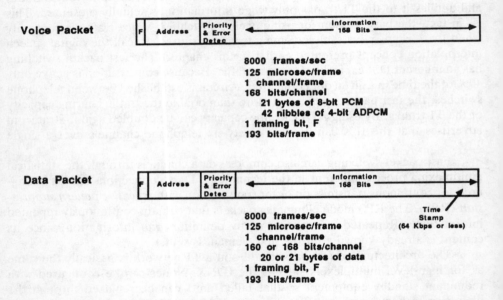

Voice Packet

8000 frames/sec
125 microsec/frame
1 channel/frame
168 bits/channel
21 bytes of 8-bit PCM
42 nibbles of 4-bit ADPCM
1 framing bit, F
193 bits/frame

Data Packet

8000 frames/sec
125 microsec/frame
1 channel/frame
160 or 168 bits/channel
20 or 21 bytes of data
1 framing bit, F
193 bits/frame

Time Stamp
(64 Kbps or less)

Figure 4.16 A comparison of T1 frame formats used in TDM and fast packet switching.

data from only one port channel, the reset operation will only drop data for a single channel. As mentioned, one sample lost from a voice channel will have insignificant results on the conversation. Data are less tolerant; network level protocols such as X.25, however, provide means of automatically recovering from the loss of one packet of data through an automatic request for retransmission. Since buffer reset should be relatively infrequent, the effect on data throughput in such a data communication channel would again be insignificant. The same argument applies to TDM, although in that case buffer reset drops all channels in the particular T1 link.

Fast packet switching and high-level multiplexing provide flexibility and efficiency, but there is a price. The price is that every node in the T1 backbone network must be equipped with a unit of the same design. The companies that market these systems have spent considerable sums in their development, and look forward to having a competitive edge in the marketplace. No two nodal devices of different manufacture are known to be compatible with each other. The features may be comparable, but the details of the channel assignment, timing recovery, and common channel signaling are different. Another difficulty can be that with the high cost of the nodes, it is not attractive to bring a branch office location into the backbone network. Network Equipment Technologies has introduced a less expensive product to attack this problem, and the other companies are likely to follow suit. Furthermore, fast packet technology is evolving and developing, and can ultimately become the backbone networking concept of public digital networks in the future.

4.1.3 Channel Coding and Modulation

This particular topic represents a transition from the true networking aspects of telecommunication technology to the domain of transmission systems. The review is brief here because these topics are well treated elsewhere (for example [Elbert, 1987], [Schwartz, 1980], [Noll, 1988], and [Glasgal, 1983]). Let us, however, identify and briefly review the critical aspects of channel coding and modulation because these can be important in the private telecommunication network environment.

4.1.3.1 Error Correction

Being binary in nature, digital communication links lend themselves to processing and computational manipulation. An area where science has made impressive gains is detection and correction of errors which arise during transmission over some distance. A practical example of error correction would be the spelling checker that this author used as he word processed this manuscript. The spelling checker will be capable of correcting a misspelling if one letter of the word is in error because there is a limited set of possible words (80,000 in the on-line dictionary) and the software is able to match up the most likely word with the same number of letters or with those different by one. An error correcting code over a communication link works in much the same way. Typical decoders are capable of correcting a substantial fraction of error bits, allowing usable data transmission over relatively noisy communication channels. We start the discussion with a brief review of the process by which errors occur on a digital communication link.

The sources of errors (explained in [Schwartz, 1980] from an engineering perspective and in [Noll, 1988] from that of a layman), are basically electrical noise, interference, and equipment anomalies. Noise is a random phenomenon, and on occasion can produce a spike of energy comparable in strength to the data. When this noise spike envelopes the signal for some instant, the detection circuitry of the receiver will be confused and may incorrectly interpret the data. The following paragraph provides a simple example.

Bipolar data transmission has the binary digit 1 coded as a positive pulse of certain amplitude (voltage strength) and a 0 coded as a negative pulse of the same amplitude. A given noise spike, for example, one that is positive going, will cause an error in reception of a binary digit ("bit") only 50% of the time. If the bit were a 1, the positive spike would enhance the signal and no error would result. A 0 bit, which in a typical transmission system is sent with a negative sense pulse, will be reversed with a noise pulse of greater amplitude in the opposite (positive) direction, causing the receiver to interpret it as a 1. This, in essence, is how bit errors occur on any communication path. Being random in nature, noise will act in an unpredictable way. We, however, typically know the mechanism which produces the noise. If the noise came from a thermal-equivalent source, such as a microwave amplifier, then the amplitude of noise pulses would be normally distributed (i.e., it would have a Gaussian density function, with a zero mean value). The only parameter that we need to specify for such a noise source is the variance, which is the square of the standard deviation of the intensity of the pulse. The variance turns out to be the relative power of the noise, which can be determined by measuring the signal-to-noise ratio with standard test equipment.

In the ideal case, only noise will produce errors in a digital communication link. Overcoming noise is usually achieved by increasing the signal strength relative to noise. The theoretical relationship between bit error rate and signal-to-noise ratio in an ideal link is presented in Figure 4.17(a). The parameter plotted along the y-axis is the probability of bit error, which increases as the signal-to-noise ratio (S/N) decreases. For a large number of bits, however, the probability of bit error is effectively equal to the ratio of bits received incorrectly to total bits, or the bit error rate (BER). A real digital link would not perform to this ideal level, so a measured performance curve would lie above, as shown in Figure 4.17(a). The difference in dB between the actual curve and the ideal curve is called the implementation margin. Contributors to this implementation margin (which is really a degradation) include nonideal modem performance and distortion produced by the transmission link, particularly that due to bandwidth limitation on an analog data channel.

This brings us to the topic of this section, namely the use of coding to reduce

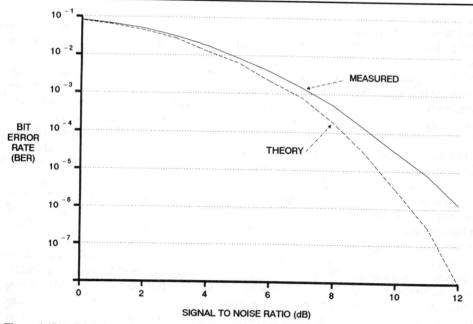

Figure 4.17(a) Digital communication link bit error rate (BER) performance.

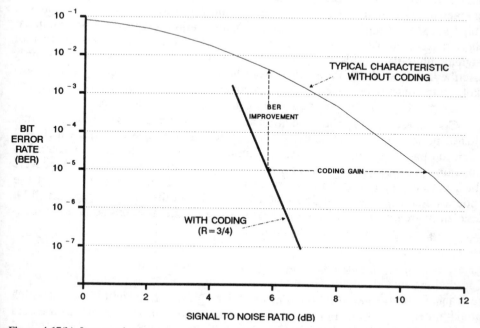

Figure 4.17(b) Impact of using error correcting coding on a digital link in the presence of noise.

the BER on a particular link. Rather than starting with the theory of coding, let us first review the benefit obtained in quantitative terms. Figure 4.17(b) presents the BER *versus* S/N performance curve just discussed. In this case, a new curve, which includes the impact of coding, has been added. We see that rather than requiring an implementation margin, coding actually subtracts margin and provides a net gain in performance. The resulting curve is actually better than the theoretical curve shown in Figure 4.17(a). For the particular example, at a bit error rate of 10^{-6}, the required S/N is approximately 3 dB less with coding. We say that there is a coding gain of 3 dB. Viewed another way, for a constant S/N, the introduction of coding reduces the BER by two orders of magnitude (i.e., from approximately 10^{-4} down to 10^{-6}).

The manner in which coding improves BER performance is highly theoretical and beyond the scope of this book. The basic idea of coding, however, is to add redundant bits to the information stream, increasing the bit rate by up to 100%. These bits convey information about the user data, allowing the receiving end to correct bit errors which have been introduced along the way by noise and other sources of interference. One of the most common techniques used in data modems and satellite communication modems is called *convolutional coding*. All coding techniques add bits and consequently increase the transmission rate over the channel. This is measured by the coding rate, R, which is the ratio of the input information bit rate to the bit rate going over the transmission channel. Typical coding schemes operate at values of R between 3/4 and 7/8. The coder usually consists of a high-speed shift register. The decoder implementation is typically more complex, where a received block of data must be compared against a set of possibilities for original input data. This is similar to the spelling checker example previously described. The sophistication arises from how a decision is made as to the most likely data that was encoded into the recovered block. Simpler decoders, which do not provide all of the potential coding gain, are also in common use.

Coding is usually applied in conjunction with other schemes to enhance transmission performance or capacity. Coding increases transmission rate, which will widen the bandwidth of the bit stream. Limited channel bandwidth would tend to constrain operation. There are bandwidth-efficient forms of modulation, however, which permit a higher speed to be passed. When this is done, we can expect the modem implementation margin to increase, but hopefully the coding gain will be sufficiently greater to overcome this factor. More information on modulation techniques is presented in the next section.

4.1.3.2 Digital Modulation Systems

The final element of the sending end of the information channel is that which modulates the signal for transmission. Modulation is the process by which the data stream, called the *baseband,* is impressed on an audio, radio, or lightwave signal, the *carrier.* Only by way of the carrier can the information be passed through the transmission link. In a telephone network, modulation is performed within the modem used to connect computers to leased telco lines. Modulation is performed in a microwave terminal between the multiplexer and the actual radio transmitter. Modulation is an important process in satellite communication. One of the advantages of satellite communication is that the link is dominated by thermal noise, which is the best understood, and for which coding has the best results. Modulation also is part of the fiber optic sending terminal, but in this case the carrier is light rather than a radio signal.

The most common classifications of modulation include amplitude modulation (AM), phase modulation (PM), and frequency modulation (FM). These are general designations which would apply to analog or digital baseband signals. (Pulse code modulation, PCM, is not modulation in this context, but rather is an A/D encoding scheme.) In the modulation process, the baseband is applied to a unique frequency, the carrier. Carrier frequencies in the audio range are appropriate for use over telephone channels. Generally, however, the carrier is at a much higher frequency, suitable for transmission over a cable or microwave system. In the case of true digital communication involving binary data, the forms of modulation are called "keying," a term which is a throwback to the days of telegraphy. *Amplitude shift keying* (ASK) amounts to bouncing the carrier power up and down or simply turning the carrier on and off in direct response to the binary data stream. ASK is seldom used by itself in digital radio systems (but radiotelegraphy using the Morse code is a true application of ASK). ASK, however, is effectively the technique for modulating the lightwave carrier in fiber optic transmission. *Frequency shift keying* (FSK) is quite popular for low-speed applications, under 4800 b/s, because the modem hardware is inexpensive and reliable. Nonetheless, FSK consumes too much bandwidth to be useful at higher rates.

The most widely accepted digital radio modulation method is *phase shift keying* (PSK). Once techniques for efficient modulation and demodulation were identified and reduced to commercial practice, PSK became the working standard. In the binary case, PSK is mathematically the same as ASK with an amplitude reversal at the transition from a binary 1 to a 0. An important feature of PSK is its strength in the presence of noise. This tends to reduce the amount of signal power needed for proper reception, thus allowing smaller antennas (in the case of satellite or microwave communication) and wider repeater spacing (in the case of cable). To enhance the bandwidth efficiency, hybrid modulation systems that combine PSK with ASK have been applied to voice band modems, which deliver 9600 b/s and higher over a dedicated telephone circuit. There is a trade-off, however, in that the inclusion of ASK raises by a substantial margin the signal power required for satisfactory reception.

The opposite of modulation is called *demodulation*. Typically, demodulation is the more complex process, and is where most of the complexity of the modem resides. Digital modems, therefore, are truly sophisticated devices. There is a trend to combine the functions of the error correcting codec within the modem. This is to be expected because the codec is usually correcting for modem and link performance. In certain cases, the codec and modem interact in some manner to pick up additional gain.

Another feature of modems is the automatic ability to compensate for distortion from the transmission link. The system for doing this is called *equalization*. It is possible to obtain analog telephone circuits which have already been equalized. When the network is used on a dial-up basis, however, any equalization must be performed by the user. Equalizing modems on both ends of the circuit initially go through the equalization routine prior to the sending of data. Wideband links which are not switched can be equalized manually by using specially designed compensation circuits. Certain types of filters can be selected for the modulator and demodulator to obtain equalization in an attempt to reduce implementation margin.

4.2 TERRESTRIAL TRANSMISSION SYSTEMS

We now address the transmission systems which provide the links among network nodes and user locations. Our emphasis is on wideband systems capable of carrying large amounts of information. Viewing Figure 4.1, the transmission system would connect from the output end of the source chain to the input end of the destination chain. In this section, we review the terrestrial transmission systems used for digital communication and the integration of digital services, including voice, data, and video.

Terrestrial transmission systems are inherently point-to-point. This means that we require a complete two-way system installed on the ground to provide each leg of the network. In comparison, a satellite transmission system is essentially point-to-multipoint, allowing us to broadcast the same information to hundreds, thousands, or even millions of locations throughout a region. A discussion of satellite transmission systems is provided at the conclusion of this chapter. In the following paragraphs, we review the salient features of terrestrial transmission systems, notably copper cable, fiber optic cable, and microwave radio. We assume that the reader already knows what these systems look like and how they work in principle. Our focus here is on their application in a private telecommunication network.

4.2.1 Coaxial Cable

Copper cable is a basic electrical transmission medium that can buried in the ground, strung between poles, or installed throughout buildings. Bandwidths measured in megahertz (thousands of kilohertz) are provided by using coaxial cable, which consists of a center conductor wire and an outer conduction sheath of braided copper strands. It is the same type of cable used in the home to connect a television set to the cable television service. In long-haul systems, a repeater amplifier is required every several miles to overcome the resistive loss of copper. Current undersea cable is also coaxial, but the distance between repeaters is extended with greater amplification to a hundred miles or more to simplify maintenance by ship. Coaxial cables using analog FDM were the main long-haul transmission medium for the AT&T long-distance system. Note that these cables were compact, each composed of a bundle of a dozen or more individual coaxial "tubes," but the FDM equipment needed to break down the thousands of voice channels was so massive that it took a skyscraper on either end to house and power it all!

Today, coaxial cable (called "coax" in the vernacular) is still an important transmission medium for short-haul links within buildings, between buildings on a campus, and even within the downtown section of a large city. A typical cable can support bandwidths from 12 kHz to a maximum of approximately 500 MHz. Above that frequency, the signal attenuation per foot is too large to be practical in medium-distance transmission applications. The attenuation per foot *versus* frequency is presented in Figure 4.18.

An excellent example of the practical use of coax is in Ethernet, which is an important LAN standard. A single cable is routed from one location to another, and a passive tap provides access to this broadcast medium. Another important use of coax in integrated networks is for terminating high-speed digital circuits. As mentioned previously, coax is ideal for distributing analog video signals to industrial and home receivers, employing the medium for one-way or two-way service.

Coax is rather simple to use because most types of electrical signal can be directly connected. In some instances, modems may need to be provided. Whether amplification will be required along the route depends on the distance to be covered as well as the baseband bandwidth. The bandwidth limitation of coax, however, places its digital capacity at a fraction of that of one fiber optic strand. Several years ago, there was a shortage of copper and the price of copper cable dramatically increased. This gave an economic push to the other point-to-point media and to satellite communication as well. The copper bubble has burst, however, and copper prices are well below their record highs. As long as bandwidth requirements are modest, such as in television channel distribution and for current LAN implementations, coaxial cable is still a viable transmission medium. Reductions in fiber optic cable manufacturing costs and improved techniques for installation are making fiber more and more attractive as a replacement for coaxial cable on the local loop.

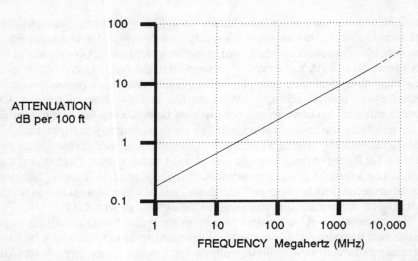

Figure 4.18 The attenuation characteristics of common copper coaxial cable (coax) as a function of signal frequency.

4.2.2 Fiber Optic Cable

Fiber optic cable systems are becoming the long-haul, point-to-point medium of choice where high capacity is in demand. The basic transmission medium is a thin strand of extremely high quality glass. Coherent light propagates in a zigzag path through the fiber by reflecting off the strand's outer surface, where there is a change of refractive index. In older fiber cables, this is produced by a cladding of glass of a different density which coats the center strand. Improvement in fiber performance has been realized by replacing the coating with a fiber strand which uses a graded index of refraction. Because the fibers are of hair thickness, many strands are combined into a single cable, which is still much thinner than a comparable copper cable. This simplifies installation in already constructed conduits and buildings.

Using light as the carrier is more attractive than electrical signals for two reasons. First, being at the high end of the electromagnetic spectrum, the potential bandwidth of a light carrier is substantially greater than possible with copper cable. Second, to monitor signals as they propagate along the fiber is extremely difficult without making an optical connection. Copper coaxial cable is fairly quiet in the sense of electromagnetic radiation, but the signal can be picked up by using a sensitive reception device located outside of the physical cable, while in close proximity to it. The issue of security is probably most important to the Department of Defense, but there may be other cases outside of government where such security issues are still vital. In any case, the bandwidth superiority of fiber is unquestionable. The fibers exhibit less attenuation than their copper counterparts. Optical amplifiers are not used at the intermediary points, but rather the modulated light is regenerated using a demodulator-modulator combination connected to another light source.

Fiber optic transmission incorporates the digital modulation process along with coherent light generation and detection. There is a break point between transmission speed ranges at approximately 100 Mb/s. Inexpensive *light-emitting diodes* (LEDs) are used as the light source below this speed, while above it lasers must be employed. The current operating capacity of a single optical fiber is approximately 2000 Mb/s. (That is two gigabits per second!) If the installed cable had 72 fiber pairs, the total capacity of 144 Gb/s would carry more than two million voice channels or 3000 digitized television channels. A comprehensive overview of fiber optic transmission systems can be found in [Smith, 1985].

We pointed out in Chapter 1 that, to be economical, a long-haul fiber optic system must be heavily loaded with revenue-bearing traffic (see Figure 1.4). Otherwise, the network would be a financial albatross. Most current applications of fiber are in the long-haul transmission portion of a private network, where the capacity is leased from a common carrier. Fiber optic transmission has a role to play within the metropolitan area and the campus environment as well. The kind of bulk loading required for financially justifying the use of fiber represents a near-term barrier to extending fiber on the local loop and into each residence. There are, however, business applications where fiber loops can be employed today, and eventually, through the telco, to the home.

An important consideration in using fiber as part of the telecommunication plant owned by the strategic unit is that of maintenance and repair. Fiber installation is usually a task for specialists familiar with the rules for pulling, bending, and connecting the cables in conduit and within buildings. Once operational, things will be fine and users may enjoy the benefits of an all-digital communication medium. Breakage of a single fiber within a multipair bundle can be restored by selecting an unused pair. This can be performed within the telephone exchanges and equipment rooms where the fiber optic cable terminates. The more serious kind of breakage is one which takes the entire cable out of service, requiring splicing. We have already mentioned the back hoe fade, wherein the buried cable is cut by a piece of earth-moving equipment.

Restoration of a cut fiber optic cable will take a considerable time due to the complexity of splicing the individual fibers back together. To splice a broken fiber, the two fibers must be perfectly aligned in much the same way as a piece of movie film is spliced. This will permit the coherent light to pass through the junction with a minimum of attenuation. The fibers are held together with light-transmitting bonding agent, which is similar to ordinary epoxy. Much has been learned in the years since fiber links have been in operation so that current splicing systems are now thought adequate for long-life operation.

4.2.3 Microwave Systems

Line-of-sight microwave systems have played a very important role in long-haul communication in North America and around the world. (See Figure 4.19.) As a point-to-point transmission medium, microwave can carry wider bandwidths than coaxial cable — and is able to leap tall buildings without even touching the ground! Not long ago, microwave towers were considered to be a sign of progress as telephone and television services were extended across wide geographical areas. Terrestrial microwave seemed to be overcome by satellite communication during the 1970s and early 1980s. Today, it would appear that microwave has little place in the world of fiber optic communication. In the context of long-haul communication in North America, Europe, and Japan, microwave's days are numbered. To paraphrase Samuel Clemens, however, the news of microwave's demise is premature.

The ability of microwave links to jump from site to site makes the technology viable for the final mile of a private network. A new breed of low-cost, digital, microwave radio has appeared on the market. Aimed at the private telecommunication network market, these products are used for short-haul links at the T1 or DS-3 level. The concept is to use the long-distance fiber optic network and local loops from the telcos, where available. There are, however, instances where the telco is unable to extend the digital backbone to every primary location. Private digital microwave systems can be erected in a relatively short time and operated from the same facility where the network node equipment is located. Total installed cost of a short-haul microwave terminal is a fraction of that of the heavy route variety used by the long-distance carriers. As an alternative to purchase, the services of a radio common carrier (mentioned in Chapter 2) could be used.

There are two complexities which must be faced when implementing a short-haul microwave link. The first is the verification of line of sight (LOS) between the two terminals. This involves determining the profile of the earth along the path, which can be plotted from a topographic map of the region. Any buildings or other obstructions must be included in the profile. An example of such a profile between two terminal sites is shown in Figure 4.20. Apparent in the figure is how the curvature of the earth is taken into account. The plot is also adjusted for refractive bending of the microwave beam as it propagates through the varying density of the atmosphere. Minimum clearance of obstacles is needed to prevent diffractive loss as well. The other complexity with which to deal is frequency clearance. The responsibility of the prospective operator of the link is to obtain frequency assignments and licenses from the Federal Communications Commission, the appropriate governmental regulatory agency in the U.S. The FCC places the burden of selecting frequencies from the available pool on the operator. Because these frequencies are used throughout the country, the new entrant must

Figure 4.19 A typical line-of-sight microwave terminal installation with tower-mounted parabolic antennas. (Photograph courtesy of Hughes Communications, Inc.)

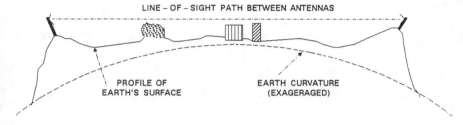

Figure 4.20 A profile of the earth used to determine if a line-of-sight path is clear between microwave terminals.

always contact current users to obtain their permission to transmit. This process is called *frequency coordination,* and can take several months and considerable expense to work out potential problems with the existing systems.

In connection with frequency assignments, the prospective operator of the microwave link has a number of choices in the way of the specific frequency band. The FCC has allocated frequencies in several different bands, each intended for particular purposes. Common carriers can use frequencies in the 4 and 6 GHz bands, which overlap the C-band frequency range used in commercial satellite communication. Private networks can employ frequency bands which are considerably higher. The higher bands experience greater rain attenuation, but this is not a problem as long as the link is shorter than approximately ten miles. Private microwave links can employ 18 or 22 GHz, and several vendors offer relatively inexpensive equipment which operates at these frequencies.

Fortunately for the private network operator, there are organizations which perform both path analysis and frequency coordination. These services could be obtained from a single vendor as a package along with the microwave equipment and installation. Another approach is to buy the equipment and to purchase path engineering and frequency coordination services from a qualified consulting company. One such organization is Alliance Telecommunications Corporation, formerly known as Spectrum Planning, Inc., of Dallas, Texas. Alliance has been in business a long time and has assembled a database which contains all of the currently operating microwave stations in North America. Alliance is thus able to determine quickly which frequencies are easy to coordinate. This, of course, reduces the time, since the smallest possible number of other operators will need to be contacted. It is advisable to check which frequencies in particular bands are more readily available before selecting microwave equipment for purchase.

4.3 SATELLITE COMMUNICATION NETWORKS

Satellite communication has valuable benefits for the private telecommunication network user. Organizations have used satellite communication effectively in the pursuit of business objectives. For example, cable television could not have developed into a $10 billion market without the satellite delivery vehicle. Other industries are discovering that satellite communication when combined with today's digital technology is capable of providing more flexibility than typical terrestrial network solutions. We must, however, proceed carefully when considering a satellite approach, since there will definitely be an effect on the current operation of a business communication environment which begins from a terrestrial perspective.

We provide a brief review of the capabilities of satellite networks. A more complete examination of the design, implementation, and operation of an entire satellite system (including the satellites, launch vehicles, and operations facilities) can be found in [Elbert, 1987]. The discussion in this chapter focuses on satellite communication for private telecommunication networks. The emphasis is on digital technology, a field which is not new to satellites. In fact, many of the multiplexing and switching concepts now popular in terrestrial networks were first tried and proved in the satellite context. Today, the facilities which once required 90-foot diameter dishes and a $2 million investment per site now are available at a cost of under $20,000. To quote former FCC commissioner Nicholas Johnson, this represents a 99% discount!

The broadcast capability of a satellite transmission is perhaps the most attractive feature. Historically, traffic found its way onto satellite because of its ability to reach many different locations on Earth and to provide a clear, reliable link. Most of the traffic on the first international satellites was for analog telephony. Subsequently, television transmission grew to the point of playing a major role in justifying the investment for spacecraft. Domestic U.S. satellites operating at C-band are now primarily used for the distribution of video signals. The introduction of Ku-band satellites in the early 1980s, however, opened the door to inexpensive interactive communication with small ground terminals. These are referred to as very small aperture terminals, VSATs, as have been discussed in previous chapters.

4.3.1 Service Capabilities

Satellite communication is being applied in a number of private telecommunication networks (such as described at the end of Chapter 1). In the following discussion, we review a number of the most common arrangements, emphasizing the transmission of digitized voice and data information. Analog video for one-way teleconferencing is also considered. At the conclusion of this section, we treat the use of international satellite links from business earth stations under an offering called IBS.

4.3.1.1 Thick Route T1

A *thick route* in the satellite context is a link which carries several channels of communication. The link therefore employs a relatively wide bandwidth within the satellite transponder. The prime example of a thick route would be one or more T1 channels used in a backbone link between major corporate locations. A typical application would be for very high speed data transmission to connect host computers for load sharing. Another example is compressed, digital, two-way video for interactive teleconferencing (i.e., the remote meeting application described at the end of Chapter 3).

In general, the thick route link operates on a point-to-point basis between pairs of earth stations. Satellite transmission is still point-to-multipoint, with one station transmitting to the satellite and the others receiving all or a specific portion of the transmission after relay from space. Multiple access, which is discussed later in this section, is achieved through either *frequency division multiple access* (FDMA) or *time division multiple access* (TDMA). In the case of FDMA, there is at least one carrier transmitted to the satellite from each earth station, and the transponder will carry several of these signals. Likewise, for a destination earth station to be connected, it must receive the carrier of the originating earth station. In a full mesh network (i.e., where every station is linked to every other) of ten earth stations, there are ten carriers and each carrier has traffic for nine other stations. In sum, ninety links are established.

The thick route type of satellite application requires relatively expensive earth stations capable of transmitting the high power levels. Also, thick route traffic is usually of a high priority and justifies the inclusion of redundant equipment within the earth station. Antenna size is typically in the range of three to seven meters, depending on the traffic requirements, satellite radiated power in the direction of the earth station, and the expected amount of attenuation due to heavy rain fall. Because of these factors, C-band can still be used effectively, even though the frequency coordination with terrestrial microwave systems can be time consuming. Using Ku-band tends to reduce siting difficulties, but both the power and the cost of the satellite transponder tend to be somewhat higher than the C-band counterpart.

The thick route concept potentially can provide a foundation for a VSAT network. The relatively large and expensive earth station for thick route can serve double duty as the hub. Earth stations tend to be fairly flexible and versatile, provided that there is an initial understanding of the range of possibilities when the station is constructed. For example, a small set of thick route earth stations could provide the first stage of a private backbone network using T1 transmission. The network arrangement would be point-to-point and equivalent to that shown in Figure 4.14. Use of high-level T1 multiplexers as nodes at each earth station site would be sensible. In fact, the same multiplexers would connect to terrestrial T1s as they were introduced or already in use. Subsequently, some of the earth stations could be made into hubs. The T1 links over the satellite might be moved to terrestrial transmission as fiber optic systems expanded to cover all necessary locations. A thick route station can be employed as a television uplink for one-way or two-way video teleconferencing.

The configuration of a thick route station with the capability of four T1 links as well as full-motion video transmission is shown in Figure 4.21. This particular block diagram indicates that comprehensive redundancy is provided for every active element in the station. The size of the antenna would be in the range of five to seven meters. It could operate either at C-band or Ku-band; to be a hub in a VSAT network, however, Ku-band would be preferable. An independent transmitting and receiving capability is provided for a full-motion television channel, complete with audio and data channels for network coordination. This type of

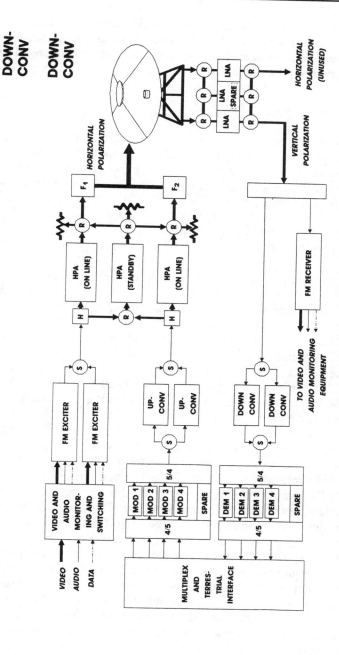

Figure 4.21 Example of a major communication earth station with the capability to provide full duplex analog video and T1 digital transmission. [Elbert 1987.]

thick route earth station could serve the changing needs of a major corporation or government agency over a number of years. The antenna and transmitters would remain in place, while the rest of the equipment could be rearranged or replaced to reflect evolving requirements.

4.3.1.2 Thin Route Data and Voice (VSAT)

We have already highlighted the VSAT network approach to extend the full range of digital services to branch offices and other remote locations. This particular technology, which draws heavily from years of experience over international and domestic satellite systems, continues to evolve. The applications for VSAT networks range from the most basic of connecting two points with a 9.6 to 56 kb/s data line to a dynamic wide area network for integrated data and voice. Our emphasis will be on the latter case, since this presents the greatest opportunity for building a flexible private network, one which is totally under the control of the strategic unit.

The configuration of a VSAT with capability for data, voice, and video communication services is presented in Figure 4.22. Typically, the terminal is divided into outdoor and indoor sections. The outdoor equipment and antenna, shown in the dotted box, can be mounted either on the roof of a building or on the ground in close proximity to the building. (Examples of such installations are shown in Figure 4.23.)The nonpenetrating roof mount is designed for low-cost installation, requiring access to the roof of the building. A ground mount is secured on a concrete pad. Either may be employed in a Ku-band VSAT network. Offset-fed parabolic reflector antennas are typically used, with the diameter in the range of 1.2 to 1.8 meters, depending on the power of the satellite and the severity of local rainstorm activity. Attached to the feed is a *solid-state power amplifier* (SSPA) capable of transmitting approximately five to ten watts. The power level of this amplifier will be the primary determinant of the traffic capacity of the terminal. In high-capacity applications (T1 and greater), a high-power *traveling wave tube amplifier* (TWTA) would be necessary. Attached to the back of the mount is an enclosure containing the rest of the outdoor unit. The antenna and outdoor unit ensemble are commonly to referred to as the radio frequency terminal (RFT).

The indoor equipment would normally be located in a communication center or telephone room, with convenient access to telecommunication interfaces. Typically, the indoor equipment is housed in a cabinet about the size of a minicomputer or a personal computer, depending on the facilities provided. The example shown in Figure 4.22 is quite comprehensive, having the capability of handling two-way voice and data as well as video reception. Many VSAT networks are built around providing only data communication services. Adding either the voice or television receiving capability at a later date is relatively convenient.

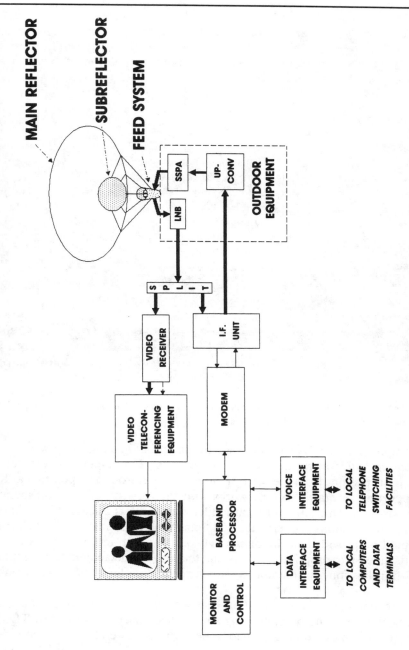

Figure 4.22 Configuration of a Very Small Aperture Terminal (VSAT) which combines two-way voice and data with one-way video. (Courtesy of Hughes Network Systems.) [Elbert 1987.]

Figure 4.23 The Ku-band hub station and several associated VSATs which are installed at the HNS Facility in Germantown, MD.

A comprehensive VSAT network environment in Figure 4.24 shows VSATs communicating with several different hub locations. The satellite at the center relays the transmissions from the earth stations, and are probably operating at Ku-band. Supporting several VSAT networks with the capacity of one transponder should be possible. (Utilization of transponder bandwidth is discussed later in this section.) The hub stations can serve the additional functions indicated in the figure. Access would be provided to the public networks, and an international gateway could be included if appropriate. We assume that any connected network employs terrestrial transmission, since a *double hop* (two satellite links) is typically undesirable.

The architecture used in almost all instances has the VSAT communicating with a particular hub station. Employing a specialized packet protocol or TDMA over the satellite is most common. These aspects are reviewed later. The interface to data communication users, however, would be a standard one, such as SDLC or X.25. Conversion between the satellite access method and the interface protocol

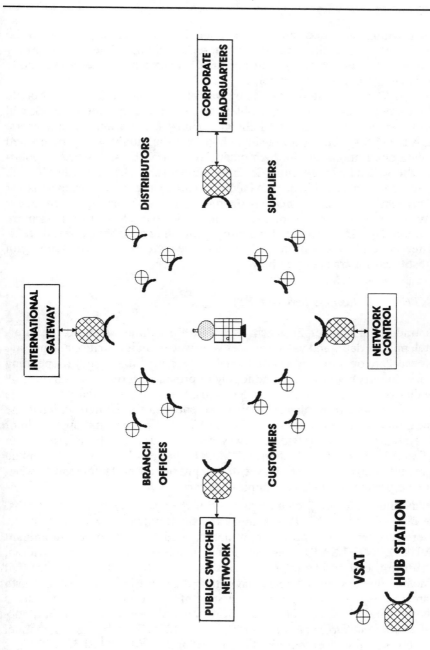

Figure 4.24 An example of an advanced satellite network using VSATs, hubs, and gateways to the terrestrial network.

is performed within the indoor equipment. Users have found that, even with the time delay over the satellite link, the end-to-end response time of a properly engineered VSAT network may be shorter than is experienced over a multinode terrestrial data communication network.

VSAT-to-VSAT communication is by way of a common hub (i.e., a double hop). This usually does not pose a problem because the corporate network will have VSATs at branch offices and hubs at regional centers and headquarters. Direct VSAT-to-VSAT links (i.e., a single hop over the satellite) may be needed for some data applications and for high-quality voice service. In this case, a point-to-point circuit would be set up with a dedicated modem at each terminating VSAT. A circuit switched service could be provided with modems controlled from a master station. This scheme is called *demand assignment* and is discussed in a subsequent section. Whether fixed assigned or demand assigned, the VSAT transmits more than one radio carrier, which is a driving force for using a higher powered transmitter. Alternatively, the size of the reflector may be increased, since antenna gain can be substituted for transmitting power.

4.3.1.3 INTELSAT Business Service (IBS)

The traditional manner in which INTELSAT satellites are used for international telephone, data, and video services is by way of an international gateway. These gateways are operated by a commercial or governmental agency (depending on the home country), which has a monopoly to provide international services. A sample listing of gateway providers, called *signatories,* for particular countries is provided in Table 4.1. In particular, the chosen entity in the United States is the Communications Satellite Corporation (COMSAT). Over the years, ownership of the earth stations used to provide gateway services has been transferred from COMSAT to AT&T, MCI, and others. COMSAT's primary business now is selling space-segment bandwidth to these U.S. gateway providers and to companies wishing to provide private business communication services.

International gateway providers must offer their services by way of tariffs, which are filed with the FCC. This is unlike the situation with regard to domestic satellite service operators, who are now deregulated (except for the dominant carrier, AT&T). INTELSAT places technical demands on these service providers, forcing them to use large and expensive earth stations. Nonetheless, the tariffed services include public switched voice and data and international television. Earth stations are thoroughly tested prior to the start of operation, and transmissions are tightly regulated. The associated technical standards are maintained in a document called the *Satellite System Operating Guide* (SSOG). Changes in channel format are not permitted, except for those coordinated with INTELSAT. This is to ensure that the overall system meets the high standards of quality which INTELSAT maintains.

Table 4.1 A Sample of Signatories of INTELSAT that will Provide Gateways from Particular Countries into the INTELSAT Satellite Networks

Country	Signatory
U.S.	COMSAT
U.K.	British Telecom
F.R.G.	Bundespost
Canada	Teleglobe
Mexico	SCT
France	France Telecom
Switzerland	PTT
Indonesia	Postel
Philippines	PhilCOMSAT
Japan	KDD
Australia	OTC

The INTELSAT Business Service (IBS) was introduced in 1984 to encourage the development of international private business communication using carrier-owned and customer-owned earth stations. This was a radical concept because earth stations previously were typically owned and operated by signatories. Some might argue that IBS would impact the "bread and butter" business of INTELSAT by removing traffic from the public networks. Consequently, not every country in the world rushed to allow their domestic users to have direct access to the INTELSAT system. The U.S. and the U.K., however, moved to facilitate the growth of IBS.

Parameters for IBS where chosen to encourage the use of relatively small earth stations, which are more like hubs than VSATs. Depending on the footprint of the satellite in the particular geographical region of service, either C-band or Ku-band earth stations can be employed. The communication links are arranged on a point-to-point basis between IBS antennas, using capacity on the Intelsat V and Intelsat V-A satellites. INTELSAT has mandated that only digital communication links be established. These links employ conventional high-speed modems, which operate at rates such as 56 kb/s, 256 kb/s, T1, and even higher in special situations.

An example of how IBS is used in a private telecommunication network is illustrated in Figure 4.25. The user in the U.S. can arrange for the circuit from New York to London with IBS providers on both ends. A terrestrial data circuit is used to connect from the customer's premises in a distant city (Washington, D.C.) to New York and the IBS earth station site. This circuit can be provided as part of the service, or the user may employ an existing domestic private network of its own. Space segment in the form of half circuits is provided by the INTELSAT signatory for the transmission path from the particular earth station to the satellite. On the other end of the link, British Telecom or the Mercury subsidiary of Cable

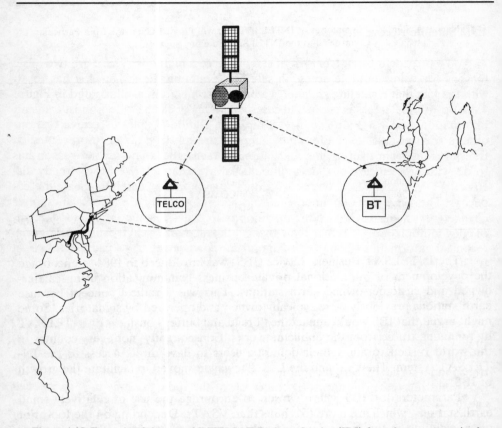

Figure 4.25 Example of the use of INTELSAT Business Service (IBS) for private communication.

and Wireless operate IBS facilities in London; either may be employed. The U.S. user may contract directly with the British IBS provider, who can then arrange for the INTELSAT half circuit and the digital access line to the distant customer's premises.

From this discussion, it should be apparent that there could be a number of ways in which IBS links might be arranged. This would depend on the domestic and foreign service providers that were used. A "turn-key" approach would be possible, where a single carrier, such as C&W, ITT, or France Telecom, did all of the coordination, obtained the requisite satellite and terrestrial capacity, turned up the circuit, and provided a single monthly bill. Some users would prefer to put the pieces together with the hope of reducing cost. Another variation would have customer premises earth stations on both ends. These could be owned by the user if qualification as an IBS carrier were arranged. Alternatively, the customer premises earth station could be provided by a carrier which was already in the IBS business.

4.3.2 Satellite Access Methods

We provide a brief overview of satellite access methods. There are two basic reasons for using multiple access in satellite communication. The first has to do with the fact that a satellite channel is a broadcast medium, as illustrated in Figure 4.4(b). Multiple access is applied in cases where many earth stations must transmit information to a common channel. Since each station is able to receive one another's transmissions over the same channel, care must be taken to prevent such transmissions from overlapping and jamming each other. The second reason has to do with efficient use of the satellite microwave channel. We have already discussed FDMA and TDMA, the two principal systems for employing satellite capacity. The other two common multiple access methods are Aloha and *single channel per carrier* (SCPC). Our intention is to provide some guidelines for employing multiple access, rather than giving a detailed technical treatise. Additional technical information along with supporting engineering data can be found in [CCIR, 1985].

4.3.2.1 Frequency Division Multiple Access (FDMA)

The most straightforward multiple access method, and the oldest, is FDMA. In fact, the satellite bandwidth is subdivided by FDMA into transponders, although the other multiple access methods are used on a subassignment basis within a given transponder. The FDMA technique is familiar because it is used in over-the-air radio and television broadcasting. Since all of the radio stations in a given area can be received at the same time by your radio receiver, they must transmit on different frequencies. In the case of satellite communication, the channels can all be received throughout the country or hemisphere, depending on the coverage footprint of the satellite antenna. Frequency assignments are typically made by the satellite operator so that *radio frequency interference* (RFI) is avoided to the greatest extent possible.

A hypothetical example of a four-node FDMA satellite network is illustrated in Figure 4.26. This network is a fully interconnected mesh, which provides a capacity of one T1 channel for each station. Each station transmits on a different frequency to the satellite. The combined spectrum of four channels is received by the satellite, and the T1 carriers are translated as a set from the uplink frequency range to the downlink range. For example, the uplink and downlink ranges at Ku-band are centered at 14,000 to 14,500 MHz and 11,700 to 12,200 MHz, respectively. Also within the satellite, the group of channels is boosted in power by a microwave amplifier (one per transponder in typical designs) to overcome the substantial loss along the earth-to-space path. The four carriers, on separate, nonoverlapping frequencies, appear in the downlink band in the same order as they arrive in the uplink.

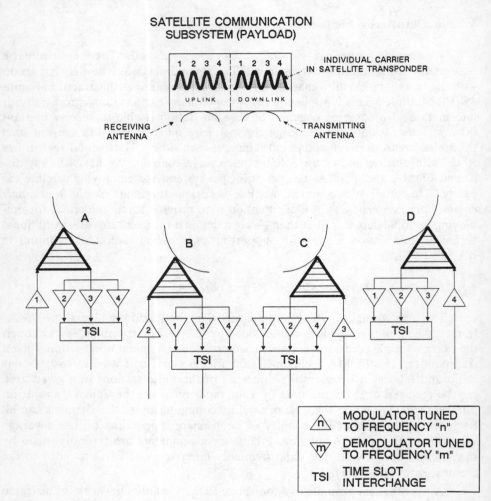

Figure 4.26 Application of Frequency Division Multiple Access (FDMA) in a digital communication satellite network.

To provide full connectivity in this mesh network, each station is equipped with three demodulators tuned to the frequency of a carrier from another station. In general, a station transmits one carrier (containing its outbound traffic) and receives $N-1$ carriers (one for each inbound link from another station). This is unlike a point-to-point terrestrial network, where a complete modulator and demodulator is required for each T1 link. The broadcast nature of the satellite allows the downlink to be received at all stations, producing an economy in transmission equipment.

Because a given carrier has all outbound information multiplexed together, the receiving station must separate out only the traffic destined to it. This is the function of the *time-slot interchange* (TSI), which is also a basic function within the DACS, described earlier in this chapter, the need for which is illustrated in Figure 4.27. There is a time frame shown for each carrier frequency (1 through 4), representing the information flows from the corresponding earth stations (A, B, C, and D). We have assumed ideal (equal) balance of traffic between the four stations. Therefore, the frame is broken into three equal blocks of data, where one block contains the traffic which is being transmitted to one destination station. This type of transmission has also been referred to as a multidestination carrier. For example, station A transmits on frequency 1 a frame with three blocks. These blocks are destined for stations B, C, and D, respectively. There would be no block for A, since this transmission had come from A. An examination of the four frequencies and their respective time frames will show that there is a duplex pair of blocks for each point-to-point link. In this example, any given receiving earth station would not have to rearrange the blocks since at no time were there simultaneous transmissions on different frequencies for the same destination. Station B, for example, receives carriers 1, 3, and 4. Note that, from Figure 4.26, it would pull the first block from frequency 1, the third from frequency 3, and the second from frequency 4.

The role of the TSI on the receiving side of each earth station should now be apparent. With appropriate programming, TSI accepts the baseband time frames for each demodulator and places them in memory buffers. Then the appropriate blocks are read out of memory in the correct sequence and sent to the user interface. The key is to make the slots match those in the uplink direction so that the user on the terrestrial side sees a full T1 time frame.

This example of using multidestination carriers takes advantage of the broadcast nature of the satellite and requires the use of a TSI in each earth station. It is certainly possible to simplify the network by transmitting a carrier for each destination (i.e., single-destination carriers). This increases the number of carriers in the example from four to twelve, which will also increase the traffic capacity of the network by the same ratio. If this capacity were used, it would be advantageous. A large fraction of FDMA networks employ single-destination carriers. The TSI will not be required because there is one downlink for each uplink, and the traffic on each carrier is dedicated to one link. The interface with the user is direct. Single-destination carriers, however, can consume satellite capacity and result in a large investment in earth station equipment, particularly modems and transmitters.

As should be apparent, point-to-point links over an FDMA satellite network could easily be blended with terrestial digital links over fiber optic and microwave systems. A high-level multiplexer can access all three types of digital links. This type of device provides flexible access to the satellite and terrestrial links, ac-

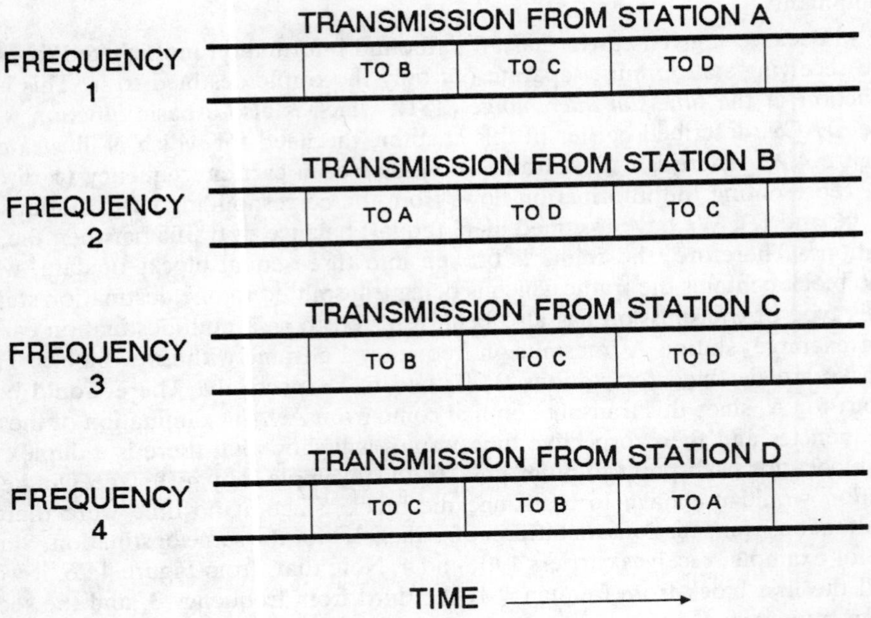

TRANSMISSION FROM STATION A

FREQUENCY 1

| TO B | TO C | TO D |

TRANSMISSION FROM STATION B

FREQUENCY 2

| TO A | TO D | TO C |

TRANSMISSION FROM STATION C

FREQUENCY 3

| TO B | TO C | TO D |

TRANSMISSION FROM STATION D

FREQUENCY 4

| TO C | TO B | TO A |

TIME ⟶

Figure 4.27 Ideal time slot assignment in a TDM-FDMA satellite network of four earth stations.

counting for changes of routing and loading in response to demand for service. Furthermore, FDMA is the principal multiple access method used in INTELSAT's IBS network.

The manner in which FDMA uses satellite capacity is illustrated in Figure 4.28, which plots the relative capacity *versus* the number of simultaneous carriers within a single transponder. If there is only one carrier, then the transponder can be operated at full capacity with no loss of output power (the main determinant of capacity). When two carriers share the power of the transponder, the usable output drops to nearly half. This is caused by nonlinearity of the power amplifier associated with each transponder within the satellite. Increasing the number of earth stations sharing the transponder causes a corresponding increase in the number of carriers. The total capacity continues to decline, but at a declining rate. After about twenty carriers, the relative capacity remains at approximately 30%. As a consequence of the loss of effective transponder power, the bandwidth becomes underutilized.

There appears to be a substantial penalty in transponder throughput, which becomes a concern because transponder capacity is a significant cost component of the network. There are, however, several aspects to transponder utilization. Capacity is replotted in Figure 4.29 on an absolute scale for 32 kb/s data channels

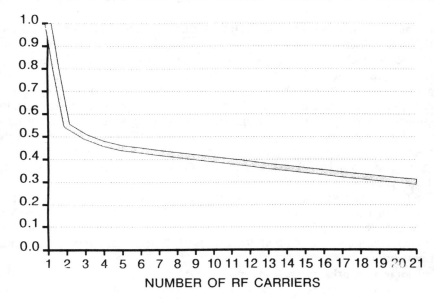

Figure 4.28 Relative capacity of a satellite transponder used in FDMA service.

(32 kB/s DATA CHANNELS)

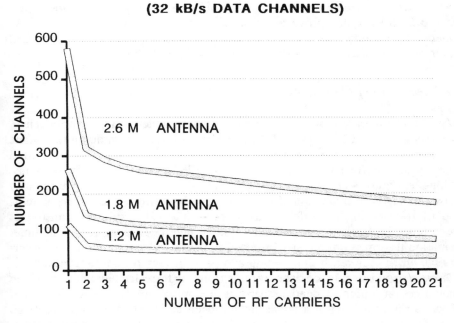

Figure 4.29 Transponder capacity versus antenna size.

that are multiplexed together using TDM. A maximum transponder capacity of approximately 600 channels is shown, assuming a domestic Ku-band satellite and a homogeneous ground segment of earth stations with antennas 2.6 meters in diameter. The factors that determine antenna size are discussed in [Elbert, 1987]. Transponder power is the main determinant of capacity, and so it is possible to trade capacity for antenna diameter. Therefore, two smaller sizes of antenna for the homogeneous earth station network are shown. Maximum capacities for the 1.8 and 1.2 meter antennas are approximately 250 and 100 channels, respectively. Behind this relationship is the physical fact that the receiving sensitivity of an antenna is proportional to the square of the diameter (i.e., proportional to the area of the reflector).

The penalty is dramatic for using smaller antennas and increasing the number of simultaneous carriers. A network of four stations with 2.4 meter antennas can provide about 280 channels in a single transponder. The capacity is reduced to only 50 channels, however, if we attempt to use 1.2 meter antennas for this four-station network.

For large networks of small antennas, FDMA is apparently an unattractive choice. The real measure of efficiency in the private telecommunication environment, however, is probably not maximum transponder loading. Rather, it is the cost of delivering the required service to the user's location. With this in mind, an inefficiently used transponder supporting a a network of very small diameter antennas (i.e., VSATs) may prove to be optimum for a particular application. As presented subsequently, the VSAT approach uses a combination of FDMA and the other multiple access methods to achieve this type of efficiency.

4.3.2.2 Time Division Multiple Access (TDMA)

As an invention to overcome FDMA's transponder loading inefficiency, TDMA was developed by COMSAT and INTELSAT for international telephone communication. TDMA has also been used on domestic satellites in Canada and the U.S., primarily for backbone or heavy route links between major locations. In looking at Figure 4.28, full transponder TDMA on a single carrier frequency achieves almost 100% utilization of the power and bandwidth of the transponder. Instead of transmitting on different frequencies, the earth stations transmit at different times in blocks of information, which have been compressed in time. Earth station transmissions are synchronized so that individual blocks do not overlap in time when they reach the satellite and pass through the same transponder. *Burst transmission* is employed, where traffic from the terrestrial interface is stored in a compression buffer. A header is placed in front of one complete block, and the block is then read out at a much higher speed and sent through a wideband modem. Transmission speed over the satellite ranges between 6 and 120 Mb/s, depending on the network requirements. The receiving downlink path contains the bursts arranged sequentially in time from all the stations (including that of the sending earth station) so that the demodulator must synchronize to each individually.

The key elements of a four-node TDMA network are shown in Figure 4.30, which is for a 60 Mb/s wideband system employing all of the capacity of one 36 MHz transponder. This is really a snapshot in time of one station's uplink. There is only a single carrier at a time; stations transmit their information in nonoverlapping bursts of digital data, containing all traffic. Each earth station has a single modem which transmits and receives the bursts that occupy the common frequency in the transponder. The digital traffic, which is arranged in a TDM frame, is assembled and disassembled with a TSI type of device. This TSI is quite different from the type used in continuous FDMA transmission or terrestrial applications, as will be clear from examining the time slot assignment for the network shown in Figure 4.31.

Stations A, B, C, and D transmit their bursts at different times, but on the same frequency. The top line shows the burst transmitted from station A, with the blocks of data for destinations B, C, and D indicated. At the front of the burst is a "header," which consists of overhead bits used for synchronization and network control. Timing of the network is maintained by the TDMA control equipment within the stations themselves. The TSI at each earth station assembles and disassembles the bursts, and arranges the traffic for connection to the users on the terrestrial side of the interface. Being a TDM system, the bursts are sent in a periodic frame sequence to provide time frames compatible with the typical user equipment. The transponder loading efficiency of single-carrier TDMA is between 95% and 99%, depending on the design of the equipment and the number of stations in the network. The upper limit is determined by the fact that time gaps must be left between bursts (guard times) to prevent overlap. In addition, the header attached to each burst absorbs some of the throughput.

The transponder loading efficiency of TDMA is impressive and was, in fact, the main motivation for TDMA trials first conducted over INTELSAT satellites in the late 1960s and early 1970s. One of the real benefits, however, is in the flexibility in traffic assignment and routing that TDMA delivers. A new station can enter an existing network by locating an empty slot in the time frame for its burst, using an automatic acquisition system contained within the TDMA terminal. The station can now communicate with the control center over the *common signaling channel* (CSC) provided in the header of every burst. Actually applying multidestination traffic to the network involves assignment of individual blocks and channels within the burst. The *network operations center* (NOC) would be connected into the CSC and could make such assignments. As we will discuss in Chapter 5, network channel assignments are determined off-line by using a com-

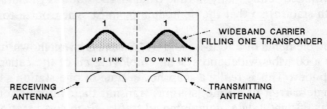

SATELLITE COMMUNICATION SUBSYSTEM (PAYLOAD)

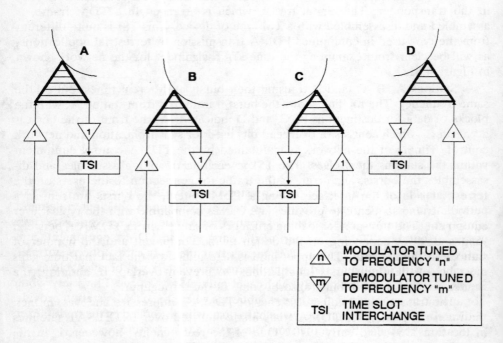

Figure 4.30 Application of Time Division Multiple Access (TDMA) in a digital communication satellite network.

puter database of some sort. This database is converted into a *network map* which identifies the time assignment of every channel to each earth station within the bursts. A new network map would be transmitted over the CSC to every terminal, including the new location that had come into the network. A subsequent command from the NOC would cause the terminals to switch to the updated network map.

A TDMA network is inherently flexible because a new network map can be introduced at any time to specify all connectivity in the full mesh network. In fact,

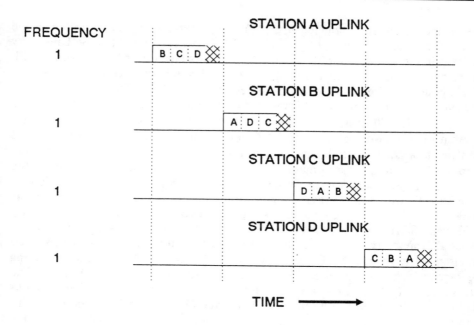

Figure 4.31 Ideal time slot assignment in TDM-TDMA satellite network.

the terminals can have several maps prestored and particular ones may be "executed" at prescribed times to adjust the network routing to anticipated demand. This type of common and centralized control for a wide area network is perhaps unique to satellite TDMA. As part of the earth station TDMA equipment, multiplexing is typically provided. This means that the port card options normally associated with intelligent and nonintelligent multiplexers will be available as elements of TDMA. Consequently, high-level multiplexers of the T1 variety would not appear to be required. The one exception is where the TDMA station provides dedicated T1 links between sites, which, in turn, will be applied to the multiplexers at locations that may be remote from the actual earth station.

4.3.2.3 Packet Satellite Access

The two principal multiple access methods in satellite communication, TDMA and FDMA, serve the needs of relatively high-capacity, thick-route networks, particularly for domestic and international common carriers. Packet satellite access, illustrated in Figure 4.32, was conceived in the early 1970s for networks of small, inexpensive earth stations. The intention was to have a network scheme to bring satellite communication down to an affordable level. The first experiments with the technology were conducted over NASA's experimental ATS satellite in the Pacific Ocean region. The specific packet protocol was developed at the University of Hawaii and the appropriate name of Aloha was given to it [Abramson, 1973]. These experiments, run on a limited budget, proved highly successful. The advent of Ku-band satellites and the keen interest in data communication applications, however, led to the commercialization of the technology in the mid-1980s.

Figure 4.32 illustrates the basic operation of a packet satellite access system, wherein a star network of several VSATs sends data messages, which are collected by a much larger hub earth station. The transponder performs its usual function of merely repeating the information after translation from the uplink range to the downlink range. Operation is very similar to TDMA in that inbound transmissions from VSATs are on the same frequency, but at different times. The hub transmits a continuous stream of outbound data to the VSATs on a different carrier frequency to prevent interference with the packets coming from the VSATs. By virtue of the broadcast capability of the satellite, the transmission from the hub can be received simultaneously by all VSATs. This outbound path is a data broadcast containing messages which are time multiplexed and addressed for individual VSATs. The terrestrial equivalent of this arrangement, shown at the right of the figure, looks like a star network with full-duplex links between each VSAT and the hub. This arrangement economizes on the transmitter power within the VSAT because the hub employs a larger, more sensitive antenna than does the VSAT.

The arrangement of the four-node network using packet satellite access is shown in Figure 4.33. A hub earth station is needed to receive inbound VSAT transmissions and to send outbound information back to the satellite. Therefore, to provide VSAT-to-VSAT data communication, a double hop must be accepted. In most applications, however, the hub would be connected to a host computer using a terrestrial circuit, thus avoiding a double hop.

This example is meant to illustrate the principle of operation for the same example of the four-node network as used previously for FDMA and TDMA. VSATs A, B, C, and D transmit on frequency 1 at a relatively low data rate, typically in the range of 9.6 kb/s to 56 kb/s. The packets being sent are assembled by a PAD located within the earth station. This is a very specialized type of PAD using a packet protocol designed exclusively for satellite operation. In most cases, the technology is proprietary, and VSATs made by different manufacturers cannot communicate among one another. Frequency 1 appears in the uplink as a relatively low-powered, narrow-band carrier, occupying a small part of the total capacity of the transponder. One inbound carrier can handle packet transmissions from a number of VSATs, typically twenty to fifty. The carrier at frequency 2 is transmitted by the hub at considerably more power to ensure good reception at each VSAT. Rather than a PAD, the hub has a time division multiplexer to send the data to the VSATs and a packet disassembler to receive from them.

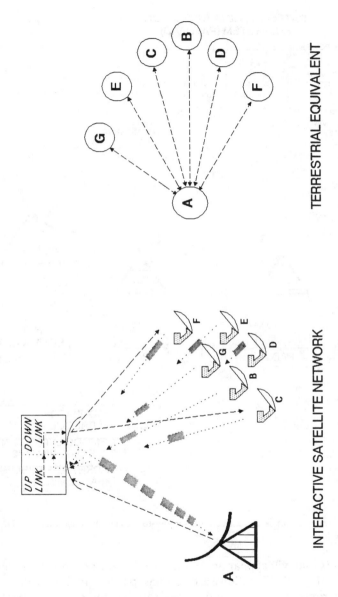

INTERACTIVE SATELLITE NETWORK   TERRESTRIAL EQUIVALENT

Figure 4.32 Packet satellite transmission medium.

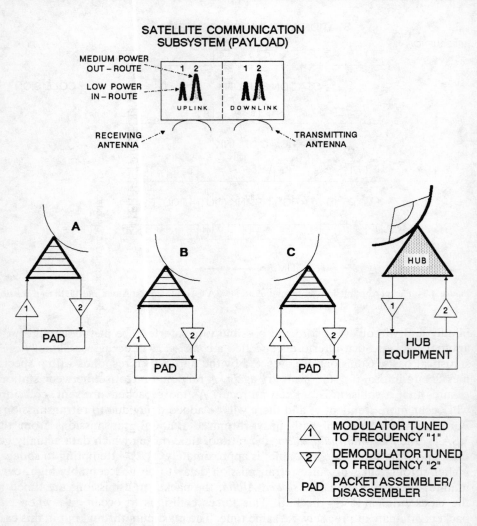

Figure 4.33 Star network using packet satellite access for remote-to-hub communication and TDM for rub-to-remote broadcast.

The actual packets that will appear on frequencies 1 and 2 are shown in Figure 4.34. We assume that the VSAT stations are sending packets to each other, requiring the hub to retransmit them back over the satellite for reception back at the VSATs. The VSATs each transmit on frequency 1 for the inbound path. Unlike TDMA, however, the packets are not assigned into specific blocks, nor timed precisely according to a prearranged plan. Rather, a VSAT with a packet transmits

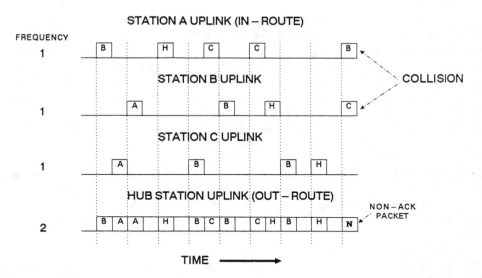

Figure 4.34 Packet transmissions from remote sites (VSATs) using slotted Aloha and TDM transmission from the hub station.

blindly and without regard to the possibility of interference at the satellite with another packet. Such interference, called a *collision,* will occur occasionally on a statistical basis. Collisions are detected by the hub, which responds with a special message to inform the stations to try again. A random time offset between stations ensures that a collision is unlikely to recur. As more packets are sent, collisions will occur more frequently, and there will be added delay due to retransmissions. For the Aloha protocol with unsynchronized random transmissions from the VSATs, the *maximum utilization,* the ratio of time during which data actually get through to the total time available, is approximately 20%. Attempting to squeeze higher throughput will cause retransmission delay to be unacceptably high, even infinite. In a variation called *slotted Aloha,* the packet transmissions are timed to the clock originated by the hub. This forces collisions to occur only when two packets originate at precisely the same time. The maximum throughput in this case is nearly double that of pure Aloha. Slotted Aloha, however, imposes the added complexity of the timing reference.

The Aloha protocols are embodied in the access technique called *carrier sense multiple access with collision detection* (CSMA-CD), which is used in local area networks on cables. Another approach is to use *hub polling,* wherein VSATs only transmit when instructed to do so. Indication from the hub over the outbound link is called a *poll.* The throughput efficiency is high with polling, but there is a significant delay because VSATs must wait their turn for every message.

The outbound path from hub to VSATs is shown at the bottom of Figure 4.33. This transmission is continuous in time, containing messages for the individual remote stations. Each block is actually a packet addressed to a station, as indicated, containing control information or actual user traffic. Typically, control information allows the network to operate in several modes for better matching the types of information transfer which users may require. In the true packet Aloha or datagram mode, the VSAT sends an addressed packet to the hub, which then sends it to the host or another VSAT, as appropriate. This requires no control other than the proper routing and addressing. A reservation mode to execute a file transfer from remote to host may also be employed. The VSAT sends a reservation request packet to the hub. In turn, the hub assigns a block of time to the remote for the bulk data transfer. The other stations in the network monitor such messages from the hub and restrain themselves from transmitting during the reservation. Alternatively, a TDMA time span is allocated within the frame for nonpacket transmissions.

Another use of TDMA in the VSAT context is primarily for voice communication between remote and hub only. Voice communication over the VSAT network requires that 8000 samples be sent per second. This typically involves compression to reduce the effective transmission from 64 kb/s to as low as 16 kb/s, depending on the features available from the VSAT. In particular, the compression algorithm called *residual excited linear prediction* (RELP) has proven very effective on VSATs provided by Hughes Network Systems. Telephone service through a VSAT of this type would be through the hub, where calls could terminate on a PBX. VSAT-to-VSAT calls could be provided by double hopping via the hub PBX. This would be acceptable for internal administrative calls, but not for true toll-quality public calling.

A typical Ku-band transponder would have several packet satellite networks in operation. These would be assigned different frequencies within the transponder bandwidth. Consequently, the FDMA mode is also being employed. Typically, the hub transmission is at 256 kb/s and simultaneously supports several VSAT subnetworks on different inbound frequencies. The hub would be equipped with several demodulators to receive packets from all inbound carriers, but only one high-speed outbound carrier would be needed.

4.3.2.4 Demand Assignment Multiple Access (DAMA)

We have seen how a satellite channel can be used as a packet switch in the sky. In this section, we evaluate the *demand assignment multiple access* (DAMA) mode, which effectively implements circuit switching among small earth stations. Direct VSAT-to-VSAT links are to be provided; therefore, either FDMA or TDMA is to be used. In the *single channel per carrier* (SCPC) DAMA system, the actual communication channels employ individual FDMA carriers. Most SCPC links in current use are not demand assigned, but rather are fixed assigned to certain frequencies within a transponder. To be demand assigned, the frequencies are not specified until the time that an actual is call set up. A VSAT requests that a circuit be established between it and a specific other VSAT. The DAMA system assigns a pair of frequencies (one for transmitting and one for receiving), and sends them over a CSC to the calling VSAT. The DAMA network also alerts the called VSAT that a circuit is to be established and of the frequency pair to be used. The called VSAT uses the same frequency pair but the transmitting and receiving senses are reversed

An example of a four-node DAMA network using SCPC transmission is shown in Figure 4.35. We have assumed that each station has the capability to support two simultaneous calls, such that it can set up two independent circuits on demand. The VSAT architecture is designed for this type of service, and each terminal has two circuit switched modems and one packet switched modem. The purpose of the satellite packet modem is to provide the CSC that is used to set up the circuit at the beginning of the call and for network monitor and control. By providing two circuit switched ports and one packet switched port, this earth station provides service which emulates the ISDN basic rate interface (e.g., 2B + D). Data may be sent via the packet satellite link, but its main purpose is to facilitate the DAMA operation of the circuit switched network over the satellite. A hub station (not shown) would process the packets in the same manner as was discussed in the previous section.

The bandwidth of the satellite transponder uplink and downlink is shown at the top of the figure. There is a total of eight frequency slots available for service within the transponder; the ninth frequency is dedicated to the packet satellite channel. Because this corresponds exactly to the number of possible carrier transmissions from the earth stations, the "switch in the sky" is nonblocking. The satellite SCPC channels are not assigned until actually used for a call. Hence, it is possible to provide fewer channels than there are modems. There would be many more earth stations with like number circuit switched modems. Most blocking would occur when one station cannot reach another because the distant station does not have an available modem. These considerations are not different from those used in traffic engineering to design telephone networks. (More information on this topic is provided in Chapter 6.)

The frequency assignments within the modulator and demodulator symbols in Figure 4.35 show one possible arrangement for traffic routing. This, of course, would change from time to time in response to the dynamics of calling over the network. Area code information, which is required to determine the destination earth station, is stripped off in the *circuit switched control unit* (CSCU) that interfaces the access lines to the modems. As in other common channel signaling systems, the area codes are sent over the CSC, rather than as in-band signals

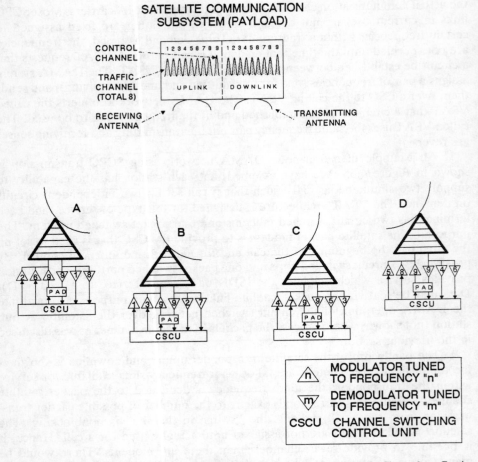

Figure 4.35 Application of Demand Assigned Multiple Access (DAMA) and Single Channel per Carrier (SCPC) Operation in a circuit-switched satellite network.

through the actual telephone channel. A PAD of the type previously described for packet satellite is used to transmit area codes and to do the communicating needed for call supervision. Tuning of circuit switched modems to assigned frequencies within the transponder is controlled by the CSCU or another element of the VSAT.

The type of information imparted by the nine carriers within the transponder is shown in Figure 4.36. Remember that this is a snapshot in time when the circuit connections are set up in a particular pattern. As calls are completed, the frequencies are released and become available for subsequent calls. In other words,

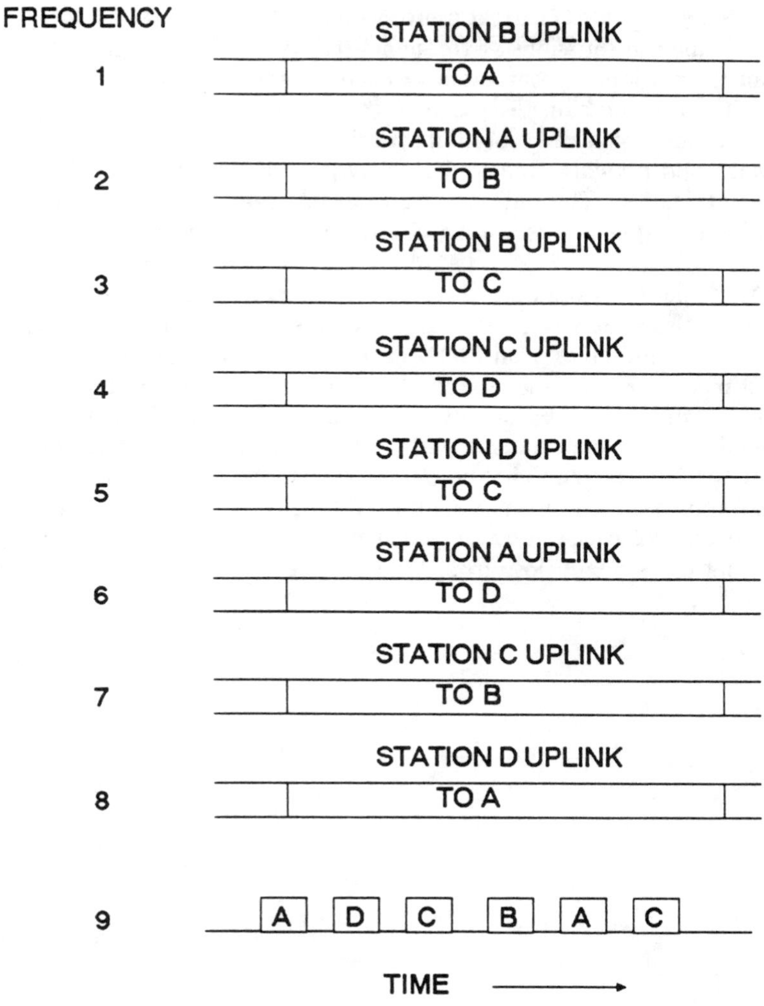

FREQUENCY

STATION B UPLINK

1   TO A

STATION A UPLINK

2   TO B

STATION B UPLINK

3   TO C

STATION C UPLINK

4   TO D

STATION D UPLINK

5   TO C

STATION A UPLINK

6   TO D

STATION C UPLINK

7   TO B

STATION D UPLINK

8   TO A

9   A   D   C   B   A   C

TIME ———→

Figure 4.36 SCPC frequency assignments in a DAMA satellite network.

the same frequency pair would be used by different stations at different times, with assignment being made at random by the central network control element. An example of a pair is frequencies three and seven, which are used to connect stations B and C for the duration of this particular call. While the call is in progress, the frequency pair is occupied, carrying voice or data traffic. In telephone service, the voice activation technique turns the carrier on and off in response to actual speech. This conserves aggregate power within the transponder, since only 35%

to 40% of the SCPC signals are on at a given time.

Notice how the transmissions on frequency 9 are in the form of packets, which originate from the stations indicated above each packet. This is the Aloha mode of packet satellite transmission discussed in the previous section. These packets contain call setup messages and other data traffic. Depending on the design of the VSAT hardware and network, the packets may pass directly between VSATs or go through a hub station. The hub may transmit on another carrier (number 10, not shown) if a central site is used for directly controlling and administering the network.

The DAMA mode of multiple access is very flexible and provides excellent quality for services such as voice. Availability of the packet satellite subnetwork enhances the system and gives it an ISDN flavor. This particular type of network has been used piecemeal, but the combination has not yet appeared on the commercial market. Perhaps, a reasonably priced DAMA system for VSATs will become available. Other experiments with packet systems like Aloha show that there is room for improvement in the flexibility and throughput of the packet satellite access approach. The ideal VSAT would include the features of satellite packet, TDMA, and DAMA, any of which could be employed to satisfy the requirements that users might pose from time to time.

Chapter 5
Network Management

A private telecommunication network is constructed from facilities and services that a strategic unit often obtains from outside sources. The management of the network resource, however, is a matter that must be taken into one's own hands. The turning of the keys (i.e., the transfer of responsibility from supplier to new owner) is only the starting point. To manage a network, the strategic unit must control all aspects: technical, operational, and financial. In this chapter, we define the network management problem and review the technologies and subsystems that make its solution possible. We describe some concrete examples of network monitor and control systems. We consider the administration of a network, viewing the system in business terms.

In Chapter 2, we reviewed a number of applications of private telecommunication networks. One of the more complex, challenging, and potentially rewarding applications is the actual management of the telecommunication network. This application brings to bear on-line data processing facilities and an integrated data communication network that is dedicated to network management. At the conclusion of this chapter, we provide concrete examples of integrated network management systems, using computers, telecommunication facilities, and people in effective ways. Vendors of network management hardware, software, and systems emphasize that these technologies turn a private telecommunication network into a strategic resource.

5.1 DEFINITION OF NETWORK MANAGEMENT

The term *network management* is a relatively new one and unfortunately its meaning is not clear. To some, the term refers to the control of specific pieces of equipment which are remote from the controlling organization. This is the most basic requirement. Network management can be taken many steps further to mean the total control, allocation, maintenance, and management of the network from a business perspective. The special facilities and personnel that take the disparate elements of the network and turn them into a functioning business form the context of the definition of network management given in this chapter.

A useful approach to this subject is to divide it into two principal categories: *network monitoring and control* (NM&C) and *administrative operations* (AO). The NM&C category deals with the technical aspects of tying disparate equipment and facilities back to an operations center of some type, allowing operations personnel to monitor the performance of critical elements of the network and to control its configuration. This is the same problem faced in dealing with a communication satellite located 22,300 miles from the surface of the earth. The administrative operations provide for running the network like an ongoing business. The requests for service from users are processed, leading to the connection of the user to the network. Henceforward, service to the user is provided, measured, and billed as appropriate. Network management from the AO viewpoint supports the maintenance of the network and the resolution of problems as they arise. The NM&C capabilities would therefore tie into the smooth running of the network. Considerably more detail on the range of NM&C and AO functions is provided in subsequent sections.

Historically, AT&T and its telephone operating subsidiaries performed true network management, and the customer (even the biggest of businesses) needed to do very little on their own. Again, divestiture and deregulation have broken telecommunication down into its pieces. Many of these pieces are provided by a wide variety of manufacturers, Western Electric does not even exist any longer, and foreign companies are now becoming a bigger factor in the U.S. telecommunication equipment market. New technologies are appearing that provide opportunities to do things through telecommunication which were not even possible before divestiture. Therefore, network management becomes the function of the strategic unit having control of the network. It is ironic that this is the situation which developing countries have previously faced. The PTT of a country like Indonesia was accustomed to buying telephone switches and transmission systems from a variety of different companies residing in industrialized countries (notably the U.S., Japan, Germany, Belgium, and Sweden). The Indonesians would then have the difficult task of interfacing these systems, maintaining operations, and building administrative capabilities around them. Today, Indonesia has one of the most advanced public telecommunication networks, due to heavy investment in modern facilities and training during the past ten years.

Today, strategic units are driven by a competitive environment that makes the difficult problem of defining new tasks even more challenging. This situation also defines the environment with regard to network management. A marketing innovation, such as the introduction by American Airlines of its frequent flier program, requires data communication facilities and software. As the new system evolves, management of the network still ought to be flexible. This reflects the PTT example cited previously because a strategic unit must purchase new equipment, and bring the conglomeration into the existing fold, while still serving the needs of the strategic unit. Today's intelligent network equipment, such as computer systems, high-level multiplexers, and VSATs, are designed with embedded NM&C capabilities. The bigger picture, however, requires a friendly interface so that people can run the network and properly serve the users. The following paragraphs elaborate the key network management dimensions of NM&C and AO.

5.1.1 Monitor and Control (M&C) Capabilities

Monitoring and control capabilities are built into essentially every device or subsystem currently on the market. Such was not always so, but M&C can now be provided on a stand-alone basis because of the widespread use of microprocessors and data communication networking, all the way down to the device or subsystem. The driving force for this was the desire of equipment buyers to reduce manpower expenses by placing equipment at unmanned sites, and using telephone and data lines to provide the control access. In the following paragraphs, we review the most common approaches for M&C within typical elements of a telecommunication facility, including the switch (voice or data), the transmission link (multiplex, fiber, microwave, and satellite), and the building itself.

5.1.1.1 Switch

A telecommunication switch is a versatile device, which dynamically processes and routes user traffic. Therefore, the M&C capability associated with the switch typically operates in the background, while the switching is conducted in the foreground. The most common type of switch is the PBX, which will be the focus of attention in this section. In the more advanced PBXs, voice and data traffic can be switched, making this class of device perhaps the most versatile.

First, we examine the voice side of M&C. A PBX must be set up initially for the requisite number of station loops and external trunks. These need to be assigned to telephone numbers with associated classes of services. The traffic management system of the PBX provides a man-machine interface in the form of a video display terminal, shown in Figure 3.5. In some PBXs, the terminal is actually a PC with internal software to perform many of the management interface tasks. On others, the terminal is "dumb," and all software and processing is performed by the PBX central processor. The basic input for adds, moves, and changes is through the traffic management terminal.

From the maintenance side, the PBX may have software within the central processor that can automatically test the loops and trunks when they are not occupied. The test measures leakage resistance and capacitance across the wire pairs. This automatic testing, in fact, was the basis for a humorous television commercial by GTE. The actor appears in a robe and says that it is 2 AM and the central office switch is now testing the lines to each customer. Having the switch identify faulty lines and station equipment before the user does so appears to be desirable. The results of such testing would be presented to the manager of the facility through hardcopy output and via the video terminal that accessed the central processor for test and maintenance purposes. Having the automatic test capability means that this terminal need not be manned all of the time.

Access to the central processor of the switch allows management to gain important information as to how users are employing the facility. The traffic system stores records for every call made, and summarizes the information for management purposes. The SMDR feature of the PBX delivers reports that provide billing information. In addition, traffic data are used to make adjustments in switch configuration. For example, if excessive call blockage were experienced on a certain tie line, an additional tie line could be added. The traffic information allows a more precise match of need so that excess capacity can be eliminated.

Information on each trunk line entered through the terminal is used to define routing of outgoing calls. *Least cost routing* (LCR) is an important capability for reducing long-distance telephone expenses. The rules for LCR must be predefined. Subsequent checks through the SMDR system will permit the verification of the assumed LCR rules.

The other types of routing devices have M&C capabilities which are consistent with their function. High-level multiplexers operate like switches, but they also operate as multiplexers used in point-to-point transmission. The configuration of the device would be entered through a video terminal connected to one of the compatible nodes, as shown in Figure 5.4. Via the terminal, it would be possible to view the current network topology, the designation of routing of channels between the various nodes. Updates to the topology could be made, and even changed automatically by the system at predetermined times. As in switch maintenance, the M&C system of the high-level multiplexer can test ports and trunks to determine their operational status.

The variety of switching and routing devices is great, and we cannot treat every one in this chapter. Suffice it to say that the ground rules for M&C at the device level have been set by the industry. The user interface defined in Figure 5.3 is now nearly standard. The more difficult task, however, is gathering these interfaces so that the overall network can be managed. The central control point for network management is called the *network operations center* (NOC).

The passing of NM&C data between PBXs and data switches is usually provided in the form of a dedicated signaling channel. A series of international standards for common channel signaling is fast being adopted by switch manufacturers, telcos, and long-distance carriers. This will eventually make possible the exchange of M&C data between switches manufactured by different companies. In addition, the structure of the M&C information will be consistent across operating organizations' boundaries.

5.1.1.2 Transmission Link

A circuit is normally composed of one or more transmission links, which individually connect between two nodes of a network. The connection point between two links may or may not have a switching capability. To determine the operability of a transmission link, a signal must be passed through the link. The integrity and fidelity of the signal would be compared, before and after transmission. Optimally, the M&C system makes its determination based on the user traffic as it traverses from node to node. For example, the integrity of a T1 data stream can be checked without examining user traffic by looking for bipolar violations, which indicate that traffic is not present. Also, the *cyclic redundancy check* (CRC) word attached to data transmissions provides a way of verifying that data have passed undisturbed through a link. The CRC approach is used for verification by the host and terminal devices at the ends of the circuits, leading to a retransmission request if the CRC indicates that errors have occurred during transmission over the network (as explained in Chapters 3 and 4).

In analog transmission, a standard test tone at an audio frequency of one kilohertz is inserted at the transmitting end and measured at the receiving end. This is used to set the effective end-to-end gain according to the transmission plan in effect. After removing the tone, the intrinsic noise power in the channel is measured by using a true power meter. The ratio of tone power to noise power is the signal-to-noise ratio, an important measure of overall quality. Other measures include frequency response and amplitude linearity. Older analog multiplex systems employ constant tones called pilot carriers, which are used to balance the signal power along the route. These systems are antiquated, but will on occasion be encountered, particularly in developing countries.

Remote testing of analog telephone links can be accomplished by using dedicated circuits and automatic test equipment. A nationwide long-distance company has reserved test channels along its microwave and cable transmission links. Through the NM&C system, test equipment is connected to a particular test channel to check the performance of individual links. For example, suppose that problems are being experienced on circuits between Chicago and Dallas, which encompass several intermediary links. At the NOC, a test channel in each link is commanded into the test mode. When one of these channels shows poor performance, the NOC technician will have isolated the fault to a particular geographic area. Then, other status information may provide further hints as to the exact cause. After remote testing is completed and the likely cause identified, the company sends a technician to the appropriate site. Without remote testing, several technicians would have to be dispatched and the time to isolate the problem — and fix it — would be many times longer.

Digital transmission quality is determined by the bit error rate (BER). In the Bell System, it is customary to specify error performance in terms of error-free seconds. A test channel with a standard data sequence allows the BER or error-free second count to be continuously monitored. Test equipment can be used to measure BER on any channel with a standard interface (i.e., RS-232, DS-0, or DS-1, for example). For trouble-shooting purposes, a series of tests employing a *transmission impairments measuring set* (TIMS) permit virtually any technical measurement on an analog circuit to be made. The TIMS would be used by skilled technicians and would not normally be connected to an operating link. Detailed evaluation of a digital link would employ specialized waveform monitors and protocol testers.

According to policies of the local and long-distance common carriers, digital transmission links are to be terminated on the customer's premises by a service unit, which the carrier can activate remotely. For DDS services at rates up to 56 kb/s, a *digital service unit* (DSU) is required. This device changes the format of the user data into the required bipolar format for transmission over the public network. A second function is that of the *channel service unit* (CSU) used at T1 rates and higher. The CSU monitors the data coming over the public network to verify their integrity (i.e., without bipolar violations), and checks that there are at least three ones out of every twenty-four bits. The latter characteristic is referred to as the *ones density restriction*. In addition, the CSU can be commanded by the carrier into loop back for testing purposes. There is a requirement that a combined DSU-CSU be provided on the user side of a DDS circuit, but only a CSU is required for T1 services. The basic CSU can be enhanced for additional testing features in conjunction with the M&C capabilities of the carrier's network.

With regard to an operating transmission link, there are simple devices which can detect if the channel is functional. An example of this is the T1 monitor made by LARSE, which detects bipolar violations and thus infers if the channel is carrying live data. The best approach, however, is actively monitoring BER with a repetitive test signal, which is multiplexed along with the data. This happens to be one of the features available with the extended super frame (ESF) format now appearing in T1 transmission services offered by AT&T and other carriers. (The T1 frame structure and the ESF were introduced in Chapter 4.)

The ideal situation for NM&C on transmission links is essentially achieved with a homogeneous environment such as is provided by the current generation of high-level multiplexers. Megaswitch and Megamux products from GDC are particularly suitable in this regard. Likewise, the IDNX multiplexers from Network Equipment Technologies and the IPX fast packet switches from Stratacom provide a consistent and functional M&C environment for devices of the same manufacturer.

A satellite network involving VSATs and a hub earth station offers a unique way to achieve a homogeneous M&C environment. The fact that links over the satellite can be used for the M&C network makes the arrangement particularly easy and versatile. One technology well adapted to network management is TDMA (described in Chapter 4). An earth station can communicate directly with any other. As with PBXs and high-level multiplexers, a versatile network topology is established because changes can be entered through the M&C system. The M&C capability of a VSAT network is perhaps more thorough because of the wide area communication coverage and nearly complete integration of functions. In particular, the availability of a common signaling channel and NOC for network control greatly simplifies the problem.

5.1.2 Network Monitoring and Control (NM&C)

We now treat the networking of M&C functions from the remote facilities back to the NOC. This class of network permits direct and romote control over telecommunication facilities. Before NM&C capabilities existed, equipment such as microwave transmitters, antiquated step-by-step switches and multiplex equipment, had to be operated by people who were physically on site. Manual patch panels and peg boards were used to bypass failed equipment or to connect spare devices. This was expensive, inefficient, and often led to extended time delays. Frequently, qualified people with the proper test equipment and tools needed to be ferried considerable distances from one site to another as the problems were isolated and corrected. Telecommunication systems are now much too complex to accommodate the manual approach.

An example of the diversity of NM&C is presented in Figure 5.1, which illustrates two means of connecting a host computer to a remote device. The circuit is routed along the path of the public telephone network, on an analog basis, from a modem through a PBX and high-level multiplexer before entering the telephone network. Within the public network, the circuit passes through central offices of the local telco and long-distance carrier, fiber optic cable paths, and destination multiplexer. A *digital access and cross-connect system* (DACS) is traversed along the route. The circuit terminates in the local telco's CENTREX equipment. This circuit represents approximately four different subnetworks and several more types of telecommunication equipment. In the alternate route, the circuit employs data communication equipment in a completely digital arrangement. A PDN is employed, wherein the data flows through the cloud of packet nodes and intervening links. This approach involves three explicit networks plus many others not shown within the PDN cloud. Providing NM&C capability across the communication pathways will depend on the interface with individual elements as shown in Figure 5.1. In addition, means of testing links between elements are vital to isolating problems which are detected by monitoring devices at nodes and terminal locations. The following paragraphs review the identifiable categories of NM&C relating to modern telecommunication facilities.

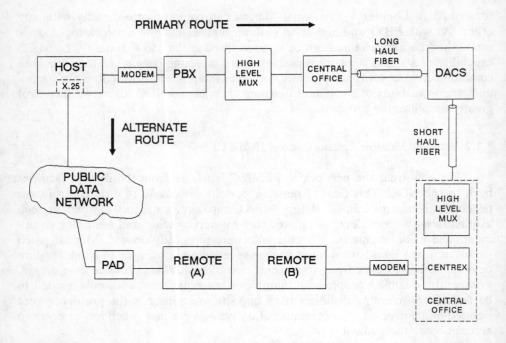

Figure 5.1 The diversity of network technologies and routing approaches is a driving force for the creation of network management systems.

5.1.2.1 Equipment M&C Interfaces

Virtually all telecommunication equipment on the domestic market today will have an M&C interface. The most basic is referred to as a *contact closure,* which is illustrated in Figure 5.2. It can transmit the discrete status (on or off, in or out, *et cetera*) of a piece of equipment or a part of a facility. This is used extensively in security systems to indicate when a door is left open or a fire alarm is set off. Detection of a problem by sensors inside the device causes the relay contacts to close, which, in turn, activates an external signaling system. Consequently, an important aspect of M&C is to carry this indication from the remote location to a central monitoring site, such as a security office or, NOC. The design and operation of a dedicated data communication network becomes a vital aspect of M&C, as will be shown.

Modern M&C goes far beyond the simple monitoring of contact closures, although these are still important for some aspects of the problem. As discussed in the previous section, intelligent telecommunication equipment provides a variety of useful information as to the quality and quantity of traffic being carried. In

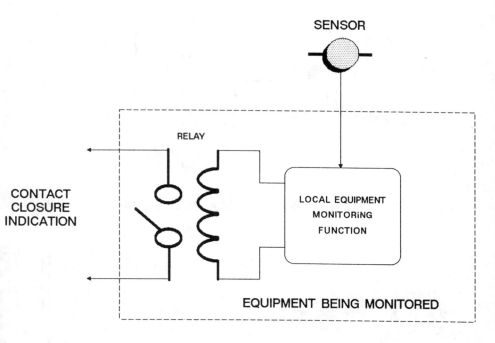

Figure 5.2 Equipment monitoring through contact closure.

telephony, the absolute and relative levels of signal and noise can be measured. Switching systems provide measures of traffic volume, routing delay, and numbers of blocked calls. Such data become important in administrative operations, as we will discuss later in this chapter. Performance of digital transmission is monitored in terms of BER, delay, throughput, *et cetera*. Another dimension is the control of the facilities being monitored. Most telecommunication devices now are run by internal microprocessors, allowing external control through the M&C interface. Aspects of the existing programming may even be changed, leading to possibilities for adapting the performance of the device and consequently that of the network as a whole.

Monitoring and control of intelligent equipment is typically exercised by a video terminal connected to the M&C interface, as illustrated in Figure 5.3(a). A data line, shown in Figure 5.3(b) can connect the terminal if the equipment is installed at a remote location where human support is not immediately available. The dial-up approach is preferred if the frequency of occurrence is low for alarm messages and control inputs. The local M&C system of the equipment detects a problem or significant change of status, and then automatically initiates a dialing sequence via the modem and telephone system. (This scenario assumes that the

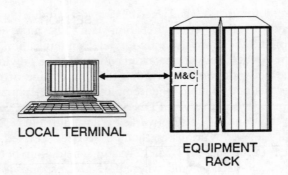

(A) EQUIPMENT M&C USING A LOCAL VIDEO DISPLAY TERMINAL

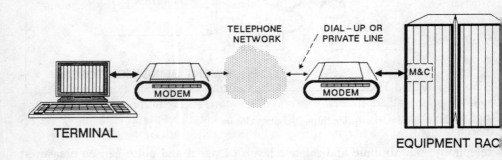

(B) REMOTE EQUIPMENT M&C USING THE PUBLIC TELEPHONE NETWOI TO REACH A DISTANT VIDEO DISPLAY TERMINAL

Figure 5.3 Equipment monitor and control of an intelligent device using a simple video display terminal.

problem is not caused by a power outage, which shuts down the equipment.) The call is received at the NOC, where the video terminal is located. Human intervention is only needed to interpret the message recorded on the terminal or page printer. A call from the terminal to the remote equipment can be initiated by an operator to poll equipment status and usage history, and to change the configuration in support of network operations. The leased line approach, either analog or DDS, is recommended for frequent M&C message interchange with the control center.

5.1.2.2 NM&C Networking

Network management and control is meaningless without a data communication environment to carry status information back to the NOC and to send control information to the remote subsystems. Even the simplest monitoring scheme using contact closures needs a reliable communication path. The more important the facility under control and the more dynamic the telecommunication network, the more important is the NM&C networking arrangement. In the previous examples, the public telephone network was employed, either leased line or dial-up. Both approaches are effective and reliable when employing the telephone network in the U.S., for example. Alternatively, a PDN could be employed. Some network managers prefer not to use public data networks because outsiders can gain access, resulting in a loss of security. Various methods have been devised, however, to increase the difficulty of accessing the NM&C system. The dial-up network is made more secure by using a call-back scheme, where the NOC responds to the incoming call from a specific remote site by hanging up and calling back the number of the remote. Good security is afforded with the leased line approach, since the circuit is hard-wired within the intervening telephone central offices. We will consider two types of NM&C environments: the *homogeneous* and the heterogeneous.

Homogeneous M&C Environment

An homogeneous M&C network environment is illustrated in Figure 5.4. Each remote element is at a different location, such as a node in a private telephone or data communication network (*nodal equipment*). By homogeneous, we mean that the elements in the private network to be monitored and controlled are essentially identical in design, or appear to be identical to the NM&C system. It is assumed that each piece of nodal equipment is "intelligent," meaning that it operates under the control of an internal microprocessor subsystem. Elements can, of course, be made to appear identical by using software programming, as will be discussed.

The function of the nodal equipment is to process, switch, or multiplex information, as indicated in Figure 4.1. Between each node is a point-to-point or point-to-multipoint transmission link. Communication traffic would therefore pass over these links to provide the telecommunication function for which the network was intended. Because the network is homogeneous, however, we can route M&C information along with the user traffic. Typically, the bandwidth required for M&C is small, with the result for throughput depending on the transmission rates over the links between nodes. In a T1 network using high-level multiplexers as nodes, the cost of carrying M&C data is almost zero, while in a low-speed data network using voice band modems, the overhead may absorb a significant but acceptable amount of the throughput. Another advantage of the homogeneous approach is that the nodal equipment at each site is tied directly to the private network, providing excellent security from intrusion.

The point of access to the M&C network is typically through a terminal connected to one of the nodes (see Figure 5.4). In backbone T1 networks using

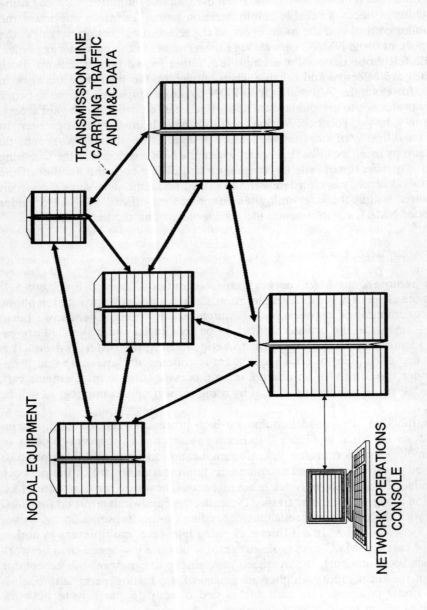

Figure 5.4 Homogeneous M&C subsystem for a common type of network equipment.

TRANSMISSION LINE CARRYING TRAFFIC AND M&C DATA

NODAL EQUIPMENT

NETWORK OPERATIONS CONSOLE

high-level multiplexers and in tandem switched PBXs, the access point may be any node in the network. The processing power of the node and the associated memory allow the terminal to configure the nodes for service and to read out the status of any element in the homogeneous network. As in Figure 5.3, the terminal could be extended from the closest node by using a dedicated or dial-up data line.

The nodal M&C scheme illustrated in Figure 5.4 is applied in many telephone and data communication network environments. Digital switches, including PBXs and tandem switches used by telcos, high-level multiplexers, and digital cross-connects, are good examples of telephone environments. For data, the IBM, DEC, Tandem, and other environments use the nodal approach. In the case of IBM, a host computer would be the control point through the front-end processor. The DECnet environment is almost a perfect match to Figure 5.4. An example of an NM&C workstation for a VSAT hub is provided in Figure 5.4(a). Any of these systems directly route M&C information over the same paths that carry user data.

Dispersed low and medium speed data communication networks, which are not integrated into a homogeneous environment, can still be managed from an M&C standpoint. A series of special products called "wrap-around" units are available on the market to provide M&C for standard modems and digital service units (DSUs). The basic arrangement is shown in Figure 5.5 for a single point-to-point data communication link over an analog telephone private line. One direction of transmission is shown for clarity, although the link is capable of duplex operation. A cable connection is made from the host (at the upper left of the figure) to the M&C unit of the data link. The modem connects to separate outputs and inputs of the M&C unit. Typically, the communication between wrap-around units and with the controller uses a proprietary networking scheme. To simplify installation of such data communication links, several companies offer modems and DSUs with built-in M&C comparable to that provided by the wrap-around units.

An example of this M&C environment is the NETCON system offered by GDC. The wrap-around units place the M&C data in a narrow slice of bandwidth below 400 Hz, so as not to interfere with that used by the conventional data modems to carry user information. Performance of the modem-to-modem link is monitored by the wrap-around units and passed to an M&C controller. An operator can command the remote wrap-around units with a video terminal or engineering work station attached to a controller. These capabilities include analog channel measurements, loop back tests, BER, and signal-to-noise ratio. To enhance operation, the connections from several data links can be "daisy-chained" together by a multidrop line hooked up to a single controller and work station. Other vendors of M&C products for data communication applications include Infotron, AT&T, Atlantic Research, Racal-Milgo, DCA, and Timeplex.

Figure 5.4(a) The monitor and control terminal of a Ku-band hub earth station. (Photograph courtesy of Hughes Network Systems.)

Heterogeneous M&C Environment

The heterogeneous M&C network environment is by nature poorly structured and difficult with which to work. A strategic unit that has been applying telecommunication over an extended period of time will need to manage a diversity of network facilities. Fortunately, every telecommunication devices have an M&C interface (or, at least, the contact closure monitoring capability). The capabilities are not so impressive as what can be done in the homogeneous case, but modern computer systems offer some hope for doing an adequate job.

The first problem in creating an effective heterogeneous environment is to work at each site by gathering the interfaces of dissimilar equipment. The objective is to link the M&C interfaces of each subsystem so that all functions can be operated by one remote control unit tied to a single NOC. If we did not do so, several individual data lines would be needed. An example of a remote telecommunication facility in a major private telecommunication network is shown in block diagram

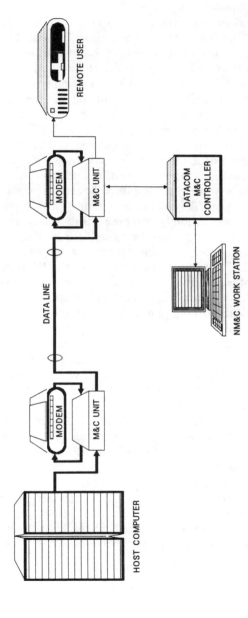

Figure 5.5 Data communication network monitor and control using wrap-around M&C units for conventional voice-band modems.

form in Figure 5.6. Connections between elements are for control and monitoring purposes (not for user communication).

Examination of the various elements at this site reveals several problems. The microwave system that terminates at this remote site has an M&C system to track the performance of every hop and each piece of equipment. Repeater shelters along the system are also monitored for alarms. If standby equipment is automatically switched to restore a link, the microwave system's M&C carries this information back to the control unit (shown within the outline of the remote telecommunication facility). In addition to microwave, the facility has a collocated earth station in a satellite network, which may be a hub of a VSAT network. Video teleconferencing is also provided, and this equipment also can be monitored. The example facility includes a piece of multiplexing or switching equipment to break down voice and data traffic coming over the terrestrial and satellite links.

The preceding example is somewhat simplified because it does not include user data communication services. This example demonstrates, however, that a difficult problem can become exceedingly complex without a reasonable M&C architecture. We suggest in Figure 5.6 that the "glue" for holding these diverse M&C interfaces together is in the form of a microcomputer. In practice, a standard PC may not be the best choice, but the internal structure of a PC lends itself well to supporting a diverse set of requirements. Unique software elements are loaded into the PC's memory, and the central processing unit responds to the external devices when they send indications. The higher powered class of PCs (e.g., AT and PS/2) are capable of *multitasking*, wherein several different computer programs can run at the same time. Each program deals with a single class of equipment, such as the microwave system or the switch. The PC is also relatively inexpensive, with most of the expense residing in the interface port cards and software. The common microcomputer on site allows the operator to have access to the integrated M&C system by way of a video terminal.

The M&C function does not stop within the PC, but is conveyed to a host computer at the NOC, where human operators can exercies control. Assuming that there will be several telecommunication facilities dispersed throughout the network, the PCs are tied to the NOC site with data communication circuits, either dedicated or dial-up. This is indicated at the bottom of Figure 5.6. As an alternative to the telephone network, the host at the NOC could be connected to a PDN with the remote PCs accessing the PDN as needed. Another approach for tying the PCs to the host is to use private telecom facilities such as the VSAT network shown in this example. Backup communication for M&C would still be available over a public network.

5.1.2.3 Alarm Monitoring

With the M&C network implemented, the NOC is able to perform the needed technical and operational tasks. Since most of the facilities are to be operated

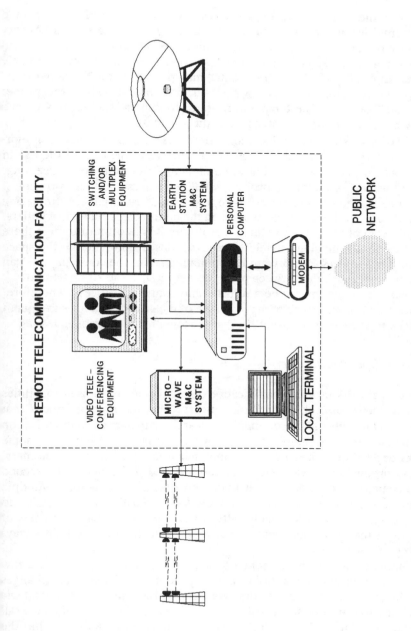

Figure 5.6 Personal computer-based monitor and control subsystem for a remote telecommunication facility.

remotely, one of the basic needs is to gain knowledge when something unexpected happens at some distant or inaccessible point in the network. Essentially every operating piece of equipment has associated with it one or more alarms. An alarm is generated when the parameters of a unit's operation go outside of the acceptable range. For example, the BER on a T1 channel may rise above the threshold value of 10^{-6}, or the power output of a microwave transmitter may drop below its proper working level of one watt. The temperature in an equipment shelter may exceed 80°, possibly because a door was left open in the summertime.

Alarm indications can be critical indicators of existing problems. Such indications can also foretell of an impending outage, such as in the case of the rise in temperature, which may lead to technical equipment failure due to overheating. People in the business of operating orbiting satellites understand these principles. The failure of a satellite or its components represents a substantial loss. Replacing satellites is difficult, involving new investment in the hundreds of millions of dollars. Obviously, you cannot reach the satellite to fix it after it is in orbit (although there were a few exceptions, involving satellites in low orbits within reach of the space shuttle). In the case of M&C in private telecommunication, the consequences can be very damaging because the services provided are vital to the strategic unit.

Interpretation of alarms can be a difficult and confusing process. Therefore, the alarm indication should come with some type of measured data for subsequent analysis. In the next section, we treat trouble-shooting using diagnostic testing.

5.1.2.4 Network Operations Center

Two forces at work in telecommunication demand quick response capabilities in the event of technical problems with a network. The first driving force is that strategic units rely heavily on telecommunication and so any loss of communication has an immediate and profound effect. As we mentioned in Chapter 1, some organizations depend on voice and data communication to reach their customers, to deliver the product (i.e., valuable information), and to be paid. The second driver is the complexity of telecommunication networks and technology. Multiple vendors provide the needed elements, composed of sophisticated switching and transmission systems, which are usually software-driven. There is a rich variety of possibilities for using telecommunication networks, and there are countless ways for things to go wrong.

The homogeneous and heterogeneous M&C environments provide the raw diagnostic information and the ability to make appropriate configuration changes to maintain the performance of a network. We strongly recommend that a strategic unit which wishes to run its own network create an NOC. An example of an overall concept is shown in Figure 5.7, where the NOC is indicated by a diamond at the hub of an M&C data communication network (shown with dotted lines). The M&C network that connects the NOC to every remote site is bidirectional, allowing

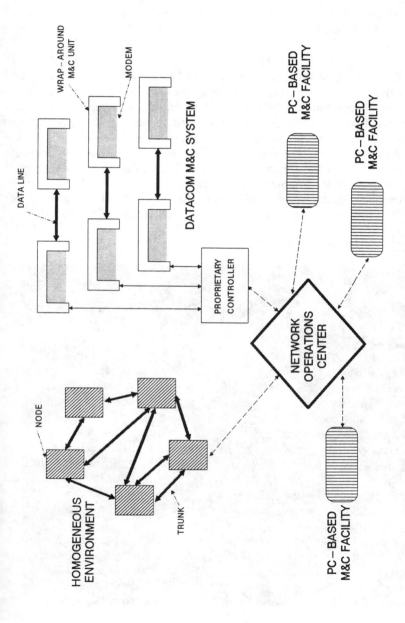

Figure 5.7 Integration of diverse M&C capabilities into a single network operations center (NOC).

status and alarm information to flow from device to NOC, and control information from NOC to device. Three different environments are shown: the homogeneous environment, the PC-based heterogeneous environment, and the data communication wrap-around system.

In the simplest approach, the NOC is a room where the individual video display terminals for each of the environments can be placed and manned. This approach gives the network operations personnel access to each of the environments. The principal disadvantage of the separate terminal approach is that operators must move from one terminal to another to respond to alarms and to perform routine maintenance. In the ideal case, there should be only one environment, and hence only one terminal would be required. As we will discuss, progress has been made in the difficult problem of integrating the various environments in a common NOC computer and operator work station. This direction is very desirable because the number of environments will tend to increase as network capabilities are expanded. A photograph of a typical NOC is shown in Figure 5.8.

Figure 5.8 Example of a network operations center (NOC). (Photograph courtesy of Contel/ASC.)

The existence of a problem on the network can be reported to the NOC in one of two ways. Traditionally, the user was the one to tell the network operator of the existence of a problem. Obviously, this approach will always exist — and work! — but its extent will be reduced by the advent of computerized M&C systems. For the fault to cause an automatic alarm to be sent over the M&C network to the appropriate NOC terminal is much more desirable. In a relatively small network, the alarm would be heard as well as seen, alerting an operator to the existence of the problem. Large networks typically have so many simultaneous alarm indications that audible alarms are inappropriate (i.e., the alarm would be ringing so often that the operators would leave it turned off anyway). Instead, the alarms are tagged with priority, indicating severity and urgency, which can be presented to the NOC on a summary display. Then a special computer system comes into the picture for automatically handling and prioritizing alarms. Color graphics are ideal for displaying alarms, where red can indicate those with the highest priority.

Incoming alarms can cause confusion if the configuration of the network is not well known. Therefore, the NOC should include a readily accessible representation of the telecommunication network. Advances in computer graphics and microcomputer work stations make possible on-line displays of the network. The work station approach allows the network to be examined, either in its entirety or in suitable segments, which can be shown on the screen by using function keys. Much effort is needed to create a computer-based status display by entering every element of the network into the database of the work station. Henceforth, updating should be a fairly easy process. The work station offered by Codex is shown in that firm's data communication environment in Figure 5.9. The attractive point about this approach is that the real-time M&C network is tied directly into the work station, where it interacts with the network representation database. The computer graphic display can therefore be used as a trouble-shooting tool because commands may be sent out to modems and other elements in the network to run various types of tests.

In the future, there appears to be a role for the area of artificial intelligence known as *expert systems*. Research by organizations such as GTE Laboratories point to applications for expert systems in telephone switching and network maintenance. IBM Japan and Sumitomo Metal Industries announced in 1988 the completion of joint development of an expert system for fault checking on an SNA network. The expert system employs over 1000 rules devised by Sumitomo's network control experts. Developing an expert system is not simple, but the potential for problems with trouble-shooting equipment and network anomalies is very great. More possibilities are presented in Chapter 9.)

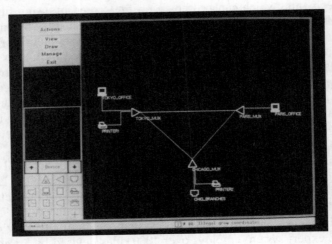

Figure 5.9 The 9850 Management Workstation of the Codex Integrated Network Management System. (Photographs courtesy of Codex Corp.)

5.1.2.5 Network Reconfiguration

The ability to alter the network configuration in response to service demand is equally important as maintaining the network's performance. The most direct purpose is to deal rapidly with problems in the network when they arise. Some examples of critical events are that particular voice circuits or T1 links may go down, multiplexers may develop bad port cards, and packet switched nodes may lose power and shut down. Through reconfiguration, however, a major point of blockage can be quickly bypassed.

A traffic measurement system is a useful feature to have in addition to the status and alarm functions. Telephone traffic is registered in terms of calls in process, those completed and not completed, and call durations. Data communication is typically broken down into packets of information, so the number of packets carried per unit time and time delay are typically the desired measures. Another important use of traffic information is to develop the billing to customers. For strategic business units, the billing may be divided much the way that telcos do for their commercial and residential subscribers. This type of information is passed to the administrative side of network management.

Traffic information can be folded back into the NOC to alter the network configuration. In most instances, the reconfiguration process employs a person who is familiar with the network and knows what can be safely changed. Some of the newer NOCs, such as that developed by GTE, can respond automatically to traffic buildup and change circuit quantities between specific nodes in the network.

Reconfiguration is a touchy process because what appears to be a desirable change can produce disastrous results. Fortunately, an on-line NOC can correct any inappropriate change that it makes. Traffic monitoring and alarms are useful in this regard. Technology can perform M&C functions, but there is no substitute for competent hands. In a diverse network, provision must be made to have people either on site or who can be called to the site. The use of administrative systems to support the human factors of network management is discussed in the next section.

5.2 ADMINISTRATIVE OPERATIONS (AO)

Administrative operations begin after the network is relinquished by the implementation team and declared to be operational. The network facilities that were put into place are capable of serving the needs of the strategic unit or units. Let us, however, clarify that telecommunication is a service business. Internal users represent a captive customer base, but they know how to give the telecommunication management headaches. (Imagine what happens if your company's chief executive is not provided with the service that he or she needs or demands.) In some organizations, the private telecommunication network is operated as a profit center, and users have options with regard to the amount of service that they will take. This heightens the need for the full spectrum of AO management.

The following definition of AO draws heavily from the work done at GTE. The background of GTE is particularly germane because they have installed comprehensive network management systems for telephone companies, state and federal government networks, and large corporate networks in North America as well as foreign countries. Network management is a living and breathing thing, not something that is installed and held fixed for all time.

The administrative side of network management deals with taking orders and initiating service for new customers along with any other changes that may be requested. Facilities which make up the network (e.g., telephone switching equipment, transmission systems, telco lines, and computers) must be maintained in inventory and assigned for the provision of service. Repair crews and hardware inventory are to be deployed as needed. Billing for services rendered and equipment installed requires constant attention, lest the financial side become out of control. Each of these areas is reviewed in the following paragraphs.

5.2.1 Service Orders

A service order is initiated by a user to request service of some type or scope. In telephone systems, a service order is placed by an individual for a telephone extension and instruments (in comparison, residential service since divestiture no longer includes the telephone instrument itself). There is an analogous service order for a computer terminal and data communication network access. There will be hundreds or even thousands of such requests in an organization, necessitating an efficient administrative system so that service is initiated in a timely manner. Once on the network, a user receives service on occasion. The service order indicates the types of services that are available to this user and which types are restricted. Ideally, the organization can enter the service order into the AO system and have it cause facilities to be appropriately configured, which almost always involves action by installers and technicians. The time frame for the installation needs to be defined, and the order must be tracked through completion.

In an automatically switched network, circuit or packet, the intelligence of the network will provide service to the user when required. The collection of usage information is also an on-line process, feeding into billing, which is an administrative function discussed later in this chapter. There will be instances where the user will want to change the manner in which service is provided. For example, the telephone instrument, terminal, or video display may need to be moved to another office or building. A new class of service, such as direct long-distance dialing or voice mail, may need to be incorporated for the particular station. These particular changes cannot be made by the user, and so he or she needs human intervention from the telecommunication organization. Adding service means enhancing the capability of the particular station, leading to increased or more expensive use. The cost must be recovered and, if the network is a profit center, there is the opportunity to increase operating margin. These financial implications can be recorded when making a change.

In a large network, members of the telecommunication staff cannot operate informally, relying on their memories. This is the same situation faced by telephone companies, or any other well run organization in a service business. A computer database system is ideal for entering and tracking service orders. The best type is the relational database which is common to all AO applications and functions. This way, basic information about a user or facility need only be entered once. As you come to be more familiar with the process and as the size of the network grows, the need for a computerized AO system becomes apparent.

5.2.2 Facilities Management

The telecommunication plant of a private network can be substantial, representing an investment of millions of dollars. *Facilities management* is the process of identifying and allocating the capacity of telecommunication equipment and transmission systems located throughout the network, including leased facilities provided by other organizations (particularly telcos). In GTE's AO system, facilities management deals with the functions of the elements of the network. For example, a PBX has capacity of so many subscriber lines and trunks, and may be capable of simultaneous voice and data. These characteristics are identified for the particular facility being inventoried. A logical representation (i.e., an idealized model) of the entire network is stored in the central computer. Thereafter, the switch capacity can be assigned pursuant to service orders entered periodically into the system. In addition, capacity can be reallocated by the NM&C system in response to network failures. Inventory of equipment from an accounting standpoint is covered in the next element of AO.

There is clearly a direct tie between facilities management and the NM&C environments of the network. The NOC has available the up-to-date configuration of the network, which is provided through the facilities management capabilities of the AO system. The status of these facilities is gathered and stored at the NOC, however, providing valuable feedback as to the availability of facilities which have not been fully committed to service. The information should flow in both directions and be made available in usable formats, such as visual displays and hard-copy reports.

5.2.3 Inventory

While facilities management relates to the functional layout of the network in terms of its elements, the inventory system deals with the constituents in a physical and financial sense. There are two classes of inventory in an operating network: installed plant and spare parts. The installed plant is subject to economic depreciation, reflecting its expected useful life. This is necessary for assessment of the cost basis for operating the network. Economic principles relating to telecommunication network will be covered in Chapter 6. Future additions to the network can be analyzed by using historical cost data gathered through the inventory system. In other words, experience with using and maintaining the equipment provides an economic baseline with which to compare prospective additions and changes. Spare parts are held in inventory for subsequent use in providing new service or effecting repairs. Major pieces of new equipment conceivably can be inventoried, although there is always the risk of technological obsolescence or physical loss.

One of the bigger headaches in network management is maintaining a precise inventory for each location. A computerized database is useful, but maintaining such a database is no simple task. This is particularly critical in the case of spare parts which may be needed to restore a failed node or user device. The time required to complete a repair will be dependent on the availability of spare parts which the inventory system should track. If none are on site, then arrangements must be made to locate the part (at some other site) and have it sent to the site of the problem to meet the repair person when he or she arrives.

5.2.4 Customer-User Directory

The creation and maintenance of user directories is an uncelebrated AO function. The directory is essential, however, as the means of serving and billing users, and as the index for locating any user on the network. Outdated directories are a hindrance to effective management of the network. In the ideal system, the directory is created from the same database that the service order process creates. If a relational database is used, any changes entered through service orders will automatically update the directory.

In PBX and CENTREX-based telephony, the directory is usually in the form of a printed document. On-line directories may appear with the growth of ISDN in the private telecommunication environment. The Minitel system in France (described in Chapter 2) provides such an on-line telephone directory to the public. One of the nice features of a well integrated directory is that the call can be placed automatically when the user selects the called party from the directory listing.

The directory function in data communication is considerably more complex than in the case of telephone for the simple reason that a machine-to-machine conversation is normally involved. If the input is incorrect, the system may not respond at all, and there is probably no human operator to assist. E-mail applications are very dependent on on-line directories because user addresses in E-mail are often initially selected by the computer.

A well integrated AO system will use the directory for a variety of functions, as mentioned previously. The underlying database again proves to be a powerful asset in telecommunication management.

5.2.5 Billing

The cost of operating the network eventually must be covered. Private telecommunication networks can be self-funding by users who pay for their use of the network's capabilities. A billing and cost accounting system should be established under AO so that network expenses could be billed to the appropriate department for better cost control and more accurate accounting. The system would process call detail records from the traffic data gathered through the NM&C environment. Bills to users would be for charges for network features, service orders or repairs, as appropriate, and equipment assigned to users through the inventory function of AO. This billing system would appear much like the administration and accounting system of a utility, such as an electric company or, obviously, a telephone company.

As we discussed in Chapter 3, telco-provided CENTREX service provides billing information under a format called *station message detail recording* (SMDR). Passing through these costs directly to the respective user is appropriate, but any internal expenses for network operations or administrative operations will not automatically be added. Therefore, the SMDR information can be obtained on magnetic tape and then input to the AO system for adjustment. Information of this type is also provided by the long-distance carriers for MTS and WATS, allowing the charges to be routed to the station or individual making the calls.

The billing function can be performed for the private telecommunication network operator by an outside service bureau which gathers the SMDR information from a variety of sources and produces cost allocation reports. For example, Telecom MIS, Inc., a subsidiary of M&SD, Inc., of Denver, CO, provides these services for more than 500 client locations. Another approach is to purchase a software package which accepts SMDR data from magnetic tape.

Tracking expenses for services purchased outside has two important attributes. First, the billing system needs flexibility so that new pricing for facilities and services purchased from the carriers can be accurately passed through. Second, as discussed in Chapter 2, the bulk long-distance packages offered by AT&T, MCI, and US Sprint are driven by the competition in that business. Cost analysis through the billing system would compare these service offerings for the actual usage through the network. Then, the most advantageous combination can be chosen. Hence, the mix would be maintained as close to optimum as the business permits.

5.2.6 Repair Administration

The administration of the network will include the management of repair operations. This function can be managed out of the same NOC where all alarms and complaints are collected. As the central point of control for all phases of the repair process, the NOC is the focus of service restoration in the event of outage.

The actual repairs would be performed by either internal staff or outside vendors under NOC supervision. An associated management function is to make assignments of resources and to set priorities for the completion of tasks. It is advantageous to use a computerized reservation system to perform all of the centralized functions.

In telecommunication repair administration, the first record of a valid network problem is called a *trouble ticket*. The standard process in telecommunication repair is that a trouble ticket is opened when a legitimate complaint is registered. This is either entered into a notebook or, in an on-line system, directly into a computer database. In the more advanced systems, such as the GTE NMCC, a trouble ticket can be opened automatically by the computer in response to an alarm received through the M&C network. The inventory and billing functions are activated when spares are expended and hours for repair services are accumulated. A trouble ticket becomes closed when the repair is complete and service is fully restored. Records of past service outages and subsequent repairs are extremely useful for tracking systemic problems. These problems could be due to a particular piece of electronic equipment or to an intermittent trunk or data line.

5.3 INTEGRATED NETWORK MANAGEMENT SYSTEMS

We use the term *integrated network management* (INM) in reference to computer-based systems that consider nearly all of the aspects of network management previously described. These aspects include network monitor and control, with either a homogeneous or heterogeneous environment, and the administrative operation of the network as a business. Our approach here is to use examples of systems that have been developed to achieve INM in real situations. These are broken down into two areas. The first is telephone management of the type common to local telcos and corporations with large user populations. The second area is data communication, where host computers serve many users in what is typically a packet switched network. These implementations overlap when considering integrated voice and data services, which is particularly the case for ISDN.

5.3.1 GTE Network Management Control Center

The Network Management Control Center (NMCC) concept was developed by GTE West to integrate and to automate the various functions needed to run a telecommunication business. In particular, NMCC was tailored to take over the NOC and AO requirements of a local operating company. The NMCC software systems run on Tandem NonStop™ computers. These were selected after it was determined that *on-line transaction processing* (OLTP) computer architecture would be optimum. Tandem, as mentioned in Chapter 3, has established itself as a leading supplier of OLTP computer systems and software. GTE states that NMCC can help increase productivity, control costs, improve service, and minimize support requirements for a private telecommunication network. To justify the financial and personnel investment for NMCC implementation, the network should be relatively large and would have a substantial user base. An example of a good candidate for NMCC would be a private telecommunication network for a state government or a corporation engaged in interstate business operations.

5.3.1.1 NMCC Overview

NMCC currently consists of a pair of software subsystems, which respectively support the monitor and control operations and the administrative operations. These subsystems correspond directly to the two broad functional areas of network management outlined at the beginning of this chapter. Figure 5.10 provides an overview of NMCC in schematic form. A central relational database for network management is maintained, containing all information as to the constituents of the network, the user base, and the deployment of each. Information that is entered or changed in one subsystem automatically updates the central database, and is therefore available to every other subsystem. This eliminates redundant data entry and improves the reliability of data created in different departments of the tele-communication management organization. Inherent in this on-line system is the ability to update the network configuration on a moment's notice. When service is requested, new telephone numbers can be assigned immediately, and facilities such as PBXs and intelligent multiplexers can be reallocated automatically. Re-location can even be done according to changes in traffic requirements. Inventory records can be maintained and updated as items are assigned and used. From an accounting standpoint, costs can be apportioned to using organizations, and other financial aspects such as depreciation are efficiently handled by the same system.

The use of a transaction-based computer architecture such as the Tandem system creates a responsive and reliable M&C environment. The computers utilize parallel processors, redundant storage media, and on-line data communication facilities to minimize the possibility of NMCC failure. This computer architecture permits expansion for new service capabilities such as ISDN. Customer service representatives, NOC operations personnel, and repair desk can all have their own terminals connected by data lines to the central computer.

5.3.1.2 Common Database Approach

Anyone who has worked with an elaborate computer system that supports a multifaceted, ongoing service business realizes the importance of a comprehensive and well constructed database. The database system is implemented by using a relational database structure, which combines information from the other subsys-tems. For example, each service order automatically updates information on billing,

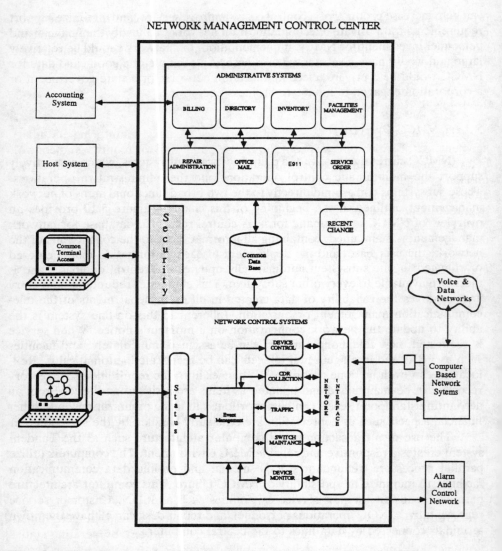

Figure 5.10 Block diagram of the GTE Network Management Control Center (NMCC).

directory, inventory, and facilities management. This reduces the possibility of data entry errors as it reduces workload. The database can be partitioned for individual subscribers, based on their specific requirements. Administrative tasks can be automated, thereby reducing personnel requirements and increasing worker productivity. With the database, the system generates complete management reports from appropriate subsystems. Facilities management reports, for example, would be used to support daily activity and to summarize use of transmission equipment. Repair reports include average clearing time and historical trends.

5.3.1.3 Network Control Subsystems

There are five NMCC subsystems to provide technical monitoring and management, all of which operate from the NOC. Capability for monitor and control of every network node is provided, including circuit and packet switches, transmission systems, multiplexers, and computers. The network configuration can be altered to add capacity where needed, or to reroute traffic around a failed or degraded node. Because the network control subsystems are integrated with the NMCC administrative operations subsystems, all departments can receive up-to-date technical data upon request from the system. The individual network control subsystems are identified in Figure 5.10 and are reviewed in the following paragraphs.

Alarm Monitoring Subsystem

The network alarm monitoring subsystem tracks the status of the entire network, enabling the NOC to pinpoint problems quickly and to provide a rapid response. The reliability of tracking alarms is enhanced by the fault tolerant nature of the computers and software. Fault and alarm messages from each network device trigger appropriate levels of response from the alarm system, including audible alarms, hard copy reports, automatic telephone calls, and graphic displays. The messages and corresponding actions by the NOC can be updated to accommodate network changes.

Traffic Administration

Optimal quantity and arrangement of network elements can be determined by using the traffic administration subsystem, which reports the operating status of the network on a real-time basis. A graphic display indicates load changes across the network, and batch reports analyze accumulated traffic data to identify trends. The traffic data would be collected from remote switching nodes, indicating actual usage of the network. Analysis of the data using standard traffic tables is also included. Thus, the design of a network by using techniques covered in Chapter 6 can be verified.

Network Interface

At the heart of the network control system is the centralized network interface, which gathers all of the monitor and control information from the remote nodes, transmission facilities, and computers. Each type of NM&C environment, covered at the beginning of this chapter, can be connected to the NMCC interface to allow an operator to communicate with and to have remote control of any intelligent device in the network from a single terminal connected to the central computer. The operator at the NOC can also create a message filter that determines which messages are to be interpreted and logged, and can define automatic or manual command files, which are canned procedures to execute a series of instructions as conditions require.

GTE has built an open architecture which is already compatible with a substantial fraction of the telecommunication equipment in current use. Furthermore, software drivers can be added when new capabilities are introduced at existing nodes, or when a whole new class of telecommunication facility is invented. In particular, GTE developed the interface to NetView/PC, IBM's integrated network management environment. NetView and NetView/PC are reviewed in the next section.

Switch Maintenance

The switch maintenance subsystem allows the NOC to control remote operation of the telephone switches on the network. Modern PBXs and switches provided by telcos are controlled by internal software, running on redundant internal processors. There may be a common signaling channel. The switch maintenance subsystem of NMCC obviates the need for operators to service the switch on site, instead relying on the Tandem M&C network to provide remote access. Functions include common event management for all network switches, regardless of manufacturer. Traffic routing plans can be checked before downloading into switches. Automatic switch testing routines are also made available. Following along similar lines, the subsystem can be set to respond automatically to alarm messages from remote switches, and thus maintain service, even when equipment and transmission difficulties are encountered.

Call Detail Recording

The on-line process of *call detail recording* (CDR), another name for SMDR, ties directly into accounting and billing. The CDR subsystem collects call detail records necessary for timely and accurate billing and cost allocation. CDR improves billing accuracy and reduces the effort required to correct bills because accurate and detailed data are available from the computer. Back-up systems are provided to process records, even in the case of M&C network or equipment failure. Extensive validation procedures and complete integration with the NMCC subsystems ensure the accuracy of the call detail records.

Internal Electronic Mail

Owing to the availability of Tandem's transaction-based computer environment, the network control subsystems include E-mail. This makes interoffice and intraoffice communications more efficient and tends to reduce costs by eliminating unnecessary paperwork. Any user with a terminal can create and route E-mail to any other defined user on the system. The user also can file, cancel, or discard incoming messages. Other features include hard-copy output and the ability to send E-mail automatically to everyone on a distribution list. These features are, of course, typical of E-mail systems (as discussed in Chapter 3).

5.3.1.4 Administrative Subsystems

There are six AO subsystems for the management of information related to running the network like a business. These include service order entry, facilities management, inventory management, directory, billing and cost accounting, and repair administration. The AO subsystems use information gained through network control subsystems and drive aspects of network control. For example, a service order under AO can drive network control to establish a new connection to the network.

Service Order Entry

Service order entry and status reporting comprise an on-line subsystem that supports service orders for adds, moves, and changes on single- and multiple-line telephones or key systems. If facilities can be controlled directly through the on-line network control subsystem, this AO subsystem will automatically initiate service. The system is flexible so that service orders can be entered in any sequence, and data access for account queries is also facilitated. Service dates can be assigned, based on availability of work groups and the number of minutes an order requires. Reassignments can be made automatically when telecommunication personnel are changed or service orders canceled. To assist with overall production management, pending-order status reports are also provided.

Facilities Management

The NMCC facilities management subsystem is used to track, allocate, and manage switching equipment and other network components such as cable pairs, data channels, and video display terminals. The allocation of network elements can be balanced as needed to improve the overall performance of the network. When a service order is initiated through the AO subsystem, the facilities management subsystem assigns a pending-order number to each element. This allows the reallocation of facilities when pending orders change. In addition, comprehensive management reports can be provided to give a clear picture of the entire network.

Repair Administration

Provision of a central point of control for all phases of the repair process is the purpose of the repair administration subsystem. Repair orders are tracked and ranked to determine proper allocation of resources. This subsystem supports all repair administration, including recording trouble reports, providing diagnostic assistance, and dispatching repair personnel. The subsystem records trouble tickets automatically or it lets an operator enter them. The repair administration subsystem provides management reports, and is integrated with the inventory and billing subsystems to ensure that repairs are conducted in a cost-effective manner.

Inventory Management

The purpose of the inventory management subsystem is to reduce excessive inventories and to reduce the time and expense of managing the individual pieces of equipment in the network. This subsystem facilitates the control of inventories by tracking all network hardware, including switch ports and cards, instruments, network transmission link options, and available spare parts. In addition, the subsystem maintains records of each item's manufacturer, warranty, initial cost, and depreciated cost. Charges to be billed such as monthly depreciation and individual feature costs are sent to the billing subsystem for processing. This subsystem sends spare parts inventory data to the repair subsystem as appropriate.

Directory Assistance

Directory assistance and directory printing are offered via the *directory* subsystem. Since this system is coupled into the service order subsystem via the common database, any directory printed by the NMCC contains the latest subscriber information. Using the OLTP capability of the Tandem computer system, uery procedures are flexible and response time is fast. Any terminal on the MCC computer network can access directory information once security clearance as been established. Private network users can print their own comprehensive or artial telephone directories.

illing and Cost Accounting

The final AO subsystem is the billing and cost accounting subsystem, with hich the cost of network access and use can be billed back to the appropriate sing department. This subsystem processes call detail records from all switches, nd generates phone bills for all network users. Charges for network features, ervice orders or repairs, or inventoried equipment can be billed to users on a line-y-line basis. Tandem's transaction monitoring facility provides complete data ntegrity by reconstituting database files automatically.

A well designed and constructed INMS such as that described would greatly implify the network management job. GTE has dealt most effectively with the elephone side of the business. Other such network management offerings are eginning to appear on the market. For example, Telwatch, Inc., of El Dorado Iills, CA, offers an INMS called NetExec 2000. We now move on to the data ommunication environment, where the focus is on the host computers and the ther nodes of what is generally a packet switched network.

.3.2 IBM NetView and NetView/PC

Host-based data communication networks employing IBM's System Network Architecture (SNA) may now employ an integrated network management envi- onment called NetView. Basically, IBM combined all the functions of the main etwork monitor and control programs and the subsystems used in SNA. The mphasis is on the technical and operational aspects of running a data network. The business side (i.e., accounting and billing), would be handled externally by n AO system using appropriate software. NetView is directed toward IBM's large nainframe network customers because of its integration with the existing hardware nd software product lines.

The majority of NetView operating software is located at a host IBM main- rame computer, making it most appropriate for large data communication systems. The basic characteristics of SNA were reviewed in Chapter 3; we only concentrate ere on the network management aspects. At the core of NetView is the on-line etwork manager called the System Service Control Point (SSCP). If there is a host or peer node in an SNA network, an SSCP is operating. SSCP effectively controls that part of the SNA network which centers on the particular host. This SSCP contains the address tables, name-to-address translation definitions, routing tables, and instructions for using the routing tables. It is critical to network operation that the SSCP establishes virtual circuit connections, selects routes, and controls the flow of user data. Network operations personnel can manage the network and its resources over the same data communication links that pass user traffic. There is a set of IBM software products to extend that control to the NOC. The functions of a number of key programs are reviewed briefly in the following paragraphs.

The command facility is the basic operator interface into the SNA network, providing the base for centralized network management. The command facility incorporates the necessary services for the operator so as to enhance control of services provided over the network. The session monitor supervises the actual user sessions and monitors the performance of the network. This includes response time monitoring. Another key element is *hardware monitoring* where the NOC can be informed of hardware and line problems on the network and respond with appropriate action from the central location. (This is NM&C discussed earlier in the chapter.) The status monitor is intended to give the operators at the NOC the ability to review the status of the network in terms of the operating facilities. The monitor also identifies the commands available to control the elements of these facilities. Other supporting NetView functions, such as help menus and a browse facility, provide assistance to the NOC in using the other features. To enhance the man-made machine interface, IBM is introducing a graphics-driven operator interface called the Integrated Work Station (IWS). An important benefit to NetView users is that the IWS will use icons and windows, eliminating the command line display previously employed.

In 1987, IBM introduced a gateway product, called NetView/PC, to use NetView in a heterogeneous M&C environment, because NetView by itself is completely dedicated to IBM-compatible computers, software, and data communication products. The gateway is through an AT-compatible PC, which is running IBM's multitasking software. Critical to the success of NetView/PC is the development of the interface to it from non-IBM M&C environments. IBM offers considerable assistance to vendors wishing to implement a NetView/PC interface with their products and services. As more of this is accomplished, NetView should become a leading M&C environment in private telecommunication.

NetView/PC is now primarily a system for reporting alarms and performance on a non-IBM segment into the IBM NetView host. Some of the companies which have announced interfaces with NetView/PC include GDC, Infotron, MCI, NET, Paradyne, Racal-Vadic, StrataCom, TelWatch, and Timeplex. For example, a vendor-specific M&C environment such as the GDC Netcon might detect a fault condition of one segment of a data link and this information would pass over the M&C network to the NETCON central computer (which could be a minicomputer or PC). The fault might be measured in terms of an excessive bit error rate. The Netcon central computer will format a message and transfer it to the PC which is running the NetView/PC software. The message is automatically reformatted and sent to the IBM NetView host over an SNA/SDLC link between the PC and host. The fault information, including the identity of the problem section and the measured data, if any, is then processed and displayed at the NOC with other network status information. Interpretation of the fault message would be up to the operator at the NOC. To effect a change in the external non-IBM network, the commands must be sent from the NETCON computer. One NETCON feature allows for statistics to be gathered for the larger network (IBM and non-IBM) and analyses to be performed to gain more insight into network configuration and performance.

IBM has incorporated a NetView interface in its new line of PBXs (formerly sold under the name of the Rolm subsidiary of IBM). The new PBX (model 9751) appears to be doing well in the corporate market. While telephone networks may not make effective use of NetView, the fact that PBXs are beginning to incorporate data switching, and ultimately ISDN, means greater prominence for NetView in general and IBM PBXs in particular. There are, however, competitors who offer software which supplants NetView on IBM SNA networks. Cincom Systems, Inc., has garnered a market for its NetMaster system software, users of which have said that NetMaster outperforms NetView in several important areas. A subsidiary of U.S. West is also marketing its product, called Net/Center.

5.3.3 Other Integrated Management Systems

There are two avenues for the development of integrated network management solutions over the digitized public telephone network. The first has been pursued by AT&T in conjunction with the nationwide common channel signaling system implemented over the past ten years. This signaling system, called Common Channel Signaling System No. 6 (SS-6), is used for routine long-distance call setup, and to provide special features like 800 service and credit card verification. The second thrust is in the international standards area for a new signaling system, called Common Channel Signaling System No. 7 (SS-7), which is based on the OSI seven-layer stack. To provide some clarity as to the usefulness of SS-7, the following discussion draws on work done in Japan by NEC Corporation, a leading manufacturer of digital telephone switching equipment.

5.3.3.1 NEC ISDN Network Management

The NEC model NEAX61E ISDN switching system employs SS-7 along with an integrated network management system suitable for use by local telcos and long-distance companies. A photograph of a typical NEAX61E installation in a telephone exchange is shown in Figure 5.11. Conceivably, the same processes could apply to private networks as well. All of the network control interfaces would be provided, (as discussed in Section 5.3.1 on NMCC). These interfaces are reviewed in the following paragraphs.

Network operation consists of modification and expansion of the network configuration. New services are provided, their usage is recorded, routing information is entered, and utilization of switching resources is recorded for analysis by other subsystems. Within *network supervision,* equipment status is maintained, providing input to overall network status. *Traffic flow* is measured and critical loading points can be determined. This process is particularly straightforward, since the M&C information flows through the same network (SS-7) as the actual call processing information. Routine tests throughout the network can be initiated and logged. As in NMCC, the ISDN *maintenance subsystem* collects alarms and trouble information. Trouble would be isolated through on-line diagnostic capabilities and recovery could be instituted as appropriate.

The final element offered through the NEC's ISDN M&C system is *network management.* Its purpose is to collect the information and facilitate overall network control.

These subsystems would overlap several of those of GTE's NMCC offering. NMCC, however, has administrative capabilities which could be melded with those just described. NEC offers an on-line computer system associated with the switch, and SS-7 for maintenance and operation of the network, consisting of switching systems, subscriber lines, transmission links, and terminals. The emphasis is on the telephone circuit switched portion of ISDN. For packet data over ISDN, there will be an evolutionary process as NEC and others resolve how such services will be measured and billed.

5.3.3.2 AT&T Unified Network Management Architecture

With a preeminent background in telecommunication network management, AT&T has an impressive base from which to enter the integrated network management picture. Their use of on-line M&C networking through SS-6 was previously mentioned. Developments in SS-7 on the domestic and international scenes are favorable to AT&T's plans for taking a significant share of the integrated network management business. The umbrella that AT&T is opening is called Unified Network Management Architecture (UNMA). The existing structure is consistent with AT&T's view of the intelligent network of the future. This would appear to be heavily telephone-based with ISDN capabilities added as appropriate. AT&T is taking an approach similar to IBM's NetView/PC offering, encouraging vendors to adopt UNMA, or at least to be capable of interfacing to it. Recently, AT&T released a set of specifications for its Network Management Protocol (NMP), which is being made available to vendors of telecommunication network products. The protocol follows the OSI model, implementing several of the layers as they relate to network management. At the conclusion of this chapter, we consider the OSI network management application for data communication environments.[ATT, 1988.]

The intelligent network concept appears to provide a broad infrastructure for network management. With the largest domestic network, AT&T could offer UNMA as a service to users. This offering would tend to implement a fundamental telecommunication strategy of binding customers of long-distance service to AT&T. Services such as WATS, virtual private networks or software-defined networks, credit card calling, and PDN offerings could eventually be using the AT&T network management infrastructure.

Figure 5.11 The NEC model NEAX61E ISDN switching system. (Photograph courtesy of NEC.)

A basic view of the intelligent network architecture is presented in Figure 5.12. At the left are shown user terminal devices (telephones and terminals), which are connected through local telco lines to the service switching point (SSP) of the long-distance carrier (i.e., AT&T). At the right, SS-7 provides the glue for controlling and monitoring usage of the long-distance network, allowing circuit switched and packet switched services to be provided, according to a wide variety of pricing packages and software definitions. Fundamental to this type of system is the database that contains directories, routing tables, telephone number translation tables, facilities information, and facilities for network maintenance. On the

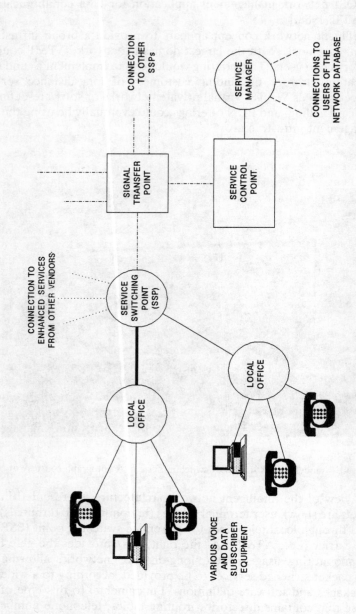

Figure 5.12 Basic capabilities of intelligent network architecture.

wer right is access for outside users to employ the network management services
AT&T. Our discussion is brief owing to the early state of UNMA's development.

3.3.3 OSI Management Standards

The International Standards Organization (ISO) is providing a series of stand-
ds for communication management. Currently at the draft level, the OSI man-
gement framework would reside at the application layer (layer 7) of the OSI
ference model. The framework will give developers of network management
stems in the data communication area a tool kit of protocols to allow exchange
management information between networks, both public and private. This sys-
m has the potential to reach all OSI and non-OSI compatible components in a
twork and to automate many routine tasks.

Defined within the management system architecture are four levels at which
twork management data are processed. Each level is responsible for aggregating
d analyzing a subset of the data. The *sensor* level retrieves raw data from the
twork in much the same way as the NM&C networks do. The raw data are
gregated, filtered, and analyzed at the next level, called the *agent*, where there
a unique agent installed in software for each type of network component. At
e *manager* level, the next level, agent output is processed from the entire network
d passed on to the *super manager* level. The difference between manager and
per manager is that, while the manager is an on-line function, the super manager
ay be executed off-line. Moving from architecture to actual data processing and
mmunication, the ISO committees have defined a set of protocols for imple-
enting INM within the OSI structure. This set of protocols is designated the
mmon management information protocols (CMIPs). In addition, there would be
set of applications to perform such functions as configuration and name man-
gement, fault management, performance management, accounting management,
d security management. Trials of CMIPs using products from a variety of vendors
ere initiated in late 1988.

The promise of a standard architecture and protocol structure for network
anagement is attractive, particularly for the operator of a private network. Hard-
are vendors and long-distance carriers may find it in their interest to maintain a
osed INM system. For those of us who must use a variety of facilities and services,
wever, some under our control and others not, the prospect of a common control
d accounting system is attractive. With regard to OSI in general and CMIP in
rticular, some progress has been made on the practical side. The network man-
gement offering from Codex, mentioned previously, includes CMIP. Also,
T&T's UNMA would also apply some of the recommendations for the OSI
chitecture and protocols. Until an OSI solution is provided, major private net-
orks may find it advantageous to use current INMs, such as GTE's NMCC or
3M's NetView to their fullest capabilities.

Chapter 6
Network Design and Economics

The design of a private telecommunication network is a complex problem for strategic units. One of the most basic goals of implementing a private telecommunication network is to reduce the cost of telecommunication for the strategic unit, so we treat the design of telecommunication networks from an economic perspective. We are not talking here about economic theory, but rather how to determine the cost of implementing and operating the network. We also consider how to make economical decisions on network design and equipment configuration.

6.1 OBJECTIVES OF ECONOMIC STUDIES

Because private telecommunication networks represent an investment or major expense, they have potentially substantial economic effect on an organization. In general, the network should be justifiable in terms of reduced cost or improved performance in the realization of the strategic unit's goals. A business unit typically examines the network's effect on the "bottom line," that is, the consequences of operating the network for the financial income statement. If the private telecommunication network is operated for a profit, it can be viewed as a business unit itself, requiring its own set of accounting statements, much like a telephone company. In fact, many of the concepts and procedures presented in this chapter evolved from the approach taken by regulated utility companies, particularly Bell Operating Companies.

Most networks are not operated for profit, however, because they are components of a much larger business. In the context of Michael Porter's *Competitive Advantage* [Porter, 1985], information technologies in general and telecommunication in particular are fundamentally part of a company's value chain. Porter defines the value chain of a strategic unit as the collection of activities that are performed to design, produce, market, deliver, and support its product. The value chain of a strategic unit reflects its history, business strategy, the manner in which the unit implements its strategy, and the underlying economics of all of the unit's activities. Telecommunication and information processing capabilities are particularly pervasive in the value chain because every activity creates or uses information. Furthermore, this information usually must be carried from one point to another, from one location to another. Chapter 8 is devoted to the topic of strategies, and further develops these concepts.

Showing direct correspondence between the investment and operating cost of the network with the economic performance of the unit is usually difficult to do in a nontelecommunication business. Nevertheless, the decision to implement, alter, or expand the network is to be evaluated in economic terms. This is typically done by a budgeting process, where the telecommunication manager is allocated capital and operating expenses to perform the requisite tasks. Alternatives would be compared and investment decisions made so as to minimize costs for the desired capability.

We will review some basic economic principles, which are needed for evaluating a new network capability, including topics covered under the general area of management finance. We have focused on the telecommunication problem, and we will use some of the specific approaches of the telephone companies. This must be qualified because telcos are regulated monopolies, and they are more able to make long-range plans. Strategic units may not have as much latitude. Any investment in facilities, however, is of its nature long-range. Twenty years ago, the long-range "planning horizon" may have extended fifteen years, while today five years is more typical. This is the result of rapidly evolving digital technology. Likewise, the external business and economic environment is more competitive and turbulent than it was fifteen years ago, particularly in the way that information technologies are employed in the business environment. The time needed to prove that a telecommunication investment will provide the necessary economic benefit (called the "prove-in time") must be short. Telecommunication managers prefer to show prove-in during one or two years.

6.1.1 Review of Economic Concepts

The following discussion is intended to review certain relevant economic concepts. This draws heavily from two fields of study: managerial finance, which is usually taught in colleges and universities as part of a business administration program, and engineering economics, which is an aspect of operations research or systems science. Those readers familiar with these concepts can skip this section and move on to the examples. The techniques of managerial finance allow us to examine a private telecommunication network as an investment in a business. A key parameter is the *return on investment* (ROI). Engineering economics is a facet of a technical field, putting dollars to network performance. The particular strength of engineering economics is in examining alternative ways of doing the same job, which is useful in deciding how to proceed with the network. Generally, the comparison is on the basis of the equivalent annual cost for running the network. A good engineering economic study thus includes managerial finance and *vice versa*. The economic concepts reviewed below apply to both ways of examining the problem. More detail can be found in [Smidt, 1970].

6.1.1.1 Investment Cost

Also called the "first cost," investment cost is the initial outlay of all installed plant, including equipment, material, engineering expenses, and installation expense. Typically, the investment cost for the network is entered into the asset side of the balance sheet of a company. In most units, funding of the investment cost is through capital budgeting procedures, where investments in different projects are compared with one another in terms of their effect on the ROI. The moneys could come through debt or out of operations. Investment decisions have long-range implications because they often commit future expenditure for debt repayment and operation. Once the investment is made, portions of it are written off annually, with the result that at the end of its useful life, the investment is essentially zero. A residual or salvage value may exist when the investment is no longer maintained.

6.1.1.2 Annual Cost

The cost of running the network may be expressed on an annual basis. This is a key result of using economic principles to evaluate the network. The annual cost includes various components, some of which relate to the writing off of the investment and others incurred during a particular year of operation. In comparing investment in facilities with use of common carrier services, all costs must be examined on an annual basis. As we will discuss, annual cost includes depreciation, the cost of money (interest or debt service), and operating expenses such as maintenance, rent, taxes, and leases.

6.1.1.3 Depreciation

Depreciation is an important component of annual expense and represents the portion of the investment which is to be written off in a given year. Those of us who have complicated income tax returns know that depreciation can be allowed as a tax deduction because the IRS recognizes it as a legitimate expense of doing business. There are many different formulas for computing depreciation and the application of a given formula is straightforward, but the concept behind depreciation is somewhat abstract.

Basically, depreciation expense provides for the use of the invested facility and corresponds to its depletion of value. For example, an automobile has a useful life of eight years. Every year, we use up one-eighth of its value so that after the eighth year the investment is nearly gone. The car still has a residual or salvage value in that a teenager may be willing to deal with the effort of maintaining what is now an old car. The original buyer, however, may find the old car to be unsuitable for its intended purpose. To run the comparison like a business, money should be set aside each year from operating revenue (let us say, the car was used as a taxi) so that at the end of life, there would be a pool of money for the replacement vehicle. The money set aside yearly is called *depreciation expense*.

6.1.1.4 Cost of Money

We mentioned that the investment expense is funded through the capital budgeting procedure. The money is often borrowed from an outside source. This capital is raised through the sale of bonds or notes at a specific rate of interest. Another source is from the sale of stock. The strategic unit typically has an average rate of interest which considers all sources used. Every year, cost of money expense is set aside to pay lenders for the use of the money that comprised the initial investment. Repayment of the investment principal itself is through depreciation expense, which sets aside the money annually. The combination of depreciation and cost of money covers the annual equivalent cost of the investment dollars which were initially put into the network.

6.1.1.5 Present Value

Another somewhat obscure but valuable economic concept is present value. The best example of present value is the home mortgage. When we wish to buy a house, we make a purchase commitment at a specific price through a sales contract. From this price is deducted the down payment, resulting in the principal. The principal is effectively the present value of the mortgage. From one year to the next, we make mortgage payments, which include repayment on the principal plus interest (i.e., cost of money). If we add all of the mortgage payments together, the sum is several times larger than the principal. The added money represents the interest expense, a component of annual cost. In a mathematical sense, the initial amount to be financed, the number of payments, and the interest rate determine the monthly payment. Conversely, the initial amount financed is uniquely determined by a given stream of monthly payments over a certain period and a fixed interest rate. Those readers familiar with financial pocket calculators know which buttons to press to make this computation in either direction.

In engineering economy and managerial finance, the initial principal to be financed is called the present value. This powerful concept allows us to bring all monetary expenditures to a common point in time with essentially equivalent financial worth to the strategic unit. Another related concept is the future value of a present amount. This simply means that a pile of money right now is as valuable as a larger pile at a future date, where the increase is determined by the period of time and the interest rate. This is the same concept as taking the principal and putting it in a bank account, drawing that same rate of interest. At the future date, the balance has been increased by the accumulated interest payments.

6.1.1.6 Apples to Apples Comparison

One of the most difficult tasks in performing economic studies of alternative telecommunication network approaches is that of achieving a valid comparison. We use the term "apples to apples" comparison to indicate the desire that the alternatives do the same job, operate over the same time period, and live by the same economic ground rules. We often find ourselves instead making an "apples to oranges" comparison. This happens when the analyst colors the comparison by accident or on purpose. A new network may be attractive because of technological features that have potential for future applications, but do not produce value today. We might be tempted to push the conclusion in that direction, rather than simply following a path of lesser technological sophistication. Of course, if the economics were equal (presenting a "wash"), the selection of more advanced technology could be the right choice.

6.1.1.7 Evaluation of Deltas

Most strategic units already have invested heavily in telecommunication facilities, and economic studies are used when considering changes or expansions. The appropriate way to proceed is to accept as given that which has already been committed. Analysis of the change (delta) proceeds almost independently of the earlier project. An advantage of using deltas is that this approach puts the alternatives under a microscope. If the fixed part of the network is constantly added (although it is a constant for the alternatives or changes), the important economic differences will tend to be less visible. Nonetheless, there is a danger in just evaluating deltas because an assumed constant aspect of the network may actually need to be changed. Reality, however, has an effective way of catching up. Recall Murphy's Law: *if anything can possibly go wrong, it will*. A good way to prevent such problems is to involve in the study people who are intimately familiar with the construction and operation of the existing network. Their review of the assumptions will usually uncover something important that may have been ignored.

Another aspect is that only future actions should be considered. The fact that an organization installed a mainframe computer at one location should not preclude the consideration of a totally new approach using smaller minicomputers in a distributed processing data communication network. The cost of disposing of the old equipment should be considered, but the emotional attachment to the older system might slow the transition to the new capability needed for a growing business. Of course, one alternative to consider would be maintaining or expanding the mainframe, and retaining a host-based environment. Analogies to telecommunication switching and transmission equipment are also appropriate.

6.1.2 Planning for the Future

In telecommunication, as in other areas of high capital investment, decisions made today have profound results over an extended period. Modern information technologies contribute to the flexibility and versatility of the telecommunication network even before money is spent. Selection of a particular technology such as packet switching is a long-term commitment because the organization thus becomes wedded to an architecture. If the technology is proprietary, only one vendor may be available to handle future requirements. While a sole vendor will have economic consequences, the choice of a technological direction is typically based more on philosophy than sound economic decision-making.

The telcos learned long ago of the importance of such planning. They have created the procedures of engineering economics for us to apply in the private telecommunication network environment. Managerial finance provides the corporate mechanism for funding the project. The telecommunication manager, however, should blend these ingredients without forgetting the significance of each. Stated another way, an impressive telecommunication network can be designed and optimized through engineering economics, but it had better benefit the strategic unit in terms of its business role.

The most effective way to make plans is to develop a set of options on paper. The options are projected into the future by using the economic principles outlined in the previous section. The introduction of the personal computer and spread-sheet programs (e.g., Lotus 123 and Microsoft Excel) have revolutionized economic analysis of complex projects such as those in telecommunication. Anyone who has worked with one of these programs understands the power that they provide. It is just as easy to set up a problem with a spread-sheet program as it is to create a manual layout on a wide piece of accounting paper. From that point, various alternatives can be easily investigated with the computer spread-sheet, by either altering the current sheet or creating a duplicate to hold the changes. Spread-sheets can be tied together to produce very complex analyses, providing the results in convenient summaries. Graphic plotting is also included for visual representation of results. The visual makes sensitivity testing as easy as determining the range of possibilities that need to be included. Spreadsheet software packages also contain

inancial analysis routines, including present value, annuity calculations (to com-
)ute monthly or annual payments on loans), statistical functions, and a wide variety
)f standard mathematical relationships.

Still more powerful computer tools are available for the designer of telecom-
nunication networks. Traffic analysis software allows us to simulate the operation
)f the network and determine its performance for telephone calls and data trans-
nission. Another class provides current tariff cost data for a wide range of available
services from local and long-distance carriers. (This topic is treated later in the
·hapter.)

6.1.3 Integration of Uses

In the context of digital private networks, the integration of the widest possible variety of uses has become an important objective. The economics should
also bear out the value of combining uses to achieve an economy of scale. The
function of integration is typically performed in a digital switch or high-level multiplexer (as discussed in Chapter 4). These devices are usually purchased and
subsequently operated by the strategic unit. Local telcos and long-distance carriers,
however, are offering such capabilities on a service basis. An economic analysis
would be performed on the result of purchasing *versus* leasing the new equipment.
High-capacity trunks using T1 links would interconnect many of the privately owned
nodes. Being able to combine annual costs of leasing trunk capacity and operating
the new equipment is vital to proper consideration of integration.

6.1.4 Break-Even Points

As mentioned previously, today's telecommunication manager wants to see
a new facility proving itself economically in one or two years. The point in time
when the costs of operating the existing network *versus* a new network are equal
is called the *break-even point*. Thereafter, the new network would produce savings.
As an example, the break-even point for an integrated T1 network might be 18
months as compared to the existing approach using leased voice private lines and
DDS multidrop lines. The study might show that over a ten year period, the new
network would reduce total telecommunication expenditures by 30%. This author
is always leery of statements that the new approach will save a specific amount,
such as $2 million. Economic analysis, however, will give a reasonably reliable
estimate of break-even and savings. Of course, we cannot predict what will happen
when common carriers add new capabilities and change pricing due to competition.
Also, a ten-year commitment to one technology poses the risk of obsolescence.

6.1.5 Business Development Leverage

Many private telecommunication networks are implemented to give the strategic unit a capability that did not previously exist. This can present an opportunity to improve the business performance of the unit from the telecommunication viewpoint. For example, a major credit card organization previously handled all transactions by paper, sending the little credit card imprint slips across the country between processing centers. A new technology was developed, based on scanning the slips for entry into a computer processing system. The transmission of the electronic images, however, required high-speed digital lines at rates between 256 kb/s and T1 (i.e., 1.544 Mb/s). This was the driving force for a new telecommunication network capability, which could not be accommodated over the telephone network. Economic considerations came into play in that the organization considered two forms of satellite communication as compared with terrestrial fiber. The terrestrial fiber approach was selected because, for this network, the cost of leased T1 was lower on an annual basis. In addition, the management thought that it needed to make a commitment to this transmission technology at an early stage of its domestic development.

6.2 DEFINING THE NETWORK

One of the toughest assignments for the network planner is defining it so that a proper economic analysis can be performed. We have already mentioned that an existing network offers the opportunity to gather useful data on usage, traffic flows, operation flexibility, reliability, and operating costs. Much of this information can be included in the analysis of new capabilities. In the absence of actual data, the analyst must make intelligent estimates. Network engineering is the process whereby traffic requirements are established and converted into physical characteristics of the facilities and services to be maintained or implemented. A variety of network topologies would be considered for satisfying the requirements and subsequently analyzed by using the economic principles.

In the following paragraphs, we review a logical process for defining the network. These concepts would be applied wholly or in part, depending on the particular situation of a strategic unit. The general sequence of steps in the network definition process is presented by the flow diagram of Figure 6.1. Many of these steps are iterative, as will be apparent from the discussion that follows.

6.2.1 Estimating Demand

The first step in estimating demand is to obtain a baseline requirement for the initial operation of the network. On the telephone side, usage records could

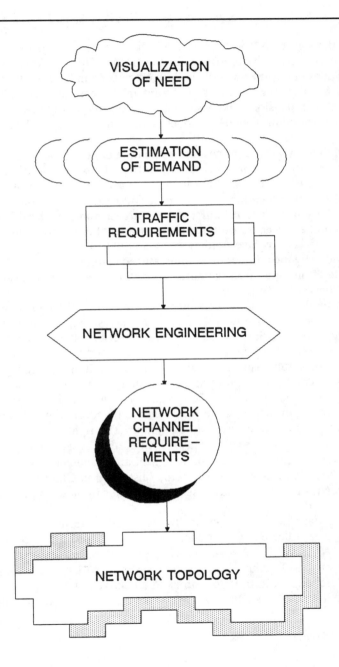

Figure 6.1 Network design process flow diagram.

be obtained through PBX and CENTREX call detail records. If they are lacking, the existing network can be used to infer the traffic. This is possible because the quantity of nodes and trunks was likely derived by trial and error. If a given switch capacity had been inadequate, at some point in the past user complaints would have led to local expansion. Likewise, blockage on frequently used trunks would have led to trunks being added to reduce it (and the corresponding complaints) to an acceptable level.

We will first review the process for estimating demand for telephone service. Each location in the network serves a quantity of user stations through local switching (PBX, CENTREX, or telco MTS). Records of usage indicate the number of calls placed at that location in a given period of time, typically one hour. In any day, the *busy hour* is that hour with the largest number of calls in progress. Besides the number of calls, we must also consider the period of the call, called the *holding time*. The average holding time is equal to the sum of holding times of all calls divided by the number of calls, measured in minutes. The calls in progress and average holding times are segmented according to local calling *versus* outside calling on particular trunk groups.

Converting the existing call data into the format described above is somewhat cumbersome. Once done, however, it is relatively simple to proceed with the network design process. Specialized software, described later, converts the data into meaningful traffic loading for a network under study.

Another important aspect of estimating current usage is the gathering of operation costs. The SMDR information provided by the switches should include service charges by line and trunk. This is the best information for making comparisons with new approaches. Other costs should be accumulated from operating budgets and the organization's accounting system. The right accounts should have been set up ahead of time so that network operating costs could be examined. Thus there is an interplay between the technical and administrative operations of the network. Network management systems (discussed in Chapter 5) incorporate various approaches for implementing such capabilities in modern private telecommunication networks.

Traffic in data networks is typically measured in terms of packets. These statistics should be collected by the nodes and hosts within the data communication environment. As in voice networks, blockage would be detected in terms of response time for users. Trunking between nodes is done with either analog or digital private lines. Hence, the number of trunks in an active network would have been matched to actual use. This complex topic is covered in more detail later.

6.2.1.1 Using Surveys

Real data from current operations are always the most credible. If such data are not available because the requirements are new or the accounting systems were

in place, user surveys become vital. The problem with any survey is that the
r tends to give unreliable input. One reason might be the user's desire to have
ra network capabilities installed so that busy signals would be less frequent.
e telecommunication manager, however, prefers to see a tolerable level of busy
als (due to blockage at switches and trunks), which indicates a better balance
ween quantities of facilities and their usage. Users also try to give the answers
t they think the surveyor wants to hear. What would happen if we could assign
ts to answers given so that users can visualize the results of their recommen-
ions?

Whether the network exists already or not, the other difficult aspect of plan-
g is to make projections into the future for growth of usage. Straight-line
umptions at fixed growth rates are relatively easy to make. Perhaps the best
proach is to perform sensitivity studies with the economic model in the form of
omputer spread-sheet, testing the effect of various growth rates over the planning
rizon. The one nice feature of traffic growth is that the statistics of telephone
ling and data communication develop an economy of scale. For example, a
ubling of traffic load does not require a doubling of the amount of facilities
eded to carry that traffic. This type of economy appears throughout the network
enever information flow is aggregated.

.2 Network Engineering

The process by which traffic requirements are converted into specifications
· network facilities is called *network engineering*. A subdivision of the field, called
ffic engineering, deals with the sizing of transmission and switching systems to
rry a known or projected quantity of throughput. The tools of the traffic engineer
e typically based on mathematical models of queuing theory, thoroughly reviewed
[Schwartz, 1987].

The analysis of traffic flows differs somewhat for the examination of circuit-
itched *versus* packet switched services. In circuit switching, the pathway is set
during an initialization process, and then held on a full-time basis for the duration
the call. As mentioned in Chapter 4, call setup time is typically a small percentage
the call duration. In packet switching, each packet is addressed and processed
dividually. Loading of the circuits between nodes is completely random in nature,
th differently addressed packets being sent next to each other. The transmission
user information is statistical, as packets are applied to the path and routed
rough the network. Of course, the node-to-node communication is continuous
ing a synchronous protocol such as HDLC.

6.2.2.1 Circuit Switched Traffic Analysis

The estimation of facility requirements in telephone networks is quite well understood, and traffic engineering has been reduced to practice. In a circuit switched network, calls follow a fairly consistent set of laws, which are derived from well defined statistics using queuing theory. With the increasing percentage of calls being used for data, there has been an introduction of some uncertainty in the analysis, but not enough to make the procedures for traffic engineering unusable. The next section deals with packet switched networks, where there is a much wider distribution of information statistics. Consequently, packet switched networks are less straightforward when predicting needs in the absence of experience with the real network and users.

The telecommunication traffic represents the activity of conducting actual communication. In telephone networks, traffic is measured in calls and call durations. Data communication traffic is typically measured in terms of the quantity of packets and their individual lengths. The telecommunication traffic is carried over switching and transmission facilities that have been sized to carry the traffic offered to the network, called *offered traffic*. We can implement various network capabilities to carry the offered traffic. Therefore, the channel capacities on links between nodes are physical characteristics, and not traffic in and of themselves. This distinction should be kept in mind because even experienced telecommunication people could confuse trunk capacity (in channels) with the traffic being carried (in calls). In the idealized case (which is not too far from the real world), the processing of calls is analogous to serving a number of incoming clients who arrive in a random fashion. Service is provided to them by the switch and the trunks that go to particular destinations. We assume that the statistics of the arrivals are known, being the distribution in time of the intervals between arrivals and the lengths of the calls when they have been set up. Traffic engineers use a Poisson distribution to model the arrival of calls, which are assumed to be independent of each other, and the call duration. This typically gives adequate results for conventional telephone applications within the switch as well as on the trunks between switches.

An example of a circuit switched telephone network is shown in Figure 6.2. This is a hypothetical arrangement with five nodes of a fully interconnected mesh, where each node has some traffic to be carried to every other node. The fully interconnected mesh network has the benefit that, in the event of an outage along one trunk group, calls can still be routed through other nodes on a tandem basis. The traffic between two points is measured in the number of calls during the busy hour. For convenience, we assume that the average call duration is three minutes. The traffic requirements could be met with public switched service (MTS or WATS) or by private lines using customer owned or controlled switching.

The nodes originate traffic and accept it from other nodes, as summarized in

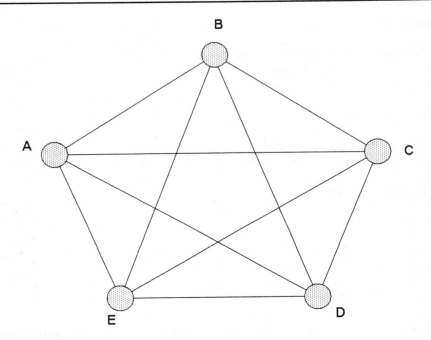

A B
E
C
D

Figure 6.2 A five-node mesh network to be used in traffic analysis example.

Table 6.1(a). The traffic indicated is only that to be carried between locations. Consequently, the diagonal of the matrix is nulled, since switching among local subscribers is not considered in this example. We use the abbreviation A-B to indicate traffic originating at A and destined for B. For example, the A-B traffic is estimated to be 44 calls, which happens to equal the B-A traffic. The traffic load is 44 × 3 minutes, or 132 call minutes. Traffic engineers prefer to use call seconds measured in hundreds of call seconds (CCS). Therefore, the traffic is equal to (132 × 60)/100 = 79.2 CCS. This conversion to CCS rounded to whole units is shown in Table 6.1(b). Traffic between two nodes being unbalanced is not uncommon. There are simply more calls originated in one direction than in the other. Table 6.1 indicates that D-B traffic amounts to 103 calls, but B-D amounts to 90 calls. This result has only to do with the particular calling pattern for users at these locations. This location has a total call activity, shown in the last column for originations and the bottom row for terminations. Location B, the one with the most traffic, originates 264 calls in the busy hour, while it accepts only 178. The total calling of the network is equal for the sum of both directions, which amounts to 1426 calls or 2566 CCS.

Calling activity places demand on the trunks between nodes. If too many calls are attempted, the trunks will be full and subsequent calls cannot be served

Table 6.1 Traffic Characteristics between Nodes for the Network Illustrated in Figure 6.2

(a) Number of calls in the busy hour (average duration of 3 minutes)

	TO: A	B	C	D	E	TOTAL
FROM: A		44	12	30	13	99
B	44		70	90	60	264
C	13	45		19	20	97
D	32	103	21		24	180
E	14	30	8	21		73
TOTAL	103	178	99	130	104	713

(b) Traffic measured in CCS

	TO: A	B	C	D	E	TOTAL
FROM: A		79	22	54	23	178
B	79		126	162	108	475
C	23	81		34	36	175
D	58	185	38		43	324
E	25	54	14	38		131
TOTAL	185	320	178	234	187	1283

(i.e., the excess calls will be "blocked"). The solution is typically to increase the size of the trunk group by adding more channels between the problem nodes. Although we cannot be certain of the relationship between the size of a trunk group and the frequency of call blockage, there are statistical relationships which have been found to be reasonably accurate for planning purposes [Schwartz, 1987]. A synthesis of one such relationship, the Erlang B equation, is presented in Figure 6.3. This plot shows the traffic carrying capacity of a trunk group containing from one to fifteen channels. For example, a trunk group consisting of five point-to-point channels between a pair of nodes can handle a traffic load of up to 80 CCS. The statistics for this case provide that with this traffic load, on the average, there will be one blocked call for every twenty attempts. The ratio 1/20, or 5%, is called the *grade of service*. The average utilization of this trunk group is 2.2 channels. Therefore, the group is less than half utilized on the average. There will be times, however, when all channels in the group are occupied with active calls, and the next attempted call will be blocked because there are no available channels.

The Erlang B equation for a particular grade of service, used to create Figure 6.4, allows us to convert from traffic load into the required number of channels in a point-to-point trunk. In this fully interconnected mesh network, the minimum trunk size is three channels, while the maximum is nine. Note that trunk A-B is equal in size to trunk B-A, which results from equal traffic loading. Trunk B-E, however, is sized for seven channels while a capacity of four channels is sufficient for trunk E-B. These directions refer to that which call originations can take, but

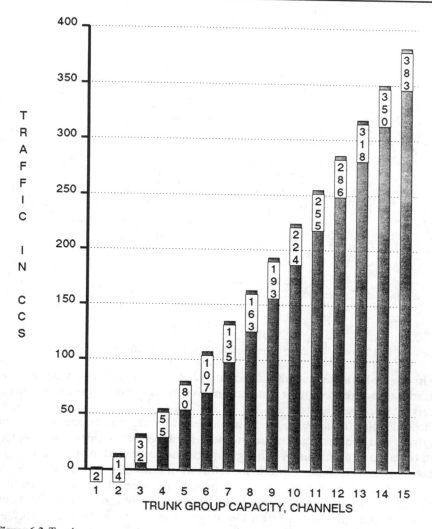

Figure 6.3 Trunk group capacity at 5% blocking probability (Erland B Equation).

the channels themselves provide full-duplex operation for normal voice and data calls. While not shown in our example, a trunk group can be configured to accept calls from either end.

This simple example should give the reader a feel for the process of traffic engineering. The formulas and tables used by the traffic engineer are beyond the scope of this book. Fortunately, the study of network design is made more convenient by the computer software tools which incorporate traffic engineering tech-

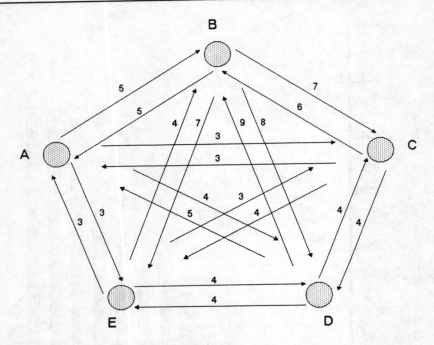

Figure 6.4 Capacities of trunk groups in the example of a five-node network.

niques. In cases where the recognized formulas such as Erlang B are not applicable, we can still resort to computer simulation. We review both approaches using computers later in this chapter.

The utilization of a trunk group increases in percentage as the size of the group grows, and this fact can be used to the advantage of the network designer. Through the process of *bundling,* trunk groups can be combined to achieve a larger cross section, which is basically done by reducing the number of major nodes in the network. Traffic will be routed on a tandem basis from nodes that currently have a lesser requirement. Also considered is the geographical distance between nodal locations. Obviously, traffic on the longest routes should be aggregated as much as possible because the cost of telecommunication generally increases with distance. (The exception to this rule is satellite communication, of course, as discussed previously.)

The process of bundling will now be applied to the simple five-node mesh network. In Figure 6.5, the connectivity between nodes has been reduced so as to increase the traffic flows on certain of the links. Several direct connections have been eliminated, requiring that traffic traverse some intermediate nodes in tandem. For example, A-E traffic must flow from A-B, through B-D, and then D-E. This increases the utilization of the B-D trunk group to give an economy of scale.

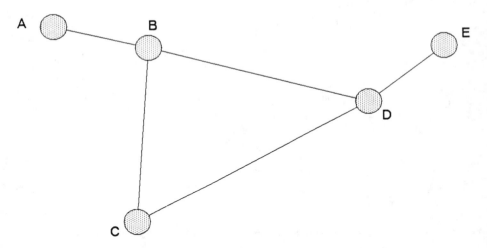

Figure 6.5 Five-node network example with traffic bundled to increase trunk utilization.

Economics is further enhanced if the A-B and D-E links are relatively short and the others are long. Combining the traffic in terms of calls and CCS, we obtain the new matrices given in Table 6.2. This traffic was applied to the utilization function (Figure 6.3) to obtain the new sizing for trunk groups, shown in Figure 6.6. Comparing the total number of channels, we can see that the bundled approach

Table 6.2 Traffic Characteristics between Nodes for the Reduced Network Illustrated in Figure 6.4

(a) Number of calls in the busy hour (average duration of 3 minutes)

	TO: A	B	C	D	E	TOTAL
FROM: A		99	0	0	0	99
B	103		82	193	0	378
C	0	58		39	0	97
D	0	179	29		117	325
E	0	0	0	73		73
TOTAL	103	237	111	305	117	972

(b) Traffic measured in CCS

	TO: A	B	C	D	E	TOTAL
FROM: A		178	0	0	0	178
B	185		148	347	0	680
C	0	104		70	0	175
D	0	322	52		211	585
E	0	0	0	131		131
TOTAL	185	427	200	549	211	1750

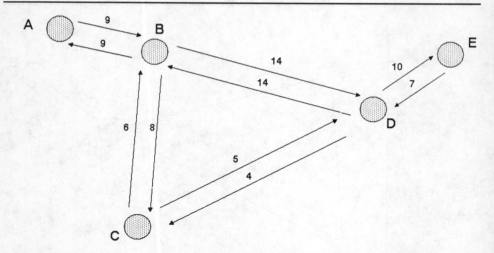

Figure 6.6 Capacities of trunk groups in the bundled five-node mesh network example.

has reduced the number of channels by a total of 20.

The bundling process is a useful network optimization technique, but it has a potentially undesirable effect on the real network. Notice how all traffic from location A must pass through B, eliminating alternative routing possibilities. A failure at node B would also remove node A from the network. Also, traffic now flows on a tandem basis between intermediary nodes. If these are full-time channels, there probably is no additional delay. If we are interested in switched services, however, the intermediary nodes will add delay (which may still be acceptable) during the call setup process.

6.2.2.2 Packet Switched Traffic Analysis

We now examine traffic engineering in the packet switched environment. Networks involving data switching and statistical multiplexers at nodes can be examined by using the same basic procedure. While circuit switched traffic can be expected to follow some well defined laws, packet switched traffic varies. Indicated in Figure 6.7 are typical characteristics of data communication transactions for a number of common applications. These include conversational services, like airline reservations systems; inquiry response, as in database access for information retrieval; data entry, employed by manufacturing companies in managing inventory; batch mode or remote job entry; and intermachine transactions, where mainframe computers share major processing tasks. For simplicity, the duration of the transaction, in minutes, is plotted in the bar graph. We have omitted other important parameters, such as transactions per unit time, data volume or rate, and statistical

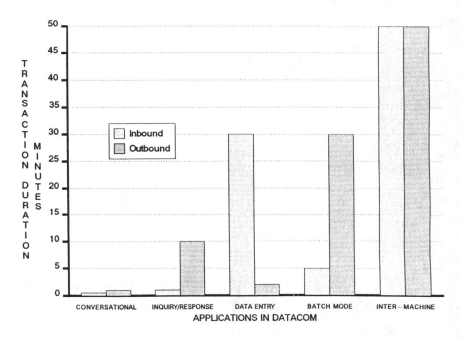

Figure 6.7 Data communication requirements for various computer applications.

properties. Even in this simple comparison, we see that distinct applications place widely different demands on network facilities. As shown in Figure 6.7, the transactions differ in the relative duration of the inbound transmission (from remote to host) *versus* that of the outbound transmission (from host to remote). Note that the inbound is shorter than the outbound in inquiry-response, while in data entry the inverse is true. This leads to uneven and unequal use of transmission and switching facilities. The intermachine (host-to-host) application could be balanced on the average, although in a particular session one direction might dominate.

In a single-purpose network, such as one developed for airline reservations in a conversational mode, the design process is simplified because a limited set of conditions can be assumed for the transactions. An integrated multipurpose network, however, probably needs to handle two or more applications with a wide variety of statistical data properties. This makes the analysis almost unwieldy. We will, however, attempt to clarify the subject.

The most important performance parameter in a number of applications indicated in Figure 6.7 is the total response time. This is the time interval between when an terminal user presses the "enter" or "transmit" key and when the requested data from the distant end is displayed on the terminal. Conversational transactions demand the shortest response time, usually in the range of one to a

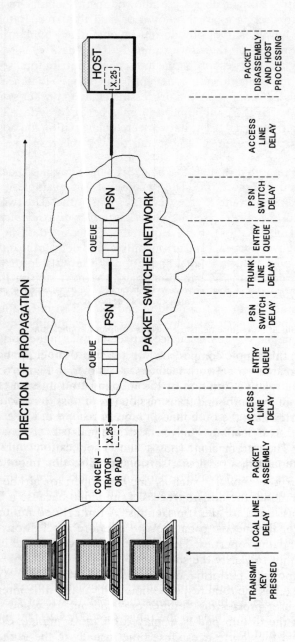

Figure 6.8 Composition of response time in an end-to-end connection of a packet switched network.

few seconds per entry. Longer delays are allowed in the inquiry-response mode because the user is expecting a larger amount of data to arrive. Batch mode and intermachine links, however, typically do not involve much human intervention, and the devices can be left on their own to hand-off the data.

We must recognize that response time has several contributors, some of which are related to the telecommunication network and others that are completely external. Network delay is derived from processing delay in packet switched nodes and transmission delay in the communication lines between nodes. These will be treated in some detail. External to the network are delays within the host computer, which must process the incoming request and deliver the response back to the network interface.

Connection of remote terminals to a host through a simplified packet switched network is illustrated in Figure 6.8. The purpose of this example is to highlight those elements which contribute to the overall response time. The two basic building blocks of the packet network are the *packet assembler-disassembler* (PAD) and the *packet switched node* (PSN) itself. Both devices accept data in blocks, add additional bits for network routing and error control, and deliver the data to another device. Storage of packets in a queue, shown on the input side of each PSN, is usually provided to deal with the situation where the next step in the process is unavailable due to previous use. The PAD is at the access point to the network and sets up user data for routing from end to end. Recall from Chapter 4 that the PAD function is typically performed by software programming within the host. The PSN, the intermediary switching device between trunks in the network, accepts the packets from the PAD or other PSNs addressed for routing. Arrival of packets into the queue is random in nature because we assume that the users are external to the network and provide no clue as to when they want to transmit information. This is analogous to the arrival of calls on a circuit switched network.

We recognize that transmission line speed is measured in kilobits per second (kb/s). Performance of PADs and PSNs is rated in packets per second, where the size of the packet will be adjusted for a particular message or network design. The possibility of several packet sizes should be recognized because of functions various packet types could play. There should be a maximum packet size so that throughput performance would not be seriously degraded by the expected BER. For example, when setting up a virtual circuit, an initial routing packet, which is shorter than a data packet, is sent through the network to establish the most efficient path. The data packets may be fixed at a certain size. A size of 1000 bits is optimal for a data packet for low and medium speed applications. A longer packet would likely contain at least one bit in error for a fixed channel BER. The relative throughput of a shorter packet length would be diminished by overhead, since the address and other process bits should be the same regardless of the size of the packet. The optimal size increases to something in the range of 4000 to 10,000 bits on circuits operating at T1 rates and higher.

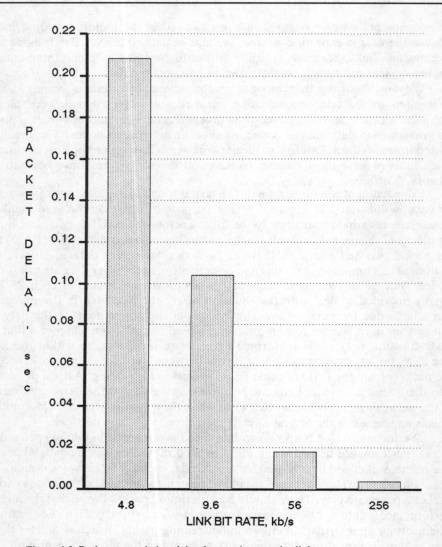

Figure 6.9 Packet transmission delay for a point-to-point link.

The most important measurement of link or network performance is the time delay in transmitting the packets through nodes. This approach is not particularly meaningful in circuit switched applications where calls are not held in a queue if all circuits are busy. Instead, a blocked call is simply terminated or dropped from the network. Packet switching is different in that if a packet cannot be sent through a PSN or a trunk, it is held in a queue (a data buffer) until capacity is available.

There are two principal sources of transmission delay: transmission time over the trunk and time spent in queues.

Transmission time is presented in Figure 6.9 for four different link transmission rates. We assume a constant packet size of 1000 bits, which includes overhead bits. This was done to have a round number for the sake of comparing several alternatives in the range of 4.8 kb/s to 256 kb/s. The packet delay plotted along the vertical dimension is simply the time taken, in seconds, for the 1000-bit packet to traverse the link. Any propagation delay, whether from terrestrial links or satellite communication, would be added to the transmission delay. Transmission delay in seconds is computed by simply dividing the packet length, 1000 bits, by the transmission rate in bits per second.

The bar chart in Figure 6.9 shows the significance of the transmission bit rate to delay. At 4800 b/s, the packet will require approximately one-fifth of a second, which is about equal to the propagation delay of a single-hop satellite link. Moving up the scale, doubling the rate will halve the delay. At 256 kb/s, the transmission delay is only 0.004 seconds (i.e., 4 ms). Interestingly at 256 kb/s, the sum of the two delay components over a satellite link is not significantly greater than the case for a terrestrial link at 9600 b/s.

Packet switched nodes can become a bottleneck in the network because they have switching capacity (in packets per second) for processing and switching packets. While this rating is straightforward enough, there are no accepted standards in the industry for node performance, and the PSNs of various vendors are difficult to compare. Nevertheless, we use the packet per second rating to make rough estimates of throughput.

The time delay encountered in a switching node can be estimated by using some of the concepts of queuing theory. A PSN is modeled by two elements: an input queue and a server. In Figure 6.8, the queue and server are shown by the horizontal bar and circle, respectively. A typical queue would be represented by the notation M/G/1, where the first symbol characterizes the arrivals to the queue (incoming packets), the second indicates the statistical distribution of the time duration of service, and last figure is the number of available servers. This particular queue type has a Poisson distribution (M for Markov process) for the rate of arrival of packets, while the service times within the PSN are specified in some particular manner (G for general) [Schwartz, 1987]. In the example that follows, the service time is assumed to be constant (indicated by D for deterministic). As another simplifying assumption, the queue is assumed to have an infinite capacity, meaning that any quantity of incoming packets can be held in line if the PSN is continually occupied in switching packets at the head of the queue.

The average waiting time for this model is presented in Figure 6.10, for a constant service time of 0.02 second. In other words, the PSN can process a packet in 20 ms or equivalently its capacity is 50 packets per second (typical values are actually in the hundreds of packets per second). The independent variable along

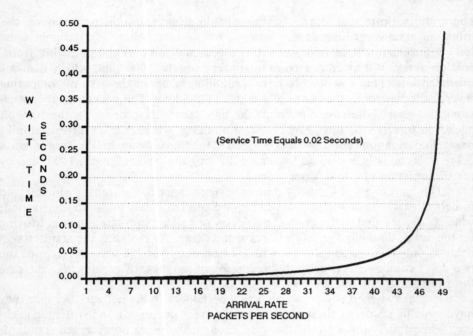

Figure 6.10 Average packet waiting time in a node, assuming the single-server M/D/1 queue model.

the x-axis is the average arrival rate of packets which are randomly applied to the queue by users. Plotted in the y-direction is the average waiting time of packets in the queue. The graph rises rapidly at an arrival rate of 49 and this is not surprising because the PSN in the model can only handle 50 packets per second on average. If packets arrived at a rate of 50 per second or greater, the delay would be infinite. The graph indicates that average queuing delay is reasonably small for arrival rates of under 25 packets per second, which is half of the rated capacity of the PSN.

This simple example provides some insight into the operation of a single server queue, which can be compared with the operation of a basic PSN. Queuing theory, however, is very difficult to apply in more typical configurations of nodes and links, primarily because of the complexity of the mathematics. The other approach, which is the one recommended in this book, is to use computer simulation models. That topic is covered later in this chapter.

6.2.3 Network Topology

Traffic planning and engineering permit us to determine the capacity of trunks between nodes in the network. We also introduced the concept of bundling for an economy of scale. These and other dimensions provide the ingredients for configuring the network topology. Selection of nodal devices and their placement is as much an art as a science. It is a form of architecture, where the designer must have a feel for the actual purposes to which the network will be put. This "soft" side of design often produces a network that provides the unit with a strategic advantage, not just a cost saving. In the following paragraphs we present some key criteria bearing on network topology. Topology is also discussed in Section 6.3.1.

6.2.3.1 Inventorying What You Have

Most networks evolve from some type of existing capability. Rather than tearing out what you have and starting over, it is vital that the origins and operation of the existing network be thoroughly understood. After all, the communications facilities were installed in response to needs. Since the same organization is likely to work with the new capability, to perform such an analysis makes sense. Separating the voice, data, and video systems of the network is much like uncovering the layers of an ancient Toltec temple, discovering older structures upon which the newer ones were built.

A rough outline of a telecommunication facilities inventory is presented in Table 6.3. Broad headings are shown, ranging from that which is on the customer's premises to elements used for processing and carrying information between major locations. The list is not exhaustive, so it is important that each organization develop its own version. All items ought to be included, whether they are owned or held on a lease basis. A good way to proceed is to enter the information into a computer database so that it can be easily updated. The network, after all, is never a static thing. Software packages are available to do the job on a stand-alone basis. For major installations, the AO software discussed in Chapter 5 greatly facilitates the process, even allowing the updating of the inventory through an on-line process as services are installed.

In cases where traffic data have not been collected, the inventory process can provide information from which current usage can be inferred and thereby estimated. For example, an existing trunk connecting PBXs at two locations is currently handling traffic satisfactorily, but measurement of that traffic cannot be made directly. We can assume a certain grade of service (e.g., 1/20), and then apply the theoretical relationship given in Figure 6.3 to infer the traffic being carried. For a trunk consisting of four channels, the estimated traffic from theory is 55 CCS. All trunks would be analyzed in this manner to yield traffic estimates for use in the design of the new network topology.

The inventory process proceeds through every location, identifying each category of equipment. This process often uncovers devices such as terminals, modems, and multiplexers, which are not needed at that location but may have utility

Table 6.3 Typical Network Elements to be Inventoried prior to Redesign of a Network

Item	Location	Description
Customer's premises		
Stations		
Terminals		
Computers		
Loop		
Cable plan		
LAN		
Switching		
Voice PBX		
Data PBX		
Packet nodes		
Multiplex		
Channel banks		
STAT MUX		
Terrestrial transmission		
Microwave		
Fiber optic		
Satellite transmission		
Earth stations		
Transponder capacity		
Shared access		

elsewhere. A simple transfer can produce important economic results because replacement costs are usually much higher than the original purchase prices. Older devices can sometimes be upgraded with new interface circuitry or software to make them compatible with a more current product line. This typically applies to switching equipment, terminals, and computers. If the equipment were old enough, however, the cost of operating and maintaining it would be uneconomic. The reasons for this include a lack of spare parts, requiring cannibalization of spare units, or burdensome support requirements, such as electrical power, air-conditioning, and on-site operating needs.

6.2.3.2 Selecting Nodes and Technology

In Chapter 4, we reviewed many of the technologies that can be applied to digital telecommunication networks. This provides the basis for specifying the devices to be placed at nodes. First, however, we must have an overall game plan or strategy for the network. Many organizations are pursuing a strategy of only buying equipment and services which vendors claim to be compatible with ISDN. Another approach restricts all buying to the product lines of one or a few vendors. Other organizations will not use satellite communication. The reasons behind these strategies should be logical, but were often emotional. A vendor-specific strategy reduces risk in terms of being sure that the network functions when it is complete. A generic approach has the organization develop a set of interface specifications to which many vendors can supply equipment and services. As we will discuss in Chapter 7, there are significant implications and risks for each of these strategies.

The economics of the terrestrial approach are dominated by the need to provide point-to-point facilities. These tend to be fixed in place and may become "sunk" investments. Specifically, it is extremely difficult to redeploy terrestrial transmission links after they are installed. We must plan many years ahead so that the sunk investment can be recovered without detrimental devaluation of the equipment before its useful life ends. In today's turbulent environment, however, such criteria are very difficult to follow. An example of an organization which can expect to use terrestrial facilities for their useful life is the local telco. Therefore, a strategic unit may find it more economical to leave investing in terrestrial transmission facilities to the telco and long-distance carriers.

Nodal equipment is different than transmission facilities because we can redeploy or upgrade it as required. A PBX or high-level multiplexer is relatively easy to disconnect from one location, then pack, ship, and install it at a new location. This is not unlike how things are done in the military, where nodal equipment is installed in vans and trailers. Microwave radio equipment is another candidate for redeployment. One of the major long-distance carriers, in fact, installs all of its link termination equipment, microwave and fiber optic, in shelters. These are mounted on trucks or railroad cars and can be placed at a site on a concrete pad. As requirements change, the shelter can be disconnected and moved to another network location. There is a trade-off in that the shelter must include a self-contained environmental control system and access for utilities. Alternatively, installation within a building can reduce costs for such support functions and minimize cabling distances.

6.2.3.3 Applying VSATs

We have already discussed the important attributes of satellite networks involving VSATs, particularly in comparison with terrestrial systems. Topology for VSAT networks is generally in a star arrangement, where links can be placed anywhere within the satellite's coverage region. Ground distances and obstacles pose no problem. The VSAT equipment is relatively compact, and it is easy to install and to redeploy as necessary. Hub equipment is different because of the size of the antenna and its support facility, which tend to be large and highly customized. A hub earth station has a usable lifetime of between ten and fifteen years, assuming that current functionality is maintained. Facility obsolescence could reduce the economic lifetime to five years, or even less in particularly unfortunate circumstances (see the discussion on SBS in Chapter 2). Upgrading an existing hub for new services is much like renovating a building. There is always the issue of whether it is more economical to upgrade or to start over.

A star network of VSATs and a hub station provide a good match to a host-based data communication environment. The host may be located in close proximity to the hub with a direct physical connection made with internal cabling. If the two are separated by a significant distance which precludes direct physical connection, a terrestrial circuit leased from the telco or other common carrier would be required. The bandwidth of this circuit is sized to support the total throughput of the star network. A dual terrestrial path for redundancy would be recommended because of the potentially vital aspects of the communication services being carried by the satellite network, and the devastating effect of a terrestrial circuit outage.

Remote users will be connected to respective VSATs which are located directly on the premises of the branch office or other remote facility. The capacity of a VSAT is set by the transmission rate over the satellite. This link, however, is shared by several VSATs using a multiple access technique such as TDMA or ALOHA. These modes are available on the same network on demand, and so the VSAT offers considerable versatility in a changing network environment. This versatility has good consequences for the strategic unit, which may alter the network as to location of sites and services provided to the sites. Sufficient bandwidth should be available over the satellite path to support a range of possible service requirements. The gross rate of 128 kb/s, for example, must be shared among the number of VSATs which access the satellite and hub on a common frequency channel. The throughput parameters of the hub and VSAT would be specified by the manufacturer of the VSAT network hardware.

A VSAT is cost justified at a given remote location if the annual cost of operating the VSAT is less than that of using the terrestrial network to reach the particular location. Also, having these Ku-band antennas on the roof of each location allows video teleconferencing to be added at any time for a modest cost. To do a complete analysis, an allocation of annual cost of the hub is made to the remote site in addition to a like fraction of the space segment lease, which is like a full-time leased line for months or even years. From an architectural standpoint, the designer needs to consider what quantities of data and voice traffic can realistically be put through the VSAT. With regard to the hub, a rule of thumb used in the industry is that a dedicated hub station can be cost-justified if there are at least 400 VSATs in the common network. A strategic unit with a requirement for much fewer than 400 remote sites can still economically employ VSATs by using a shared-hub facility offered by a common carrier. For example, Hughes Communications offers shared hub services through compatible teleport operations in New York and Los Angeles. The equipment employed is provided by Hughes Network Systems, which is a leading supplier of VSAT and hub hardware and software. Similar arrangements are possible from AT&T, Contel/ASC, GTE, and

Scientific Atlanta.

6.2.3.4 Including Per-Call Services

Per-call or switched services obtained from common carriers are attractive because of the near elimination of user investment. In Chapter 2, we discussed various MTS, WATS, and other bulk pricing schemes currently available. If conventional analog voice access lines from the local telco were adequate, the switched service would be a logical choice for much of the needed telephone service, as well as use of PDNs where appropriate. Dedicated T1 access lines from the customer location to the long-distance carrier, however, offer economy of scale and the prospect of getting the deepest discount for bulk long-distance calling. PBX switching and DACS facilities can be used in conjunction with T1 lines and the bulk calling packages, leading to hybrid network arrangements. (This is a hybrid of dedicated links and switched public services, which differs from the hybrid of terrestrial and satellite communications.) The economics would then include investment aspects as well as direct service charges paid on a call-by-call basis. An attribute of per-call services is that a full accounting by line is typically provided, as are calling statistics for making estimates of facilities requirements.

As discussed in Chapter 3, PBX equipment and CENTREX service include a feature called *least-cost routing* (LCR). While not always economical, the simplest thing to do is to make a commitment to one long-distance carrier and to let equal access (i.e., dialing 1 before the area code for every call) do the rest. Instead, several arrangements can be made with the carriers for MTS, WATS, Megacom, Prism, and the switch used to make selection on a call-by-call basis of minimum cost service. This approach could also bring in private lines and bypass circuits, which would normally be the first choice in LCR, since the user pays for them even if not used.

6.3 ESTIMATING NETWORK INVESTMENT COST

While all organizations use public networks on a service basis, it is fundamental that a private telecommunication network involves investment in owned facilities. The capital for this investment must be obtained at the time of installation. Afterward, the operation of the facility will involve annual costs (covered in the next section). In this section, we consider only the investment cost aspects. A concept called *life cycle cost* was introduced by the U.S. government to evaluate more properly major purchases of equipment by the U.S. Department of Defense. The intent of the concept is to consider the initial cost of a new system and the cost of operating and maintaining the system throughout its useful life. It would be wise to follow this same course when comparing alternatives. A low initial investment cost could conceivably have with it a high cost of eventual operation in the network. Therefore, investment costs should not be considered in a vacuum. For example, one arrangement of new facilities may require that additional personnel be hired, trained, and supported. The cost of doing this could wipe out the cost saving from replacing a leased service with the privately owned facility.

6.3.1 Configuration Diagrams

A preliminary layout of the network, indicating location and type of all nodal equipment and facilities, is needed before investment costs can be assessed. This process follows from the study and definition of network topology, employing the principles previously described. It is usually best to involve network engineering personnel in the layout of investment costs associated with the facilities because each definable element is usually an element of cost as well. Where possible, the technical performance and capacity of an element should be tied to its investment cost. For example, a faster packet switch node would normally cost more than a standard model. Also, as more channels are added to a multiplexer, the total cost of the installation increases. This particular aspect is usually the easiest to project because the added services require additional port cards and redundancy.

As a working example of a network layout, consider the topology in Figure 6.11. Three classes of nodes are shown: a major node with the capability for extremely high capacity; a regional center with significant transmission requirements; and remote sites, which can be branch offices served by the closest regional node. Major nodes A, B, and C would probably have large computing centers and telemarketing operations, placing heavy demands for voice and data transmission. Connecting these nodes are high capacity DS-3 (45 Mb/s) links, each capable of 28 T1 channels. With anticipated growth and the integration of all services on a digital basis, the DS-3s were thought justified. (They may not be available for the particular cities where the nodes are located.) The links to regional centers (T through Y) have a smaller cross section and are adequately served with one or more T1 channels. Finally, one or two 56 kb/s digital circuits are used to connect the remote sites to the regional centers. For some remotes, even this amount of capacity can be excessive, and dial up telephone service using voice band modems may be adequate. The digital integration of the network was maintained, however, for illustrative purposes. Perhaps, requirements will evolve to the point where the 56 kb/s lines will need to be replaced with T1s in the future.

A hypothetical set of transmission requirements for major node A is presented in Table 6.4. Across the top are the connectivities indicated in Figure 6.10 (i.e., A-B is the DS-3 link between A and B; A-T is the T1 link between A and T). The table presents the number of links needed to meet the transmission requirements derived during the traffic engineering process. We show that the A-B link,

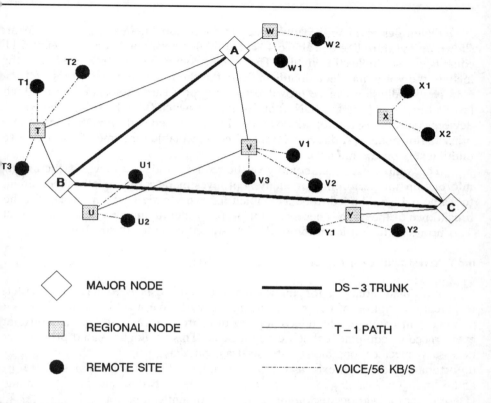

◇	MAJOR NODE	▬▬▬▬	DS – 3 TRUNK
▢	REGIONAL NODE	────────	T – 1 PATH
●	REMOTE SITE	------------	VOICE/56 KB/S

Figure 6.11 Example of a private network topology showing nodes and interconnecting links.

Table 6.4 Transmission Requirements for Major Node A

	A-B	A-C	A-T	A-U	A-W
DS-3 links	1	1			
T-1 channels	19	22			
T-1 links			2	3	2
Voice circuits			38	44	25
56 kb/s circuits			4	7	5
9.6 kb/s circuits			12	5	15

while consisting of a DS-3 capable of twenty-eight T1s, is only required to carry nineteen T1s in support of expected traffic. The use of a lower number can result in a saving in port cards and other supporting facilities. Between A and C, the requirement is for twenty-two T1s. (Note that link B-C is not shown, since it does not pass through location A.) Each T1 contains 24 DS-0 channels for PCM voice and multiplexed data.

Quantities of T1 links between A and the regional nodes at T, U, and W are shown on the third line of Table 6.4. The solid line between A and T (Figure 6.11) consists of two duplex T1 circuits. The specific communication services, however, include the voice and data circuits indicated in the remaining three lines of Table 6.4. The subdivision can be provided by using port cards associated with a high-level multiplexer (like the IDNX) or fast packet switch (like the IPX). These same devices could be used to break down the T1s associated with the DS-3 links to the other major nodes. (A direct DS-3 interface is available for some of the high-level multiplexers on the market).

This simple example should give the reader an idea of how to structure the information for performing an analysis of investment costs. A tally sheet similar to Table 6.4 would need to be assembled for every node in the network. Further breakdown of the node is necessary for properly identifying every element of cost. This brings us to consideration of the investment costs for these elements.

6.3.2 Investment Cost Elements

The ideal situation for analyzing investment costs is to have a complete technical description of every node in the network. We would also want to have a homespun catalogue of telecommunication products and services, describing every piece of equipment that we might need. This catalogue would provide the necessary specifications along with accurate purchase prices. Such a catalogue, unfortunately, does not exist as yet. Telecommunication managers must therefore amass their own catalogue as appropriate for the network under development. (Descriptions of various technologies and equipment can be found in Chapter 4.) More detailed information for a wide variety of products and services is presented in the telecommunication reference series by Datapro (a subsidiary of McGraw-Hill), which is updated regularly through reprints from the literature. New equipment and service introductions and discussions of applications can be found in such periodicals as *Network World,* published by IDG; *Communications Week,* published by CMP; *Telecommunications,* published by Horizon House; and *Business Communications Review,* published by BCR.

Equipment pricing information is considerably more difficult to obtain than technical performance characteristics. The industry is extremely competitive, and few vendors provide published price lists. Purchases are typically made through a bidding process (as reviewed in Chapter 7). Therefore, collecting cost information before the actual procurement of facilities is difficult. We must nevertheless attempt to collect this information. The key is to know as much as possible concerning the items before attempting to learn about the costs. Some vendors will be able to supply budgetary quotations, which are provided for planning purposes, but the vendor is under no obligation to honor budgetary prices. Nonetheless, a vendor who later makes a bid at a significantly higher price may be subject to considerable orn and bad will (which, indeed, makes a difference).

The homespun catalogue forms the basis for all estimating of investment sts. A good approach is to begin with as much accurate data as you have, and e educated guesses for the rest of the elements. Then, as better data are obtained, u can incorporate them. An important property of these overall estimates is that y inaccuracy in a few items will be averaged out in the totals. High estimates ll balance low estimates, unless the estimator is biasing the numbers in a par- ular direction. Estimating low is foolish because, if the project is approved, there ay not be enough funding available for its completion. On the other hand, a high timate may make the project seem unattractive.

Telecommunication equipment, such as switches and multiplexers which ser- e several access lines or channels, has investment cost which can be divided into o categories. There is a fixed cost associated with the basic device, and then an cremental component associated with each unit of access or service. The fixed rtion covers the basic chassis or "box" to house the devices, provide power, d administer the system (i.e., storage, central processing and software control, ning, monitor and control, *et cetera*). Individual channel units, port cards, or terface modules are needed for each channel or access line, and are priced cording to the function performed and the amount of capacity (bandwidth or ocessing rate) provided. The box may be very simple (e.g., a metal frame and power supply), in which case most of the expense is in the channel modules. ore complex systems, like a PBX or fast packet switch, have very expensive xes and the line units are less of a factor in the total cost. It is therefore dangerous generalize. In some systems, the box has a limited number of channel units that can support. There then must be multiple boxes, which can be very expensive ms.

A graphical representation of fixed plus incremental investment costs is shown Figure 6.12. Total utilization along the x-axis indicates the quantity of elements capacity (voice channels, T1s) which the particular facility includes. The first it requires a commitment to the basic box or major subassembly, as indicated the first point on the curve. As utilization increases, so does investment cost. this example, there is a fixed element of cost capable of carrying up to ninety- channels. A ninety-seventh channel would require an additional fixed element. is step occurs at each multiple of ninety-six channels. The curve connecting these ints is normally a straight line, ending at a point where elements of utilization n no longer be added due to the constraints of the particular type of equipment. e can examine the economics of this type of facility by using the measure of cost r unit of utilization, shown in Figure 6.13. Assume that utilization is in terms DS-0 channels. The measure is obtained by simply dividing the total investment st by the number of channels provided. The curve is plotted as shown, where it ymptotically approaches a constant minimum. This corresponds to the cost per ment because the box cost has been spread over such a large number of units

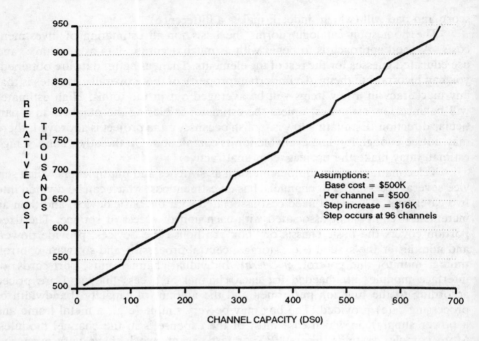

Figure 6.12 Total investment cost of a typical digital transmission facility.

that it is an insignificant fraction of the cost per channel. At some point, the capacity of the facility is reached, and another box must be added. This places us almost back at the start of the curve in Figure 6.13, although some economy of scale is possible through other sharing within the facility.

6.3.3 Use of Spread-Sheet Model

As we mentioned earlier in this chapter, spread-sheet software on a personal computer is a very effective way of analyzing the cost of a private telecommunication network. With the complexity of modern networks and the need to consider many alternatives, the spread-sheet is probably the only reasonable way to approach the task. A well constructed model is a powerful tool, one that can be easily modified as the new things are learned, both during the study and later when the network is actually in place. With enough computer memory and a fast processor, the catalogue can be incorporated as a subset of the model. Then, the channel requirements and node configurations can be folded into appropriate sections of the overall model.

A good spread-sheet model is constructed in sections which reflect the network definition and economic analysis processes. The model begins with a listing

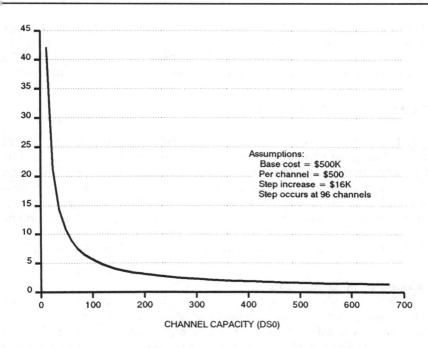

₃ure 6.13 Investment cost per channel of a typical telecommunication facility.

locations and their associated telecommunication requirements. Nodes are de-
₌ed along with the required capacities. Variables are used for elements to be part
sensitivity studies. The next section contains investment cost elements, which
late back to the requirements in the previous section. Annual costs, calculated
th formulas that account for the various categories (e.g., depreciation, O&M,
₌es), are developed in a part of the spread-sheet which can be associated with
₌ investment elements. The model includes an estimation of revenues that can
expected from users of the network. Finally, the net cost or income is presented
a summary section of the spread-sheet, providing a bottom line for comparing
₌ernatives. Most spread-sheet software packages make it easy to perform "what-
₌' studies, where tables can be generated for a range of traffic loadings or revenue
₌ssibilities.

There are specialized software products for medium sized computers and PCs
assist in the economic side of network analysis, some of which are discussed
₌er in this chapter. This author, however, believes that there is no substitute for
straightforward spread-sheet model which the analyst has constructed on his or
₌r own. The learning process is crucial for gaining a full understanding of the
₌terplay between function and cost. Such understanding greatly diminishes the
₌ance for significant errors because the analyst can continuously check validity

of calculations. We refer to these checks for validity as "sanity checks." In the next section, we consider the evaluation of the network on an annual cost basis. This makes use of the principles covered at the beginning of the chapter. Once again, the spread-sheet model is invaluable for performing this type of study.

6.4 ESTIMATING ANNUAL COST

Several components of annual cost for a telecommunication facility are identified at the beginning of this chapter. These components include depreciation, cost of money, operations and maintenance (O&M) expense, taxes, and leases. The principles of engineering economics and management finance are followed when computing the total cost on an annual basis. Also, an organization may purchase communication capabilities as a service from a facilities-based common carrier, or by sharing facilities with another owner of a private telecommunication network. We consider these various components in the following paragraphs.

6.4.1 Components of Annual Cost

Practitioners of engineering economics within telecommunication companies developed what were called "annualizing factors," wherein we multiply the investment cost to yield the annual cost. For each category of facility or equipment, such as microwave radio, satellite earth stations, or switching systems, a particular annualizing factor was computed, based on applicable economic factors. For example, an earth station was assumed to have a lifetime of fifteen years. The depreciation and cost of money were calculated and estimates were made for O&M expenses. The annualizing factor was then the ratio of total annual costs to the initial investment of the installed facility. This factor would be applied to other projects which employed the same type of equipment or facility configuration.

The annualizing factor approach made it simple to convert from investment cost to annual cost. Computing the annualizing factor, however, has become more difficult due to rapidly changing business and economic conditions. Investment decisions depend on a host of factors, and a simple change in assumptions greatly influences the choice. Therefore, annualizing factors that make fixed assumptions are no longer appropriate in today's environment. The components that the annualizing factor comprises are still valid concepts, and the analysis follows the same course by using computer models which more accurately reflect the possible situations. Once again, the spread-sheet model is invaluable in evaluating economics of telecommunication networks. The components of annual cost are reviewed in more detail in the following paragraphs.

6.4.1.1 Depreciation of Equipment and Facilities

Depreciation is a way to account for the using up of physical facilities, corresponding to its depletion of value to the user. Depreciation is a real cost of providing telecommunication service. Fortunately, this cost is recognized by the taxing authorities as an allowable business expense for tax deduction purposes. Every year, the owner of the network must set aside moneys for the replacement of facilities; at the end of life, enough funds should have accumulated for that purpose or to refund the capital to the original source. One of the serious problems in an inflationary economy is that organizations fail to set aside enough money through annual depreciation charges so that true investment capital gets used up over the life of the enterprise.

The simplest way to allocate depreciation is to assume a certain economic lifetime for the facility and then to follow a straight line with equal installments set aside each year. This approach is probably realistic from an accounting and technical sense. Another approach is called *accelerated depreciation,* where greater sums are set aside in the early years with the annual contribution tapering down toward the end of life. The sum of depreciation charges over the life is equal in either approach. Accelerated depreciation is popular when determining tax liability because it tends to show greater profit from a business asset during earlier years. For planning purposes, the straight-line approach is usually acceptable, which, for a life of N years, sets aside a fraction of $1/N$ of the initial investment as the depreciation allocation. For conservatism, these studies commonly assume no residual value.

The area where we have flexibility in computing depreciation is the assumed economic lifetime. Service providers are faced with the dilemma of determining this life. If it is assumed to be long, such as fifteen years, the depreciation charges will be low, allowing the service to be sold at a lower price. Decent revenues may not continue that long, however, because the facility may become obsolete in only half that time due to technological change or competition. A shorter lifetime of five years would reduce the risk of obsolescence, but the depreciation charge would be three times larger than for the fifteen-year life. The service may be uncompetitively priced so that losses would result from a lack of customers or from selling the service under cost. If we were successful in selling or proving a facility in a shorter period than the true economic lifetime, there would be substantial savings in annual charges during the later years.

6.4.1.2 Cost of Money

The cost of money is basically the rate of interest that the service provider experiences in financing the overall business operation, and which is allocated to the particular facility under study. Telecommunication companies finance capital purchases through the sale of stock and bonds, as well as through other forms of indebtedness (commercial paper, bank loans, et cetera). There would be an overall cost of money, which averages the charges associated with the various sources. The cost of money for a major utility, such as Pacific Bell, should be considerably different from that of a small company in a very competitive business; that is, companies with more assurance of a future tend to pay less for borrowings because of reduced risk for the lender.

A more accepted method for considering cost of money is to examine the return on investment which the new facility could produce. The cost of money then becomes a "hurdle" rate, which the proposed investment must satisfy. For example, financial guidelines for the strategic unit may require return on investment of 15% average per year on a new investment in a switch. If the company had a cost of money in the neighborhood of 10%, this investment would probably be attractive. A company with of cost of money at or above 15%, however, would be well advised to examine other approaches for meeting communication needs. Even the company with the lower hurdle might have trouble generating the projected usage and could experience a true rate of return under 10%. A careful review of a variety of options is always advisable by using rate of return as a comparison index. Also, for strategic units which employ private telecommunication networks as an ingredient of an overall business, use of return on investment will not be convenient because revenues may not be tied directly to network performance and cost.

6.4.1.3 Taxes and Insurance

Often overlooked are annual expenses such as those for taxes and insurance. A network operated as part of a profit-making business will earn revenues which are subject to income taxes. If the investment and cost associated with the network form an expense of doing business, however, and do not derive a profit, then it is likely that annual charges will tend to reduce taxable income. Property and *ad valorem* (on the value of facilities) taxes may be levied by the state or local government on the land and buildings, and for the use of telecommunication facilities. These taxes would not have been due in the absence of the network.

Organizations purchase insurance to cover various forms of liability. In satellite communication, a transponder which has been purchased on a condominium basis from a satellite operator will normally be insured against loss. This is done because a failed transponder usually cannot be fixed from the ground, unless spare equipment can be switched by command, or if the failure is healed somehow. The loss represents millions of dollars for the cost of a replacement transponder. Insurance premiums are a significant part of annual cost for the transponder. (The cost of operating the transponder once the initial cost has been covered, however, is relatively low in comparison to other telecommunication facility investments.) Terrestrial facilities will also be insured because major calamities can cause substantial destruction and loss of value. The advantage here is that the land under the facility may be recovered and equipment can be repaired or replaced by manual labor.

Insurance coverage can deal with losses other than physical loss of a facility. Liability insurance is needed to cover employees and management from law suits which could result from a claim against the organization. For example, a microwave tower could fall and injure an employee or a passerby. The tower would have been installed as part of a private network. Another type of insurance would deal with the situation where an employee acted outside of his or her authority, causing some economic hardship to a third party. The liability insurance would protect the company from a law suit arising from such situations. Annual premiums would not be high, but should nevertheless be considered in the economic evaluation.

6.4.1.4 Operations and Maintenance

The last major element of annual cost associated with investment in telecommunication facilities is for operations and maintenance. Quite simply, operational expenses are for the running of the network, including such items as personnel, utilities, administrative expenses, and other forms of logistical and financial support. In modern networks with remote M&C systems, the majority of operational tasks can be handled from a centralized NOC, which would be staffed around the clock. Local operations personnel at selected sites could still be an asset in dealing with particular needs when they arise.

The maintenance organization and support is there to keep facilities in proper working order. There are two general categories: *preventive maintenance* and *corrective maintenance*. Preventive maintenance can be planned in advance and will probably reduce the incidence of failure. For example, replacement of air filters on ventilation systems, lubrication of motors, cleaning of receptacles and connectors, and refilling supplies are typical preventive maintenance activities. Corrective maintenance, however, is required when a facility is operating in a degraded mode or has failed. Activities such as trouble-shooting, repair of broken cables, replacement of bad components or modules, and repair of electronic equipment would be corrective in nature. Maintenance expenses are directly related to the characteristics of the equipment and the facility in which the equipment is installed. Obviously, facilities which are designed so as not to need a lot of on-site preventive maintenance and are very reliable will also be relatively inexpensive in terms of annual maintenance costs.

Included are the cost of maintenance of equipment in terms of labor, on-site and shop repairs, spare parts, and other direct support. A competent engineering function is required in the maintenance aspect of network operation because much of the support for corrective maintenance is technical in nature, such as the design and application of test equipment. Trouble-shooting is an important element of maintenance engineering because the cost of downtime to the strategic unit can be much higher than that of including experienced engineering personnel in the maintenance organization.

The best way to determine annual expenses for O&M is to use historical data which the organization has accumulated from ongoing operations. Any new network capability would then be evaluated on an incremental basis. Sometimes, existing staff are able to cope with the new facility. More often, however, the in-house capability would have to be expanded. Computing annual charges for O&M is extremely difficult in the absence of real experience with the network or a portion of it. The cost of hiring, training, and sustaining a competent staff factors into the equation. The reader is advised that O&M expenses form a substantial element of annual cost and are to be carefully considered.

There are cases where the organization does not wish to perform maintenance for the network. For example, the skills necessary to repair computer equipment or VSAT earth stations may be beyond the scope of in-house staff, and test equipment may be prohibitively expensive for the small expected use. In these cases, maintenance services can be obtained on a contract basis from an outside vendor. The supplier of the equipment is the most likely candidate in this case. Were the supplier unwilling or unable (or too expensive) to do the job, then a third party maintenance organization could be located. The supplier of the equipment may be able to recommend a third party maintenance services contractor.

The cost of O&M services is driven almost directly by the speed with which action is expected after a problem is detected in the network. An organization which expects the network to be up and operating essentially 100% of the time will pay dearly in terms of annual charges for maintenance. There may be no way to avoid this situation. The fundamental way to eliminate downtime is to provide substantial quantities of redundancy and alternate routing capabilities in the facility or network. Nonetheless, if some downtime can be tolerated, the cost of deploying repair crews and test equipment decreases considerably.

The way to include O&M expense in the economic study is to identify the costs for each new site or for each new capability added to an existing site. Costs should be estimated for each year of operation of the network where personnel expenses were the most significant. If maintenance services were to be provided by an outside vendor, the anticipated cost of these contracts should be added yearly. Network planners should make the "operations and maintenance" line a permanent aspect of any economic spread-sheet model.

6.4.1.5 Long-Term Leases and Service Contracts

The remaining two elements of annual cost do not relate to owned telecommunication facilities installed for the network. Rather, they deal with facilities and capabilities in which other organizations invest and offer to a broader market or group of users. A strategic unit can take advantage of such investment made by others by contracting for the use of facilities, rather than their ownership. This advantage applies to any capability that is provided by another organization and which is paid through a lease or service agreement. Private line services are obtained from telcos and long-distance carriers on either a leased or service basis. The user usually does not take possession of a facility to employ its service capability. The seller takes full responsibility for financing, operating, and maintaining the network facilities needed to serve the organization. There is a distinction between leases and services, but very often this distinction is blurred. A number of common examples are given in the following paragraphs.

Leasing of equipment, buildings, or other facilities used in private telecommunication amounts to entering into to a long term commitment through a contractual arrangement. The lessor provides the facility and may maintain it as part of the package. The lessee then has the use of the facility under certain terms and conditions for the prescribed period. Once the lease is entered, the lessee has access to and control of the property, while the lessor retains actual ownership. There are advantages for each party. The lessee gets the use of the facility but need not worry about its disposal at the termination of the lease. The lessor receives periodic payments from the lessee, yet the lessor is in a position to count the facility as an asset within the business that it operates. For example, the lessor can use the depreciation charges to offset tax liability (in years past, the investment tax credit was also available).

Both sides can achieve economies because a portion of the capability of a larger system can be committed to revenue service through a lease. Again, the lessor obtains a customer who contributes to the annual expense of running the facility, while the lessee gets the use of the facility, possibly for a fraction of the cost. This synergy is a substantial reason for leasing to be popular. Indeed, providing access to facilities on a service basis gives the same result, as discussed later in this section. It is generally recognized, however, that a lease remains in force for one or more years.

The lease payments made by the lessee to the lessor are usually computed in the same manner as the annual cost of the facility. The contributors to the cost depend on particulars, such as who pays the insurance and taxes and whether maintenance is performed by the lessor. It is not uncommon for lease rates actually to be less than the associated annual cost. Such a situation occurs where facilities are plentiful and lessors are grateful to get the business irrespective of the terms.

This could be so because an installed facility might not be easily redeployed where more attractive rates would be available. Under this condition, users find it more attractive to lease than to buy.

Lessors comprise various entities who own telecommunication facilities. Common carriers may include lease offerings, but more likely candidates are organizations, which are actually not in the telecommunication business. As mentioned, users with excess facilities or capacity can lease them out to others. Third party leasing companies will allow the user to purchase the facility, sell it to the leasing company, and then lease it back. This is commonly called a "sale/lease-back" transaction. The leasing company is simply performing a financial service, allowing the user to free up its capital for other purposes. If the lease is for the life of the facility, the transaction is not really that different from an outright purchase. (In fact, accounting practice requires that the organization count the leased property as an asset.)

Use of telecommunication facilities on a service basis is another way to obtain some of the capabilities needed to develop a private telecommunication network. The distinction between a service and a lease is a subtle one, usually based on the fact that the customer is provided service from a facility which is operated and maintained by the offerer. Sometimes, the seller may change the facilities which provide the service so long as the result is the same. A service arrangement can be from "month to month," meaning that the customer can cancel the service at any time by giving advance notice to the provider. The provider cannot terminate the arrangement at will, however, as the user can. This arrangement gives the user maximum economic flexibility because another network or approach can be substituted to suit changing needs or to take advantage of better pricing elsewhere. Chapter 7 discusses how this approach can be used to provide interim capacity when a new facility is under construction.

Generally, leases are the domain of the specialized common carrier or leasing organization, while services are supplied by providers with ample sources and extensive facilities. The largest common carrier, AT&T, can afford to offer its capabilities on a service basis because of the size of its customer base and the extent of its network. The RBHCs and major independent local telcos are similarly positioned. Lease arrangements from specialists, particularly satellite communication providers, can still be attractive because the pricing is fixed for an extended period of time.

6.4.1.6 Per-Call Charges

Factoring in per-call charges can be a complicated problem in a private telecommunication network which also uses owned or leased facilities. Offerings which are charged on a per-call or incremental basis include switched telephone services from the local telcos and long-distance carriers, data communication services from the public data networks, and video transmission on terrestrial and satellite networks. Call accounting information is typically available from SMDR reports provided by the service company or PBX and from bills for data communication transmissions. Analysis of future circuit switched service requirements would depend on the user being able to state requirements in terms of telephone calls (i.e., CCS) with the proper breakdown identified between local and long-distance. Data communication requirements would be stated in terms of segments of a fixed number of eight-bit bytes or session holding time. Video transmissions for private broadcasting or teleconferencing are charged by the hour or fraction thereof, with the minimum duration typically being twenty minutes.

In Chapter 2, we discussed the offerings of public telecommunication network operators, particularly the telephone organizations. This discussion highlighted the various bulk calling packages which provide substantial discounts over standard message telephone service (MTS). We mentioned that the terms and conditions, including the pricing, of these packages are changing on a monthly basis. To consider this factor properly, employing a prepackaged database is usually preferable. We discuss some of the on-line services and software packages in a subsequent section.

Per-call services can be compared with using private line terrestrial services and satellite communication. The rate charged for a typical telephone call or data packet is composed of several elements. These consider the allocation of the charge to the various service providers involved in completing the connection. Prior to divestiture, AT&T provided and charged for the end-to-end connection for a telephone call. Today, the local telcos on each end collect a per-call charge, which is set by the regulatory agencies involved (FCC and state PUCs). This rate is expressed in cents per minute, and compensates the telco for the use of its switching and local transmission.

In the middle is the long-distance leg provided by an interexchange carrier such as AT&T, MCI, or U.S. Sprint. The marketplace for long-distance service is very competitive, and rates have tended to go downward during the recent past. AT&T's rates are regulated by the FCC, but the company has been allowed to reduce rates in response to competition. To further this, the FCC is expected to introduce price caps as a substitute for rate of return on investment as the criterion for regulation of AT&T's long-distance business. Most of the action in switched telephone service, therefore, has been in the long-distance segment, while the local connection has been either stable or increasing in price. As a consequence, as the price per call has come down, the relative contribution of the local connection charge has not decreased significantly. (See Figure 1.16.)

Public data networks have benefited from what may be called a subsidy in that the contribution of the local connection has been essentially zero. The telco's customers have been able to make local "free" calls to reach the PDN's access line. This situation is attractive to everyone except the local telco. The telco is already paid by the telephone subscriber for the access line by virtue of the monthly telephone line charge. There was, however, a position such that the telcos should get a bigger share for allowing the PDNs to earn revenues through the local access to their nodes. The FCC had announced its intention to levy access charges on the PDNs so that revenue would be transferred to them. In April, 1988, however, the FCC cancelled this plan due to strong objections from data communication users which was voiced after notice of the proposed rule change was announced by the commission.

In many private networks, the telecommunication manager will need to compare per-call services with the use of dedicated facilities. Any time that individual units of usage, such as calls or messages, are to be evaluated, estimation of overall usage of the facility becomes necessary. Whether the facility is owned by a common carrier or the strategic unit, the aggregate traffic loading must be determined. If the facility is new, the estimate will probably be very rough. It is not uncommon for such estimates to be off by a factor of two or more. The expected usage of a facility being expanded from an existing service base would be more certain.

Figure 6.13 points up the challenge faced by every organization that invests in telecommunication facilities. Relatively low transmission costs are achieved only if the facility can be loaded with traffic. In the steep part of the curve, doubling utilization will halve the cost per increment of use. After reasonable utilization is reached, the incremental cost of adding service is essentially constant. This process does not continue, however, because at some point another major facility expansion is required. Expansion pushes the cost per channel back up, but not as high as the initial level on the left end of the curve. Curves of this sort are easy to compute after the facility's costs have been determined.

This discussion of facility loading was presented in terms of channels. Per-call expenses involve the loading of the channels with traffic measured in CCS, packets per second, or service hours. The conversion between per-call and dedicated channels involves the traffic engineering principles previously discussed. Recall that as facilities grow larger (in terms of the number of channels per trunk), the average traffic per channel increases. This is shown in Figure 6.3. As a consequence, the economics of telecommunication facilities is doubly dominated by size and utilization of the facility. Doubling the capacity in channels will, under certain conditions, cut the cost per channel in half, and this doubling will more than double the per-call capacity, due to traffic engineering principles. Considerable economies would accrue to the facility operator if these two benefits were properly managed.

As a final consideration in per-call evaluation, let us review the way that the cost of using public and private facilities is built up. The pie chart of Figure 6.14 illustrates in general terms the rough allocation of per minute charge for making

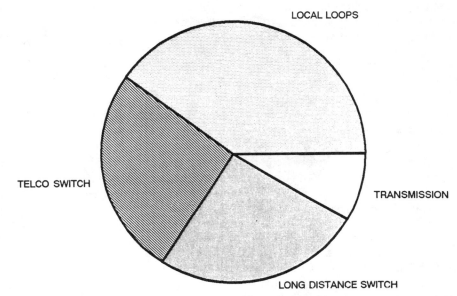

Figure 6.14 Typical cost allocation of switched per-call terrestrial service.

a long-distance telephone call. The whole pie could represent a $ 0.30 minute between, for example, New York and Los Angeles. Because of the way in which local telcos operate facilities and how they are regulated by government agencies, the bulk of the revenue is allocated to the use of local facilities on both ends. The long-distance carrier receives about one-third of the total call revenue, but notice that only a small portion is actually related to the use of the transmission system. Notice also that if the cost of transmission were to drop to zero (the promise of fiber optic transmission), the total cost of the long-distance call would decrease only slightly.

A similar pie chart for a private VSAT network is shown in Figure 6.15. The link of the network comprises a single VSAT earth station located on the customer's premises, a portion of bandwidth of a satellite transponder to carry the transmissions, a portion of a hub station to terminate the call at the main location, and a local loop from the hub to complete the call at an "off net" location. The largest contributor to the cost is the VSAT itself, which must be justified by the level of traffic for that particular location. The satellite and hub station are assumed to be efficiently utilized. The cost is converted from channel cost to per-call cost by using the traffic and economic principles previously discussed. For a well utilized circuit switched network, the per-call cost of using the VSAT is roughly equal or slightly less than that for the public network.

This type of comparison, while not trivial to compile, sheds light on the

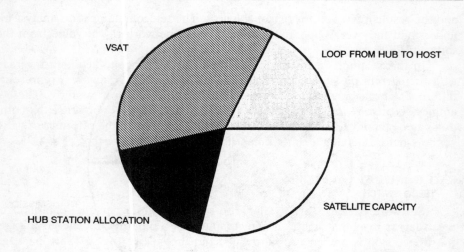

Figure 6.15 Typical cost allocation of switched per-call VSAT networks.

economic performance of public and private networks. The effect of traffic loading on economics cannot be overemphasized. Utilization involves not only putting services on the network, but also finding users who are willing to allocate funds for such usage. So most networks must be marketed to users. In commercial networks, marketing expenses represent a substantial cost of doing business. Another substantial cost, reviewed previously, is operations and maintenance (O&M). We may add users to the network, but if their service is not of a consistently high quality, they will find another way of meeting their telecommunication needs.

6.5 SPECIALIZED NETWORK ANALYSIS SOFTWARE

Many of the more difficult tasks associated with determining the technical and economic performance of a planned network have been greatly simplified by the availability of powerful network analysis software tools. A typical network is so complex that pencil and paper can no longer do an adequate job. The strategic unit cannot afford to maintain the type of staff that telcos employed to perform engineering economics studies. The software tools discussed allow the design of the network from the ground up, using purchased facilities and those used on a service basis. When the network is operational, many of the tools can help optimize network performance as requirements change. The process of setting up the problem for use by the software package is a useful learning experience for the telecommunication analyst. This is because the proper input data must be assembled and the results reviewed for reasonableness. The telecommunication analyst must understand the overall operation and use of the network to get value from these software tools.

In this section, we review the categories of software to acquaint the reader with the current possibilities. We also review two specific packages, representing different approaches, which may both be needed to perform a study. These examples of software packages on the market are neither better nor worse than others that can be employed. Readers are encouraged to evaluate the offerings of various vendors before making a commitment to a particular approach.

6.5.1 General Classes of Software

Software packages are available to run either on a personal computer or a much larger machine such as a minicomputer or a mainframe. The advantage of the PC version is low cost of hardware, and the cost of the software package is usually also lower. Because of limitations on storage capacity and execution speed of PCs, the package may only provide a fraction of the capability of the mainframe version. In many small and medium sized networks, however, the PC version can be entirely adequate. The version for a minicomputer or mainframe could handle the largest size of network and would run the program as fast as possible.

There are two general classes of network analysis software packages: those based on computer simulation models and those based on mathematical analytical models. Simulation models employ Monte Carlo techniques, wherein events (calls or messages) are actually created randomly within the computer. These events exercise the network elements, and statistics are collected throughout the simulation period. The trick here is to have a proper definition of the network elements (switching capabilities, delays, *et cetera*) and to understand the characteristics of the expected traffic. The mathematical analytical modeling approach uses formulas and tables derived from queuing theory. An example of such a derived relationship is the Erlang B formula, and the data presented in Figures 6.3 and 6.10. As mentioned, the selection of either class will depend on the type of problem being investigated. Most packages that emphasize the economic aspects of the network, working with tariffs and facilities costs, use the mathematical approach to reduce demands on the computer system. Simulation packages are essentially technical in nature, allowing the user to determine the traffic-carrying capacity or delay of a network. Simulations tend to employ the full capacity of the computer in the process of generating events and statistics, leaving little room for other elaborate functions. In such cases, the economics would be evaluated by another approach, possibly using the other class of software.

Several packages are on the market in both categories, although most emphasis is on the mathematical-economic approach. (A current listing and comparison can be found in [Van Norman, 1988].) While many examples are available on the commercial market for a variety of computers, some of the systems are only made available on a consulting basis. We will consider here two commercially available packages, representing both general categories.

6.5.2 Communication Network Simulation (COMNET)

The Communications Network Simulation (COMNET) package from CACI, Inc., of La Jolla, California, is a performance analysis tool for telecommunication networks. Based on a description of a network and its routing algorithms, COMNET simulates the operation of the network and provides measures of network performance. The package is user-friendly so that he or she need not understand the internal structure of the software and its programming. The necessary network descriptions are entered by the user with a menu-driven screen editor.

The first release of the program is called COMNET II.5, which refers to the underlying system used to develop the package. There is another more elaborate package called SIMSCRIP II.5, a powerful discrete-event computer simulation language, which is familiar to scientists and engineers involved in computer simulation projects in diverse areas inside and outside of telecommunication. SIMSCRIP has been in existence for two decades and is a well recognized high-level language. Using it to create simulations is an involved process, however, typically beyond the scope of users who are not computer scientists. The availability of the COMNET software with the ability to run it on an AT-class personal computer is indeed remarkable.

The COMNET II.5 package consists of two programs: COMNETIN and COMNET. COMNETIN is the menu-driven screen editor that is used to create and to modify a network description. COMNET is the telecommunication network simulation program written in the SIMSCRIP language, which takes as its primary input the network description created by the user with COMNETIN. The network is defined by the following major categories of information:

- Nodes (switching elements);
- Classes of service (routing and priority);
- Link or circuit groups (trunks);
- Traffic load (voice or data calls, data messages);
- Routing tables (used by nodes);
- Packet switching operation.

With the appropriate description (for which the COMNETIN requires the user to enter input in response to menu questions), COMNET is able to simulate any wide-area, point-to-point network that uses circuit switching, packet switching, or both. Packet switched networks with virtual circuit or datagram operation can be modeled. Alternate routing, least-cost routing, and adaptive shortest-path routing algorithms are built in. A typical simulation literally creates hundreds of calls within the computer as if users were actually employing the network input model. At the conclusion of the run, COMNET produces several reports summarizing network utilization and performance. Throughput results, blocking probabilities, and trunk utilization are measured and summarized for quantitative evaluation of the capability of the specific network configuration that was used as input to the model.

To use the package, we must be capable of adequately specifying the network. The most straightforward type of network is that used in telephone communication, consisting of PBXs and dedicated trunk groups. Of course, this type of network can be analyzed by using the Erlang B approach, but COMNET makes it possible to run and rerun the model after alterations of the configuration have been made. We can also consider some of the detailed properties of the switching nodes, which are rather intractable with the mathematics of queuing theory.

The package excels when considering packet switched networks with various types of nodes and interconnecting trunks, and with different packet processing times. As we discussed previously, data communication network traffic analysis is still limited in its effectiveness. Particularly, the formulas derived from queuing theory have many approximations, and usually do not yield accurate results for typical situations. If the data messages can be specified in any way, simulation is probably the best hope for predicting results with reasonable accuracy.

The COMNET package allows the user to experiment with packet and node characteristics, data rates, and quantity of transmission links. Also, protocols can be tested at the link and network levels. Any of these characteristics that are known in advance can be used as input data in the modeling process. This capability is a significant step forward in the use of scientific method in data communication network design.

Much can be learned about the characteristics of telecommunication networks by using the package. There is the ability to display the network graphically and to observe the passage of calls and packets through the nodes and trunks. This is called the "animation" mode. On a color monitor, animation provides a clear way to see what happens in the network as calls are processed. An example of this animation display is shown in Figure 6.16. Blockage is displayed on heavily used trunks (i.e., which are undersized for the demand). This feature permits the user to see the operation which is otherwise invisible when examining the output in the form of printed reports. One of the interesting properties of discrete time simulations such as COMNET is that the model must be initialized with calls ("events") entering the network for a while, until a steady state is reached. Afterward, the statistics can be gathered, representing the period of actual measurement of performance.

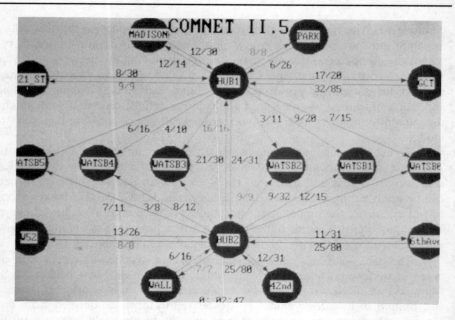

Figure 6.16 COMNET II.5 video display showing animation of a network simulation.

The COMNET II.5 package represents a step forward in telecommunication network analysis. The package will prove useful for predicting network throughput where actual data are not available. Conversely, the simulation can be calibrated against a real network, making the results of subsequent studies more credible. The output is in terms of technical performance (i.e., time delay, throughput, blocking probability, *et cetera*). Economic aspects of networks are evaluated with other tools, such as the MIND series discussed in the next section.

6.5.3 Modular Interactive Network Designer (MIND)

Contel Business Networks of Great Neck, New York, offers the MIND series of network design and accounting tools, employing detailed databases of costs and supporting mathematical analytical models. With the MIND series, the user can generate and analyze alternative network designs, apply various tariffs, determine costs, and obtain quantitative answers to specific questions. Another company which offers similar services is Telco Research, Nashville, Tennessee. Similarly, vendors of network equipment such as Hughes Network Systems and Timeplex provide software packages tailored for their particular market and architecture.

The strength of the MIND series is in the completeness of the cost database, comprising the tariffed offerings of leading long-distance carriers, local exchange carriers such as telcos, and specialized carriers including satellite operators. The tools can be run on PCs, mainframes, and minicomputers, and Contel also makes a significant portion of the software available through a dial-up service. The network design tools employ menu-driven and graphic interfaces on the video display terminal or PC. The specific packages are named MIND DATA, MIND VOICE, MIND PRICER, and MIND PACKET. Most of the emphasis is on data communication networks using packet switched, circuit switched, and private-line service offerings of common carriers. Studies of telephone networks can also be performed by accessing the appropriate product and database.

6.5.3.1 MIND DATA

A set of software packages under the general name of MIND DATA provides an interactive system for design, analysis and maintenance of multipoint private-line data networks. The type of architecture assumed is host-based, where remote points are connected via point-to-point links and multidrop data lines. MIND DATA automatically creates least-cost layouts for centralized networks, such as SNA, and considers operation under varying conditions. Various computerized network databases are created to facilitate continued maintenance of network configuration and cost accounting. While the program runs to full capacity on a mainframe, a PC version of the package is available.

MIND DATA is divided into two subgroups: Multipoint Line Simulator (MLSS) and Topological Design (TOPO). The purpose of MLSS is to allow the study of response time and throughput performance of the network. Data communication traffic loading, line protocol, and other operating characteristics are modeled so that response time can be computed. The model allows the software to predict response time in the absence of this information from an actual network in operation. If such data are available, they can be used as predefined input. TOPO is used to configure least-cost networks using the centralized architecture. The user can consider alternative topology, conduct network capacity planning studies, and engage in "what if" exercises in support of telecommunication budget analysis and management.

MIND DATA includes current tariff libraries to evaluate alternatives for the private-line links which establish network connections. The system contains AT&T interstate tariffs, BOC and independent telco interstate special access tariffs, and BOC and AT&T intrastate private-line and access tariffs. Tariffs include voice-grade analog private lines, DDS service offerings, and T1 transmission services. These data can be accessed through the time-sharing service of Contel, or input can be loaded directly into the user's mainframe computer if justified.

6.5.3.2 MIND VOICE

MIND VOICE is a software-based system used to analyze telephone traffic originating at a single location and to design the line configuration that will carry this traffic for the lowest possible cost. The user selects the desired grade of service as defined for voice networks. The package evaluates bulk discount offerings for switched services such as WATS, as well as various types of private lines. A variety of common carriers is contained in the system and tariff database.

6.5.3.3 MIND PRICER

The MIND PRICER is available as an on-line service from Contel's host computer. This on-line service is particularly advantageous because the database is kept up to date without the user being concerned about changes in tariffs, which occur at almost any time, as discussed previously. The service provides summary data and detailed cost reports covering interstate tariffs for AT&T, MCI, US Sprint, and other major carriers. Also available are special access tariffs and intrastate tariffs for voice-grade private lines, DDS, and T1 services. The information is generated on the basis of area codes and local exchanges for use in point-to-point and multidrop lines. Access to Contel's host is on a dial-up basis via the Telenet PDN.

A user may purchase a copy of the database to be loaded into a privately owned mainframe computer. Called the MIND Tariff Database, the computer-readable package provides the same tariff information that is described in the previous paragraph. Since the data will become stale due to changes in competitive positions and prices of vendors, updates are provided at three-month intervals. Obviously, a user must determine if the cost of purchasing the MIND Tariff Database and use of a mainframe are justified in comparison with simply using MIND PRICER on a dial-up basis from Contel.

6.5.3.4 MIND PACKET

Host-based and peer-to-peer packet networks can be analyzed by using the MIND PACKET package on an in-house computer. This package operates at the network level, considering the nodes and configured transmission links. MIND DATA is used to analyze the particular point-to-point links and multidrop lines which are components of the overall network. MIND PACKET, however, is used as an aid to derive a low-cost network configuration in terms of topology and line capacity assignment to meet user requirements. Extensive use of computer graphics provides visual displays of the network topology and presents curves and bar charts which depict throughput and delay performance. The system uses mathematical analytical models to determine these performance parameters for particular links. Through an iterative process, a minimum configuration is determined in terms of the number and deployment of trunks.

6.6 PARTICULAR ECONOMIC STUDIES

The concepts, theory, analytical techniques, and available software provide the user with the tools needed to consider the monetary side of network development and operation. We will review the types of problems and studies that typically use economic principles. The following is only a sampling, and it is not to be considered exhaustive. We provide an explanation for each type and a methodology for attacking the problem. The actual financial study of specific examples is left to the reader as an exercise. Perhaps the reader has a few real problems which probably provide the best possible learning experience as opposed to simply looking at numerical examples.

6.6.1 Lease *versus* Buy

Telecommunication managers are repeatedly faced with the choice of purchasing facilities for internal operation *versus* paying for the use of the facility from another entity which will operate it. The economics of making a purchase was discussed previously, where investment and annual costs are evaluated. A lease is only an annual (or, more typically, monthly) cost, which the user carries for some predetermined period of time. The equivalence of purchase and lease comes while examining the useful life for depreciation of the purchased facility as compared to the term of the lease. If the same time period can be used in the comparison, all we need to compute for the investment case is the equivalent annual cost. If the cost of the lease is lower than the annual cost of the invested facility, the lease is economically preferable. The investment is preferable, however, if its equivalent annual cost is lower than the lease rate.

The real challenge of doing a lease *versus* buy study is to include all of the costs associated with each. The case when buying must include all annual expenses associated with the investment. In particular, the facility must be maintained by the user, which involves either internal staff or the use of a third party maintenance organization. Utilities must often be paid, building space provided, and other expenses such as insurance can be incurred when buying a facility. A lease may not include all operating costs. You can see that proper accounting for all elements is crucial to this type of comparison because we must be sure that it is on an "apples-to-apples" basis.

Throughout this book we have discussed various reasons for building a private telecommunication network. Possibly leasing may diminish some of the benefits which the network is expected to deliver. For example, a leased facility probably cannot be redeployed in the event of a change in requirements. This means that the strategic unit may be forced to pay for facilities which it does not need. A flexible lease arrangement with convenient termination provisions, however, will allow the organization to "get out from under" the burden of paying for the facility before it has reached the end of its useful life. These factors should be considered as part of the economic comparison of lease *versus* buy. For a very specialized telecommunication capability such as teleconferencing or mobile position location, leasing the necessary facilities for a limited term may not be possible.

6.6.2 Bulk Commitment *versus* Piecemeal

Volume discounts exist in telecommunication as in other fields of business. Therefore, a strategic unit can achieve economies by making a bulk commitment for telecommunication facilities or services. This reduces the vendor's risk and allows it, in turn, to make bulk commitments. The telecommunication buyer could obtain the facilities, or their use under a lease or service contract, at a price which is very close to true cost. An important benefit to the strategic unit is that ample telecommunication capacity will be available at all times over the intended period, which could be five or more years, even if capacity for telecommunication service becomes very tight for the general public.

Purchasing only that increment of service which is actually needed at one time, however, minimizes up-front capital expenditure and the risk of overbuying. A supplier which must sell on a piecemeal basis would not receive full compensation at the beginning of life of the facility. The appropriate manner to price on a piecemeal basis would be to increase the margin of price over cost to allow for fallow capacity while use is growing. We generally refer to this situation as a "ramp up" of customer use and obligation. The availability of fallow capacity presents an opportunity to users with temporary requirements for services over the vendor's facilities. Obviously, the vendor would want to fill the capacity, but at the right price. In the absence of such commitment, the vendor may be willing to accept any revenue, even at rates below cost. This is much like the spot market in commodities such as oil and gas. Nonetheless, if capacity becomes tight, spot market prices can go sky high. The implication of this is that the user may have to pay a substantial premium just to "stay on the air" during the period of undercapacity.

6.6.3 Reconfiguration of the Network

The whole field of digital integration using T1 and other technologies has as its purpose the combining of various needs into a common facility. This would achieve an economy of scale for a reduction in overall telecommunication expenditure. Ample bandwidth and switching capability provides for smooth operation with minimum delay of call setup or message transmission time. Programmable devices, such as packet nodes and intelligent multiplexers, deliver flexibility in terms of the ability to add features and capacity as needed. Reconfiguration of this type of network is easy. Many changes can be made by remote control from the NOC without a physical change or local operator assistance. From an economic standpoint, the strategic unit will have this flexibility without experiencing substantial costs when changes are to be made.

In the more common, less flexible network, any change is a major undertaking, bringing with it great expense and hardship. Links and nodes to be reconfigured may need to be shut down or have their capacity impaired during the transition. (This is like what happens when a freeway is being expanded or new roads are added.) All traffic is slowed or in some cases stopped. An operating telecommunication facility which is filled to capacity may need to be kept in operation as new capability is added. Perhaps the old facility cannot be taken down because important traffic is being passed. Too often, the new facility is added to carry new services, and the old one continues as a relic of the past. What we then have is a hodgepodge of networks with different purposes and performance. Costs are added incrementally and no economy of scale is achieved. The telecommunication manager is always in the mode of "putting out fires" and not in a position to plan for gradual expansion.

As we have discussed, the modern digital telecommunication network is the way to control cost and to provide flexibility for future changes and expansion. The big challenge is to be able to select the proper technology and to make the right level of initial commitment to capacity. Typically, the all-digital network can be put into service at a cost which is no greater than that of the predecessor system. Of course, a detailed cost comparison should be performed to select the proper topology and capacity. The facilities should be designed with an eye toward expansion at a later date. If this can be done, and it is possible, more economic benefits will accrue to the strategic unit.

6.6.4 Retiring a Network

If a network is operational at the time that a major overhaul is planned, the strategic unit must plan for a phased retirement of the capabilities that it no longer needs. The economics of network retirement are determined by how the system was acquired in the first place. An ideal situation exists if the network is an assembly of service offerings from common carriers. We simply give notice that the facilities are no longer needed. Month-to-month leases permit the user to terminate the use with little or no financial penalty.

While such termination is possible with many local and long-distance services as well as standard private lines, dedicated facilities normally involve termination liabilities and costs when shut down. Long-term leases are what the name implies: the lessor would typically be entitled to a termination payment. In the worst case, this payment would be equal to the sum of the rest of the payments under the lease. A more fortunate situation for the lessee is where the payments are capped at, for example, six months of liability.

When the facilities are owned by the strategic unit, shutting them down involves a potential economic loss if other usage is not possible. Therefore, equipment that is used on a temporary basis should be movable. Buildings and cable facilities, however, are difficult to redeploy. Several years of usable life remaining for an idle facility imposes an economic write-off of the residual value. The write-off is computed by taking the cost of the facility minus the part that has already been depreciated on a yearly basis. Also referred to as the "book cost," the remaining value will be lost if the facility cannot be sold or used.

6.6.5 Switching Costs

The subject of network retirement is germane to this discussion of the cost of switching from one network situation to another. In this case, the existing capability is to be retired as a new one is purchased. The economic study would consider the loss associated with retirement as well as the cost of implementation of the new network. Sometimes, the cost of switching is so high that a change cannot be justified on an economic basis. In the marketing of telecommunication equipment and services, vendors understand that a user requires assistance when migrating a network from old technology to new. This assistance can take many forms. For example, temporary equipment can be lent to the user to support the network while a major overhaul is underway. The vendor would see the cost of assistance as being small in comparison to the projected revenues to be obtained from the user community. The user, conversely, has difficulties justifying a major change in service arrangement unless the transition is smooth, in terms of both continuity of service and economics.

Chapter 7
Buying Hardware and Services

The previous chapters covered applications, technology, and economics. Now, we deal with the real-world task of actually going into the marketplace and procuring the constituents of the network. We consider the ordering of the equipment components which the strategic unit will own and operate itself. We also consider the facilities which will be leased from others or paid for as a service (i.e., from telcos and long-distance carriers). Along with the equipment and facilities, we discuss the organization of the people within the strategic unit upon whom will fall the responsibility for operating the network.

7.1 MANAGEMENT OF IMPLEMENTATION AND OPERATION

The private telecommunication network delivers communication services which a strategic unit employs in its day-to-day affairs. No organization installs a network for the sheer accomplishment of it. Rather, it is a means to an end, or even a necessity. The network, however, must be assembled in an intelligent and professional way. Once it is up and running, there is a requirement to keep the network in working order. From time to time, changes in configuration are needed to respond to constantly changing requirements. This particular issue is often foremost in the minds of telecommunication managers. Therefore, implementation and operation are vital network functions. These rely upon experienced people, rather than hardware and software.

7.1.1 Implementation

Implementation is the technical side of telecommunication and the process places tough demands on an organization. Historically, these functions were delegated to the telcos. The new environment of divestiture and deregulation, however, presents the opportunity and challenge of performing these function within

---

the strategic unit. In Chapter 6, we presented techniques for configuring a network to meet specific needs and requirements. Implementation is simply the process of actually constructing the network using this design as a framework. Some facilities will be owned by the unit; other portions will be leased or taken as a service. The split between owned *versus* leased may be a financial decision, but it has ramifications for the implementation phase and the operation phase as well.

The division of responsibility for network implementation among various elements of the organization is illustrated in Figure 7.1. This is a conceptual drawing which identifies key groups that play an important role in the implementation phase. At the center is the core technical capability, consisting of a planning and engineering group within the company. Outside consultants could be employed by companies without a consistent demand for such ability. The first job of the core technical capability would be to define the network, which was outlined in the previous chapter. This consists of estimating user demand, setting traffic requirements, engineering the network to establish the overall configuration, and fixing the network topology (see Figure 6.1). Although other groups begin to perform more of the tasks, the core technical group supports procurement, and then stays in the picture as the network is implemented. The organization and functions of the core technical capability are elaborated later in this chapter.

Moving outward in Figure 7.1, the defined network is processed for the buying of the associated facilities and services. Later in this chapter, we will describe the *request for proposal* (RFP) process. The actual buying is carried out by the *administration-materiel* group. This name is a combination of two important functions. Administration is a set of generic functions within a strategic unit which follows through on some of the more routine tasks. As we mentioned previously, telecommunication is often an administrative function. Here, we refer to other facilitating functions, including personnel (people may be hired), finance (budgets must be established and checks have to be written), training (new people need to be trained up to speed and those already on the job may require retraining for the new facilities), and legal (contracts are to be written and negotiated).

The materiel group is crucially involved in network implementation. Another name for materiel is *purchasing,* wherein purchase orders which include specifications for facilities and services are released. Often, the other administrative functions as well as the core technical group support the materiel group so that the procurement process is conducted intelligently and effectively. The procurement officer within the materiel group is the interface between the strategic unit and the vendors. The potential benefit of using purchasing professionals derives from the continuity and integrity which they can bring. Of course, abuses by procurement officers have received much attention in the press, but a properly managed staff will usually result in significant cost savings from the running of a competitive procurement. A rule of thumb is that a procurement group should be able to save at least 10%. The strategic unit will also tend to get the most from the vendors as they push their capabilities in order to win the job.

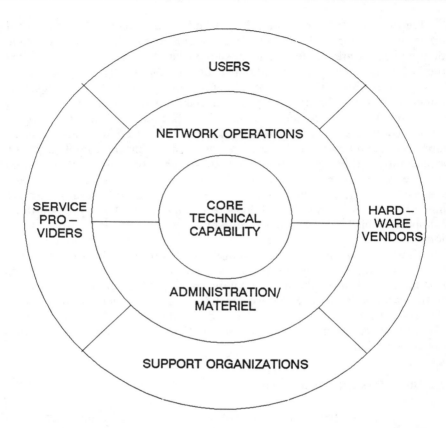

Figure 7.1 The division of responsibility in telecommunication procurement.

The network operations group, which we assume includes the maintenance function, will ultimately assume responsibility for running and maintaining the network. This group would likely overlap the core technical capability, but we can fairly say that network operation is different from engineering. An appropriate blend of capability is needed during each phase of implementation because the view taken by each group is necessarily different. For example, the engineers adhere to time schedules for implementation and require that technical specifications be met. Vendors are also expected to complete the project within the previously negotiated price. The operations side is more concerned that the facilities be maintained in working order with a minimum of downtime. While the network must work as designed, the operations group may be willing to compromise technical performance for reliability. You can see that it is important to consider the two sides properly; a network for which this is done effectively will provide more service.

On the outer edge of Figure 7.1 we find the elements that come into direct contact with the facilities which the inner elements are developing. Service providers and hardware vendors actually supply the elements of the network. Construction of a major network without involving several of these organizations is virtually impossible. Therefore, coordination among them is vital to success. The strategic unit needs a project or program management group, which usually operates from the core technical element. The facilities and services supplied by vendors would meet appropriate specifications, be installed according to an appropriate schedule, and undergo a test program to ensure compliance with the specifications. Other support organizations (shown at the bottom of the outer ring) would be brought in at various times to provide necessary assistance. These could be specialists from consulting companies doing such tasks as radio licensing, architecture and engineering for buildings, and equipment calibration. Another form of support is third party maintenance, which comes after the network is in operation.

The last component of the responsibility picture contains the ultimate users of the network. They, too, are important to a successful implementation. Proper coordination will make the users a willing participant. Users need to understand that the introduction of a new network will involve transition from what is available right now to something that will be new and, ideally, better. If users can be involved early with devices such as surveys and preimplementation trials, the transition will tend to be smoother.

7.1.2 Operations and Maintenance

The ongoing operation of a network can be viewed as more administrative in nature, using people who are good at keeping things working. On the maintenance side, technical support is needed to deal with problems, some of which will demand a thorough understanding of the underlying technologies. Maintenance engineering is therefore a highly technical skill, requiring advanced education, possibly at the university level. Operations is really a different discipline from engineering. It deals with the function of a facility, rather than its inner workings. Take the analogy of an automobile. The driver-operator needs a basic understanding of running and controlling the vehicle, with perhaps some knowledge of maintenance procedures (refilling the tank, washing the vehicle, and when to take it in for routine maintenance). The operator needs to know very little about how a car works. Of course, some operators take pride in understanding about the performance characteristics and inner design, but these details are not necessary in normal operation. The same applies for telecommunication facilities.

We discuss operations here, in this chapter on buying network facilities, because of the importance of considering the operation of the network when first putting it in place. Too often, the engineers and buyers of the network think about the initial cost of the network and its technical performance. When the network goes into operation, the environment is completely different from construction. A common problem with operations people is that they have some difficulty quantifying their needs in terms of operability and maintainability. Therefore, the engineering people need to work closely with their counterparts in operations during the network design and implementation phase.

In Chapter 5, we reviewed network management, which comprises several aspects of network operations and maintenance. In particular, we presented the case for integrated network management using specialized facilities that permit efficient control of equipment at network locations. The network management capability allows the operations side of the house to provide services in response to user needs, even as they change. Employing the video display terminals at the network operations center, operations personnel are the point of interface for users to report trouble on the network. Also, the monitor and control system enables operations personnel to respond quickly to problems on the network, thereby reducing or eliminating downtime. The number of people needed to operate the NOC is inversely related to the sophistication of the network management capability, which has been designed into the network during the implementation phase. This is where the marriage of operations and engineering can prove vital, as previously discussed. The expense of supporting customer service in the field, however, is directly proportional to the size of the user population.

Maintenance of network facilities ties in with operations as an ongoing necessity. Technicians should be available to trouble shoot problems and to repair failures. To support technical personnel, they need a stock of tools and repair parts. These must be deployed around the network to reduce the delay when problems need to be corrected. Very often, the maintenance function is subcontracted to a third party maintenance organization. A sufficiently large network can justify building an internal maintenance staff. Organizations such as common carriers, public utilities, and airlines, can afford to have extensive facilities around the country and employ an in-house maintenance staff.

Another area is the repair of equipment which has been removed from an operating facility. Typically, failed units can be returned to the original vendor for repair or replacement. This simplifies matters, but could entail delays. Also, vendors have a tendency to lose interest in what they consider to be obsolete product lines, even though the user is still getting value out of the devices. A depot maintenance facility as part of the organization could be justified in such cases. This topic is hotly debated because some organizations have not had a lot of success with the depot concept. If the in-house staff has the technical ability to repair network equipment, however, and if there is enough usage of the staff to be economically justified, a depot represents a potentially good investment. A depot would be a necessity if the facilities were highly specialized, meaning that no outside organization was familiar enough to perform the repair role.

The previous discussion should give the reader a feel for the need for operations and maintenance, which are obtained either from internal sources or from outside vendors. A strategic unit can probably not satisfy all requirements from internal sources. Later, we will discuss some of the ways that these services can be obtained from the outside.

7.2 SOURCES OF EQUIPMENT FOR PURCHASE

The U.S. economy in general and the telecommunication equipment marketplace in particular are incredibly diverse. Literally thousands of manufacturing companies make products for the telecommunication market with some of the largest corporations having a substantial stake in that business. These firms include such domestic giants as IBM and AT&T. In Japan, we have the trading companies like Sumitomo, with NEC; Mitsubishi, with MELCO; and several other company names that have become household words. In Europe, there is Alcatel, Siemens, and Philips. The foregoing is but a sampling of the major corporations. These companies produce most, but not all, of the technology which comprises modern digital networks.

Some of the most useful and interesting technology is made available by companies which did not even exist five to ten years ago. Another important factor is the manner in which these technologies are combined or orchestrated to create a sophisticated network resource. This is the role of the system implementer and architect. We will review the types of companies which develop, manufacture, and market the basic and more advanced equipment. One of the most difficult problems for the telecommunication manager is to keep aware of the new developments, and then to be able to resolve which are useful and appropriate.

Classification of equipment suppliers according to size and diversity of product lines is shown in Figure 7.2. This concept provides a framework for the rest of our discussion on sources of equipment. Across the top we indicate broad categories of application of telecommunication facilities. Voice and facsimile are on the left side of the range, employing conventional telephone networks. As we move toward the right, we encounter more specialized and demanding applications. Data transmission can be handled like voice, but data networking is an independent universe. Finally, video applications for strategic units are represented at the extreme right of the figure because of the large bandwidths and specialized facilities that are necessary.

Major corporations involved in telecommunication equipment manufacturing (discussed in the next subsection), have entered essentially all of the markets. Nonetheless, there are large manufacturing corporations which are important suppliers of equipment within large market segments. At a more modest level, many moderately sized companies have established themselves as viable suppliers of

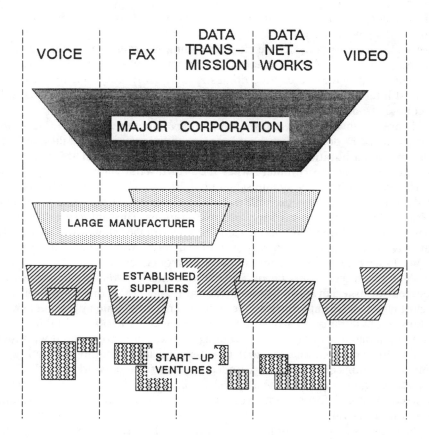

Figure 7.2 Classification of equipment suppliers according to size and diversity of product lines.

specific equipment or software. Finally, start-up ventures pursue niches, where new technology can be used to solve specific problems. These various classifications are discussed in the following paragraphs.

7.2.1 Major Corporations

Major telecommunication corporations in North America, Japan, and Europe are highly visible in the world, having built up their reputations over several decades. Each found a major market for their products, principally in the public telephone and data networks of their respective countries. With this kind of scale, such corporations could afford to develop the technology and manufacturing capability needed to pursue *megamarkets*. The corporate giants also have the financial might to deal with the rapid changes now being experienced. In particular, the major Japanese corporations are tied to leading international trading companies with affiliated banks. The Europeans often have the full support of their respective governments, providing a captive domestic market. In addition, the government will assist international marketing by providing low-cost financing, free consulting, and counter-trade provisions.

There are two big advantages for the developer of a private telecommunication network in dealing with a major corporation: confidence and comfort. All of the major corporations have the technology base, financial strength, and devotion to reputation to complete what they start. They have the wherewithal to fix mistakes, despite unavoidable schedule delays, cost overruns, and approaches that do not work. There have been instances when a large installation did not perform to the customer's and supplier's expectations. The faulty system was simply removed and replaced with more established products at no cost to the buyer. Generally speaking, a major corporation will not "walk away" from a failure without doing everything possible to correct errors.

A major corporation probably produces most of the elements needed to implement a large network. Starting with wire and cable, the major corporation can supply a wide variety of subscriber equipment, telephone and packet switches, transmission systems and earth stations, and even construction of buildings. When unfamiliar capabilities are needed, teaming relationships are formed with another experienced major corporation that has the requisite technology or service. For example, a major corporation experienced in terrestrial networks could team with an aerospace company to pursue a hybrid satellite-terrestrial project, or a hardware manufacturer could form a joint venture with a computer software developer to design a more advanced product line.

We have already mentioned IBM and AT&T as two of the major manufacturers of telecommunication equipment. The three "generals," GE, GM, and GTE, also have important positions in the telecommunication equipment manufacturing business. These corporations offer a range of products for the transmission and switching of both voice and data services. IBM is clearly the leader in computing and the associated data communication area. IBM's purchase of Rolm and marketing arrangements with companies like Network Equipment Technologies (NET) give them weight in the marketplace for integrated voice and data networks. IBM is still not a place for "one-stop shopping" where a strategic unit can have all of its needs met on an economical basis. IBM, however, certainly stands to gain with the expansion of all types of telecommunication networks simply because they have a variety of attractive products to be sold individually or as an integrated system.

AT&T is the nation's principal long-distance company and a major manufacturer of a wide variety of telecommunication equipment through the AT&T

Technologies group. The former Western Electric Company base gives AT&T credibility as an important equipment manufacturer and system integrator. AT&T continues to have the largest market share in central office switches, multiplexers, and radio equipment. Prior to divestiture, Western Electric was the primary manufacturer for the BOCs. The situation today is different because the BOCs run competitive procurements to reduce their cost bases. AT&T is virtually one-stop shopping for the telcos, although there are some gaps, primarily due to new technologies introduced by more specialized suppliers. AT&T's cost of production and distribution being higher than the overall market tends to make the company less competitive in many instances. AT&T, however, has been responding to competitive pressures by pursuing international markets, where major buyers in the Middle East appreciated being able to procure an entire telecommunication system engineered by the world's best known supplier. Domestically, several of AT&T's products enjoy a loyal following from BOCs and major U.S. corporations. Customers who see benefits from focusing on long-term aspects of quality and confidence first and price second are generally attracted to AT&T's product lines.

As we mentioned in Chapter 3, AT&T also continues to be the leader in key systems, although the competition from Japanese manufacturers is hurting their market share. In the PBX area, AT&T has been eclipsed by Northern Telecom, Inc. (NTI) of Canada. Computers represent an area where AT&T has made a series of strong pushes, and some of its networking products are gaining acceptance in certain segments. AT&T now teams with information processing organizations like Electronic Data Systems (EDS) and Computer Sciences Corporation (CSC) to pursue large integrated data network projects.

The major corporations in the U.S., Japan, and Europe have an important advantage over other types of organizations. The major corporations have the financial and technical resources to create new technology, to bring it into production, and then to market it without much help from the outside world. A lot has been said and written about a lack of creativity and that major corporations have their bureaucratic tendencies. Nevertheless, they continue to innovate within a number of areas which demand their kind of capacity and experience. IBM introduced SNA and its associated products, and later caused the microcomputer market to explode with their PCs. Likewise, AT&T has brought digital communication to the point where it has become the standard of switching and transmission for public and private networks. The major corporations have fostered the development of new products by purchasing smaller companies and providing the support needed to expand the market and to keep the technology competitive.

There is an old adage that nobody ever lost his or her job recommending a purchase from IBM or AT&T. The products and services of these major corporations will usually do the job that is expected. Again, the buyer need not be concerned if the company will be around during the implementation phase and through the life of the facility. Always going to these sources can be easy (and expensive), but it raises the concern that developments made by other companies may be overlooked. The choice depends on the needs of the strategic unit. Each time a major expansion is planned, the best options could and should be examined. Certainly, the *Big Blue* or *Ma Bell* may have the appropriate solution to the problem, in which case the business is rightfully theirs. Interestingly, the major corporations are in the best position to go after the very largest jobs, such as providing the Department of Defense or a "Fortune 10" corporation with a network. The largest buyers tend to prefer a major corporation because of the standardization that it forces across the network. As with military organizations such as the U.S. Army, this standardization simplifies training and reassignment of personnel around the network and the country. The successful bidder for a meganetwork must have the financial and technical resources to complete the job.

On the other end of the scale, smaller companies may need to depend on the major corporations to satisfy their needs. The basis for this is that a smaller buyer does not have an in-house capability to perform the necessary network engineering and implementation. AT&T and IBM both maintain direct sales forces, which support small and medium sized business customers. Appropriately sized products, such as key systems and minicomputers, can interface well with public network offerings of AT&T and other carriers. Thus, the small buyer would tend to develop economical strategies for buying appropriate equipment from the major corporations.

7.2.2 Large Manufacturers

Strategic units with substantial needs can, however, benefit from implementing networks on their own. They would select equipment from a variety of sources, including the major corporations. Next, we will review other categories of equipment suppliers which deal in more specific technologies and capabilities; there are gaps in the product lines of the major corporations. Due to the diversity and strength of the economy, there continues to be innovation by the other manufacturers that produce sophisticated equipment, which even AT&T and the BOCs purchase for their networks.

Large manufacturers have many of the properties of the major corporations; they do not, however, present themselves to the marketplace as providing one-stop shopping. The product lines are focused on a specific application. Rockwell (formerly Collins Radio) is perhaps the leading supplier of transmission equipment for microwave radio and fiber optic cables. While not the low price leader, Rockwell does have the time-tempered "built like a battleship" reputation of quality and product support. At one time, Collins attempted to become a one-stop shopping source, which offered transmission, switching, and computer systems. Due to domestic and international competition, however, Rockwell has refocused on its core strengths.

Large switches and PBXs are usually designed to do conventional things in telephone applications. If specifications were simple, brief, and easy to understand, there would be little problem in considering a variety of sources. The reality is that switch procurement is second only in difficulty to computer system procurement. This is compounded by users' desire to have ISDN capabilities built in or available in the future. We are left with claims of compatibility from the vendors. This tends to force users to rely upon the reputation of the switch supplier, rather than pure specifications. An exception to this situation is that of the telcos, which have the time and staff to make a concerted effort to test and to evaluate a variety of switches before committing to a particular design or supplier. Most strategic units do not have this luxury.

Northern Telecom, Inc. (NTI) is an important manufacturer and a leader in switching systems. In many ways, NTI picked up where the old Collins Radio left off. We mentioned that NTI made inroads into AT&T's central office switching business, and has become the number two provider in the domestic market. NTI has several other important product lines which meet a broad range of telco and strategic unit needs. These products include microwave radio, multiplexers, DACS, and fiber optic terminals, repeater systems, and packet switches. With continued, well managed growth, NTI could replace AT&T as the major North American corporation in telecommunication facilities. For the time being, however, NTI is one of the better candidates as a source of equipment supply to the telcos and larger private telecommunication users. On a smaller scale, NTI's Meridian SL-1 PBX line has a large established base in medium sized companies. Interestingly, NTI is one of the broad suppliers with its own packet switch offering. They are building on this experience with their recently introduced Meridian DNS product line.

With respect to data communication, larger manufacturers like Digital Equipment Corporation (DEC) and Hewlett Packard (HP) are important sources of equipment and know-how. They are solid organizations on which users can depend much like the major corporations. Typically, the customer must define the requirements so that a network can be configured. A problem in data communication is that specialists working for the user or buyer are familiar with only a limited range of architectures. This is a common situation with regard to IBM, which is not a particular problem if the strategic unit is committed to IBM for other reasons. Working with systems such as DECnet, system engineers and programmers have gained important practical knowledge on effectively employing the networks, even doing things that the vendor did not intend. Breaking out of one computer vendor and going to another, however, is rather difficult and risky.

As indicated in Figure 7.2, a large manufacturer would have a fairly diverse product line covering a number of categories of network applications. A strategic unit could build a network from the product line offerings of a few large manufacturers. For the selected manufacturers to have already been working with each other would be highly desirable; alternatively, compatibility can be demonstrated during a live test over part of the network. This is where the concept of industry standards plays an important role.

7.2.3 Established Suppliers

Moving down the hierarchy of suppliers in Figure 7.2, we encounter established companies that are active in particular market segments. A buyer rarely needs to worry about one of these companies being able to deliver what it has promised because that particular product is normally available off the shelf. Furthermore, because of an established business base, this type of company normally does not have financial problems which can hamper its performance under a delivery contract. Examples of such companies include the following: Timeplex, which is a subsidiary of Unisys, manufacturers T1 switches and multiplexers; Tandem Computers, which has a strong niche in on-line transaction processing systems; Codex, a subsidiary of Motorola with a strong position in STAT MUX–based data communication networks; Hughes Network Systems, the company that first introduced VSAT networks and consequently has a major market share in that technology; and Farinon, a manufacturer of high quality terrestrial microwave equipment for use in common carrier and private network applications.

Some of the established suppliers protect their positions in the market through proprietary products or software. A clear example of such an established supplier in the microcomputer industry is Apple. Its Macintosh computer is protected by patents and copyrights, making it extremely difficult for other manufacturers to "clone" their products. Any software must be written especially for the Macintosh. Telecommunication is similar in the way that proprietary products and software can play an important role in protecting a company from competition, and for buyers to play one supplier against another becomes difficult. The IDNX from NET is functionally similar to other high-level multiplexers, but the common signaling channel and frame format are unique to NET's products, forcing buyers to stick with the proprietary node equipment. Another established supplier, Timeplex, maintains the standard D4 frame format, and thus can easily interface with standard multiplex and transmission equipment. All of the functionality and options of the IDNX, however, are currently not provided.

To employ the established suppliers effectively, the strategic unit needs a core technical capability to design and to implement the network from various elements. This role could be played by a system integration contractor (as discussed in Chapter 2). The specialized manufacturers simply market their products to the larger company, which arranges for final installation. This is much like the way that a building contractor subcontracts portions of the job and purchases major subsystems such as air conditioning and heating. The company that manufactures and sells, for example, security systems will not normally build houses, but they can provide the security system to the larger contractor. Similarly, a system integrator such as Contel or EDS performs all of the design and procurement integration roles, including equipment selection, integration, and final testing.

7.2.4 Start-Up Ventures

The last classification of equipment supplier is the start-up venture, which develops a new technology and brings it to market. As we mentioned earlier, start-up ventures can introduce revolutionary concepts into an otherwise staid technical environment. Consider the company which first offered radio links in the mid-1960s to interconnect personnel in the field with the local telco. Threatened by telco restriction, the company successfully sought government authority to proceed with its business. This was Carterphone, providing the first impetus for deregulation of the telecommunication industry. Many of the established suppliers and large manufacturers were once start-up ventures. Hewlett-Packard started in a garage, where the founders assembled their first pieces of commercial test equipment. The environment in the U.S. has nurtured the start-up venture, and even today we see a great deal of activity and investment money pursuing new technology in telecommunication.

Start-up ventures are typically formed by experienced professionals, who formerly worked for established companies. Usually there is a unique idea for a market or a technology that may have originated in a laboratory. The core team of technical and business talent obtains the needed start-up funding from one of a variety of sources. Foremost are venture capital firms which pool money from wealthy investors and corporations. The start-up could also be funded by a major corporation that wishes to employ the technology after it has been proved. Of course, a major corporation can build an internal team to pursue the technology, in much the way that IBM developed its highly successful IBM-PC. Many such internal start-ups are run in nonconventional ways, along the lines of a "skunk works" [Peters, 1985].

A potential customer can consider the fact that the start-up venture convinced investors that the particular idea was viable and the team had the wherewithal to succeed. There is substantial risk when dealing with a start-up, however, because the particular item may never actually be brought to market. Also, the product may not perform as advertised. Even if the product is delivered and works as specified, there is the risk that problems encountered in the future will not be solved due to growth pains at the supplier. Start-up ventures more often than not are dissolved through any of a number of means, such as a merger or even bankruptcy. Some of these risks can be mitigated, as we will discuss later in this chapter.

The plus side is that a start-up venture can provide the key element needed to implement a vital network capability. This may give the strategic unit a competitive advantage (as discussed in Chapter 8). Many large corporations which rely upon telecommunication technology have turned critical problems over to new companies and even start-up ventures. In the aerospace industry, for a large project to require the development of a key item, such as a microwave antenna or infrared sensor, by a small and young company is not uncommon. You can imagine the leverage in such a situation, where a system which could be vital to the nation's defense is dependent on the successful delivery of the much smaller item. This scenario is perhaps less common in private telecommunication networks, but there are instances where a small company has precisely what is needed to complete an important network function.

To base an entire network upon the technology and manufacturing capability of a start-up venture would be ill-advised. Figure 7.2 indicates that start-up ventures do not provide complete coverage of the telecommunication picture. The product lines and network options offered by the more established companies, however, have gaps. The start-up venture attempts to fill a gap, and thus to obtain for itself a valuable and possibly secure market. There is potential synergy between the start-up venture and the strategic unit wishing to pursue a new business in the changing environment.

7.3 OBTAINING TELECOMMUNICATION SERVICES

In theory, a private telecommunication network could be implemented with privately owned and operated facilities. This would give the strategic unit essentially total control of resources. The expense of this probably would be prohibitive, however, because of the large number of locations that might need to be reached. (The possible exception is a satellite network.) It becomes essential that a significant portion of the network be obtained on a service basis from one or more outside organizations which specializes in the appropriate field. (This is illustrated in Figure 1.2, which has been reprinted here as Figure 7.3.) The private network sits atop the general digital infrastructure provided by telcos and long-distance companies, but other groups offer additional support capabilities. We review the reasons for doing so, along with the general classifications of such sources in the following paragraphs. (The general discussion of public telecommunication network services can be found in Chapter 2.)

7.3.1 Reasons for Obtaining Services

Perhaps the main reason for purchasing network capabilities on a service basis is to be able to share a much larger network. Except in commercial radio

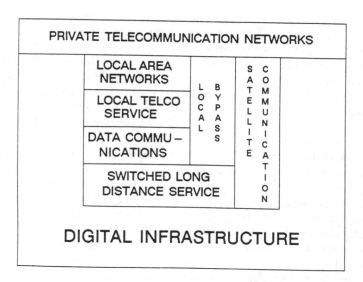

Figure 7.3 The new digital infrastructure and private telecommunication networks.

and television broadcasting, extending a private network to every possible location is really not feasible. This expansion would consider the strategic unit's internal sites as well as locations of customers, suppliers, and others who support the ongoing operations. The public telephone and data networks are established for this purpose. Of course, they are operated for a profit, and we expect to be paying at a rate which provides ample margin above cost (at least, that is the general idea).

Costs of use of the larger network by the strategic unit could still be significantly lower than if the private network were considered as an alternative. This cost saving derives from the economy of scale that is achieved when greater capacity is installed for network nodes and transmission facilities than the unit can justify. The service provider can aggregate many users in the form of the general public, other strategic units, and government agencies for better loading of the facilities. If this aggregation is successful, the service provider can achieve its need for profit margin and the particular strategic unit will get its network capability at a reasonable saving. Recall from Chapter 6 that telecommunication traffic can be loaded more efficiently on facilities with larger capacity, also giving an economy of scale.

The other side of sharing the network is that the strategic unit gives up some control and autonomy. In the past, the manual nature of the network meant that all requests for changes and assistance had to be processed on a person-to-person basis. This led to delays, confusion, and arguing. The advent of digital networks with remote monitor and control capabilities, however, now puts the user back in command, at least for a number of important functions. Virtual private network offerings from long-distance carriers employ specialized programming of switches and databases in central computers. Customer-controlled reconfiguration of T1 networks provides the type of control that a private T1 backbone will give. Another promising approach is ISDN, wherein the D channel accesses the telco's common channel signaling system. The telcos, long-distance companies, and regulatory bodies will implement the types of control that users want and the technology is capable of delivering.

To bring this control together, the strategic unit needs to implement more advance network management systems than are generally found in today's private networks. We reviewed several concepts for network management in Chapter 5. In particular, approaches based on a flexible architecture and a centralized network management computer system (such as NMCC) will play an important role in rationalizing the diverse aspects of the network.

Using experienced telecommunication service organizations has the benefit of allowing the strategic unit to delegate much of the operational responsibility for the network. Remember, implementing and operating a private telecommunication network is a difficult and tiring job. It involves many people working together, around the region or country, and typically around the clock. To be able to delegate the more tedious aspects of network implementation and operation to specialists would be comforting, but we would not want to pay too much for that kind of assistance. Nonetheless, we would do well to be able to call on others to deal with the unavoidable problems as they occurred throughout the network. We have mentioned that operations and maintenance form a significant part of the annual cost of the the network. Also, having experienced people available to support every site where a critical problem can occur is probably prohibitive. Even the largest equipment manufacturers use third party maintenance organizations to assist them in the field with their installed customer base.

Using the facilities of an established and ongoing public network tends to reduce the risk of technological change. What might appear today to be a vital new technology could become tomorrow an expensive albatross. This is a concern in proprietary network technologies, particularly in data networks. Staying with standard products, interfaces, and protocols is usually best. There are instances, however, where proprietary systems are the only means to do a particular job. The service provider may be willing to take the risk of technological obsolescence because that organization can shift the use of the facility to another part of the larger network.

Looking at the issue from another perspective, the operator of the larger network may implement a new technology before it is available on a purchase basis. Long-distance networks are now implementing the signaling system number 7, which will provide many new features for strategic units with heavy use of the public network. The area of Open Network Architecture is intended to give access to the public network for third party information service providers. These companies (videotex, electronic banking, *et cetera*) would find it less difficult to establish their customer bases.

7.4 CONTRACTING WITH MANUFACTURERS AND CARRIERS

Because of the expense and commitment involved in purchasing as well as leasing equipment for the implementation of a private telecommunication network, most relationships between buyer and seller are formalized through a legal contract. Basically, the contract is a document or series of documents which define the obligations of the parties (typically two, but more can enter into it) to each other. The seller is to deliver a certain capability according to a certain schedule and the buyer is obligated to pay certain sums when due. If this were the extent of it, a contract could be written in two or three sentences. Most legal contracts for major procurements, however, are many pages long, some even consisting of several volumes of contract documents.

A legal contract is the normal format for the agreement between buyer and vendor, where the vendor is not affiliated with the buyer. This is the normal way of doing business where a significant commitment is made by the parties to each other. For small jobs and to purchase catalogue items, the buying organization may issue a *purchase order* (PO) to the vendor, authorizing the work and delivery of products or services. The PO includes standard contract *terms and conditions* (T&Cs) of the buyer along with some kind of brief specification or statement of work. By using the standard PO, the acquisition can be done in a relatively short time as most vendors will honor the PO and proceed with the associated effort. The vendor, however, may not accept the PO due to an irregularity in the supporting documentation or a dissatisfaction with some of the T&Cs.

When work is done by one group for another within the same company, the format is usually quite different. While no contractual documentation is needed, the parties must have a clear understanding of the tasks to be performed. This is fulfilled with the work authorization or memorandum of agreement containing basically the same information as a contract, but in abbreviated form.

We will outline the somewhat standard process, whereby the strategic unit enters into a contract to realize a portion of the network. This process considers the definition of the work to be performed, the selection of the vendor, and the negotiation of the actual contract. Those familiar with the government procurement process would recognize these general steps. In our case, the government will take many months, even years, to follow the trail. This slowness is because there are many safeguards built in the process to promote fair competition among suppliers. In the commercial world, there is less concern about being fair and more interest in getting the job done as effectively and economically as possible. Many of the principles used in government procurements, however, will still help realize these objectives. An overview of the process is presented by the flow chart of Figure 7.4.

7.4.1 Generation of System Specifications

Competition can be used to the advantage of the strategic unit if there is sufficient time during the implementation cycle. Like the government, public utilities must use competitive procurements to satisfy their respective regulatory agencies. Major corporations also use competitive procurements to get the best possible deal. To do so properly, the characteristics of the network facilities or services must be spelled out in a written document called a *specification*. This is the role of the systems engineer, who is familiar with the requirements of the network as well as the capabilities of a variety of equipment from a number of sources. A well written specification is an engineer's delight — it takes a lot of know-how and understanding of the requirements and technology to do the job right. The key to a successful specification is to define the facility so that a number of different vendors can do the job, providing that the facility so specified will still perform as required in the network.

Successful specification is relatively easy to do with standardized systems like microwave radio links and D4 channel banks. Several manufacturers and system integrators can provide these facilities. Specifications are currently published by the various suppliers. Within the strategic unit, system engineering personnel develop an essentially standard specification based on those published by competent vendors. Experienced consultants and system integrators maintain libraries of equipment and system specifications for this purpose. This standard is then used in the network design activity (as discussed in the Chapter 6). Subsequently, the standard specifications for all of the various facilities to be purchased are used in the procurement process. The theory is that the vendor which can meet the standard specification at the lowest overall price is most likely to get the job. In the absence of a standard specification, the vendors may submit their own. The internal core technical staff will then have the formidable task of trying to unravel these specifications, which do not often facilitate direct comparisons among the offerings of competitors. As indicated at the top of Figure 7.4, the development of the RFP is an iterative process, where buyer and vendor exchange information useful to each other.

Buyers of telecommunication facilities and services should work hard at defining their requirements in terms of standard specifications. This would encourage vendors to make their most competitive bid for the job. Vendors that do not have the absolute lowest cost of production, however, pursue the strategy of having a unique, proprietary technology which is capable of doing much more than the industry standard. Sometimes, several functions are built into the same box. The buyer could then specify a combined facility, where potential vendors would bid

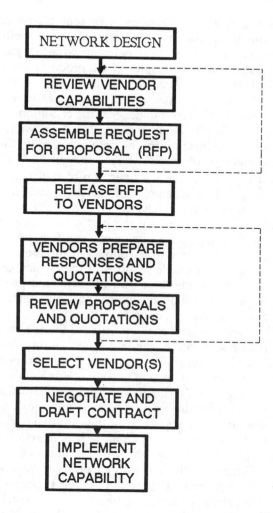

Figure 7.4 The flow of preparatory activities which should proceed the implementation of a major telecommunication capability.

either an integrated unit or a string of standard units which performed the same functions. Another approach taken by vendors is to include "bells and whistles," which are innovative features not found in standard industry products. These features may not necessarily have application in the network. The engineering group must weed out the frills, which are unnecessary to the unit's strategy, so that the best purchases can be made.

The development of specifications often requires additional study and experimentation prior to the issuance of the specification. Vendors may loan equipment to prospective buyers for short or even extended periods of time. Some organizations can afford to buy samples of the products and services available on the market, and then conduct tests akin to laboratory experiments. Real or simulated traffic can be run over a test bed of some type. This testing tends to show the performance in scientific terms under realistic conditions. Historically, this was how the Bell System always evaluated products, which would have been designed to meet the rigorous specifications of the network. Today, the RBHCs are still in a position to use this type of process. Recall that BELLCORE, which now does the specifying for the RBHCs, is a former arm of Bell Labs. Both BELLCORE and the RBHCs typically test samples of switching and transmission equipment before making a network-wide purchase.

The specification drafted before the particular vendor has been selected is usually modified prior to the actual signing of a contract, which allows the strategic unit to add to the specification during contract negotiation. The reason to do this is to make sure that any desirable features of the particular facility design from the selected vendor are incorporated into the delivered product. Otherwise, the vendor might "forget" to include capabilities that had been proposed, but which were not part of the original specification.

There will be features and characteristics of the facility or service which are not amenable to inclusion in the standard specification. Very difficult to specify are such characteristics as reliability, maintainability, and operability (collectively called the *ilities*). This does not prevent buyers such as federal government agencies and major corporations from trying to specify these characteristics in quantitative terms. Nevertheless, while numbers can be invented, they do not always relate to the real world. The *ilities* are important to the eventual operation of the network and so they must be considered. One approach is to require that the vendor be involved in the actual operational phase of the network through a warranty or service contract. Otherwise, dealing with experienced vendors who have a track record of successful delivery and operation of their products is usually best.

There are basically two schools of thought as to the complexity of technical specifications. First, there is the system specification, which defines the functional requirements for the facility or device to be purchased. This type of document is usually fairly easy to assemble because it is truly a direct statement of what the buyer is trying to accomplish. The system specification goes point-by-point through each feature desired, giving only the detail needed to complete the definition of the function and intended use. System specifications are usually readable, defining terms as it proceeds and providing only the numerical specifications that are meaningful in the context of the application.

The other type of specification is called a *detailed specification* or *unit specification*. Here, the characteristics of each significant element of the facility or device are defined in detail. This type of information is usually found in specification sheets provided by vendors. The documentation probably can only be read by an engineer who is already familiar with the design and performance of the facility. Each item in the specification bears on a particular characteristic of the equipment, requiring that the vendor perform the function in a certain way. The detailed specification usually does not leave the vendor much room to deviate from the written requirement, even though another approach will still provide the desired function. (The system specification would allow this, however.) When reading a detailed specification, we may wonder what the buyer really wants. This is not a problem with the system specification because the functional requirements are spelled out.

The optimal specification is one which spells out the system and functional requirements so that vendors can respond with the most effective and economical concept. There still needs to be a section of the document that gives certain detailed specifications which are important to the ultimate operation of the facility. For example, the functional requirements clearly explain the applications that will be performed by the network facility, the monitor and control capabilities that are desired, and the physical locations to be employed. The buyer, however, will need to specify in detail the interface characteristics at the user devices, the power supply requirements, and aspects dealing with traffic such as transmission and switching speeds. The details will, of course, depend on the type of facility being procured.

The actual writing of a good specification is both a science and an art. There is a certain skill to *"specsmanship,"* but it is usually gained with having experienced the entire process a few times. This process consists of first understanding the technical requirements for the facility, then reviewing vendor publications to see what is already available on the market. The next step is to draft the specification by using the available information as a guide. When completed, the draft specification should be circulated for review. Letting one or more vendors review the draft is desirable as they can usually provide valuable suggestions. After the specification is released to vendors along with the request for proposal, there may be additional questions from vendors before they can respond. This activity also tends to improve both the specification and its writer's skills. Once a vendor is selected, the process of negotiating the final specification is where the exact characteristics of the facility are clarified. When the equipment is delivered and put into operation, the relationship between specified and actual performance will be revealed. Engineers who see how a facility actually operates gain an important feel for specsmanship. They are more able to compress the specifications down to the essentials on future projects. All of these statements on specmanship point to the advantage of employing competent consultants and system integrators.

7.4.2 Statement of Work

The *statement of work* (SOW), which is typically attached to the specification, provides definition of tasks to be performed by the vendor. For example, the vendor is to deliver certain quantities of equipment according to a schedule. The vendor will install and test the equipment using a test plan, which may be part of the specification or an entirely separate document. The SOW is usually operational in nature, detailing the steps in the process of manufacturing, delivering, installing, testing, and even maintaining the facility. The SOW spans the period from when the contract is signed until all tasks under the contract have been performed.

One of the more important elements of the SOW is the listing of the equipment and services to be provided as part of the contract. The specification defines the characteristics and features of the equipment, but the SOW states the quantities. As we mentioned, the timing of deliveries is usually provided in the form of a delivery schedule.

Testing of the facility is an important feature which is defined within the SOW. For complex electronic systems like packet switches and earth stations, testing is performed at several points in the contract schedule. The major elements would be thoroughly tested at the factory before they are shipped to the sites. At the site, the equipment is installed and tested again before it is declared to be ready for delivery to the buyer. Another test period checks network functions and exercises the components before it is "cut over" to provide service. As part of the SOW, support from the vendor would be required during these on site testing activities as failures are more likely to occur during such periods.

7.4.3 Contractual Terms and Conditions

The contract terms and conditions (T&Cs) define the responsibilities and obligations of the buyer and the vendor to each other. Fundamentally, the vendor must deliver the capability specified, and the buyer must pay sums of money when due. If there were no more than this to contracting, we probably would not need lawyers. The T&Cs, however, are often quite lengthy and complicated. Under normal conditions, where everything is implemented and works as expected and the buyer pays on time, the T&Cs are of little importance and, in fact, dispensible. In parts of the U.S. and some other countries, contracting is not practiced simply because of the value placed on a man's or woman's word. Seeing as this is a special cultural case, we can generalize by saying that the T&Cs are vital to a successful procurement.

Preparation of the T&Cs is usually in the hands of professional contracting officers who often are lawyers. Using a lawyer to draft a contract is not a bad idea. A lawyer tries to anticipate the kinds of disputes that have happened in the past. Also, the contract may be all that survives over an extended period of time, and can be the only source of reliable information as to the obligations of the parties. This is in the nature of things, particularly if a dispute leads to a suit which is tried in court by a judge and possibly a jury.

The previous discussion is only intended to give some background about the need for preparing a contract with well thought out T&Cs. The types of issues considered are really ways of dealing with "what ifs" and worst-case conditions that can arise during the activity, which, in the case of a service or lease, may persist for many years. Standard contract forms can be used when products or services are purchased from a catalogue or "off the shelf." When development of a new capability is involved, the contract is usually customized for the particular project.

A review of the types of T&Cs normally found in a contract is beyond the scope of this book. An important concept, however, is that both the buyer and vendor need protection from wrongful acts of the other party. For example, personnel who work for the vendor company may damage the property of the buyer. This would probably be covered by liability insurance which the vendor could purchase prior to contract. Another aspect is termination of the contract. This could be caused by the buyer or vendor under certain identified conditions. For example, the vendor could terminate the contract and recover the equipment which had been sold if the buyer failed to make payments when due. The buyer, conversely, may be able to terminate the contract and not pay anything further, or not pay at all, if the vendor fails to deliver the hardware or cannot provide the service designated in the specification and statement of work.

The T&Cs usually are not drafted until a vendor is selected by the buyer. Preparation of the contract is part of the negotiation process, where the parties define their responsibilities to each other. The legal profession is highly skilled in this regard. Business and technical management, however, must maintain close involvement in contract preparation and negotiation. This provides a focus and realism so that the resulting contract is practical, without forcing either party into a situation which damages what would otherwise be a good working relationship.

Equipment purchases, facility construction and installation services, and leases are typically governed by contracts. In telecommunication services from common carriers, the obligations of the buyer and vendor are delineated in a tariff, which is filed with the FCC. Under the tariff, the service characteristics and specifications are provided in detail, as are the prices for the basic service and any available options. Terms and conditions are usually part of the tariff, although the user may also have a contract. The use of tariffs simplifies matters because all a user need do is order the service as per the standard conditions. The carrier then bills the user during the prescribed periods of time. From a marketing standpoint, the tariff publicizes the service characteristics and prices as it informs the buyer of the availability of the service and its terms. The tariff, however, prevents the seller from maintaining confidentiality of pricing and other terms.

7.4.4 RFP Release to Industry

A request for proposal is a complete package of documentation, which a buyer uses to invite prospective vendors to make a proposal or offer to supply facilities or services. In many countries, the proposal and request for proposal are called a *tender* and *request for tender* (RFT), respectively. The main elements of the RFP are the specification and the statement of work, which have been described in previous paragraphs. Instructions to bidders would be included in a cover letter or special section of the RFP, telling prospective vendors about the timetable for the procurement and the appropriate format for proposals. Other contractual T&Cs may be included to give the vendors an idea of how the buyer intends to conduct its legal affairs. Each vendor, however, may do business in a different way and therefore has its own form of contracting. Very powerful buyers like the federal government and major corporations can often impose their own forms of contractual T&Cs, which will be included with the RFP.

The vendors submit proposals which provide documentation to show that they are uniquely capable of performing the required tasks. As we stated earlier, a well constructed specification will allow several qualified vendors to bid on the project. The RFP should indicate an outline or format for the proposal responses. This will cause the vendors to provide comparable material, which is essential to a valid comparison of responses on an apples-to-apples basis. Usually, separate chapters, or volumes for large proposals, are provided for technical characteristics, project management and vendor experience, test plans, and pricing. The pricing chapter or volume should be separate from the other parts of the proposal so that the buyer's technical evaluators would not be swayed by the prices of the various aspects of the intended purchase. While maintaining objectivity is hard, it is nevertheless a valued goal in the evaluation of proposals.

The buyer should give the vendors sufficient time to prepare proposals. In a major project involving hundreds of thousands or millions of dollars, allowing the vendors several months to respond is not uncommon. Large government or commercial contracts for major networks would normally involve a proposal preparation period of as much as a year. The cost of preparing a proposal can be a significant fraction of the cost of the system after it is installed.

To prepare proposals, to evaluate them and then to negotiate the contract takes considerable time. Therefore, the buyer should set a time schedule for the procurement process well ahead of the date when the network capability is needed. The vendors will use this schedule to develop their own timetable for manufacture (if applicable), installation, and test of the system. Obviously, the buyer must do a lot of planning in advance of the date when operations are to begin.

7.4.5 Evaluation of Proposals

As part of the RFP, vendors are given a deadline when their proposals are due. Vendors usually do not submit their proposals until the due date so that confidentiality is maintained to the last possible minute. This is very important to vendors, since price is a major factor in determining who wins the contract. The buyer must protect the vendors' information because it is generally considered proprietary. A competition may last only long enough for the buyer to open up the proposals, compare them on a side-by-side basis, and then name the winner. The buyer, however, may run several rounds of proposals and proposal updates so that the best possible resources are defined and committed. If there are several rounds, the buyer may inform the vendors that there will be a "best and final offer," after which the decision will be made as to which vendor is successful.

Proposals can be evaluated in a methodical way by an evaluation team of experienced people from a variety of areas within the strategic unit or by a team of consultants. Normally, the proposals would be broken down into technical and financial portions, and appropriate teams assigned to each. An approach used in very large procurements is for the evaluators to prepare a list of criteria for evaluation prior to reading the proposals. These criteria need not be quantitative, but rather they can identify the major features of the facilities to be procured and the factors which are most important to the strategic unit. For example, the overall score could be weighted 25% for price, 30% for delivery schedule, 20% for use of proven equipment, and 25% for technical performance.

The financial evaluation team will need to set up evaluation matrices in the form of spread sheets. Financial sections of proposals usually show payments for facilities and services made over time. There may be lump sums and progress payments during construction. Typically, vendors use the customer's money to finance construction. This is, after all, only a question of how the interest is paid, but, ultimately, the buyer must pay it all, whether the funds are provided by the seller or buyer. Spread sheets are ideal for evaluation, just as they are for studying the economics of the network before it is implemented. In fact, the same spread-sheet format used during network design is desirable to use for evaluating the financial implications of the proposals to clarify which offer is most attractive in a financial sense.

When doing evaluations, we must remember that the strategic unit will eventually take over responsibility for the facility after it has been turned over by the seller. In some way, the evaluation team needs to consider the experience and reputation of the vendors during the evaluation. All other things being equal, the more experienced vendor is usually preferred. This is the ideal situation. Often, the less experienced vendor makes his proposal more attractive with a lower price. The buyer must then sort out the apparent advantage of reduced construction or operation cost, as shown on paper, with what might happen in the future if the young vendor runs into difficulty. Later in this chapter, we will treat the subject of how to deal with the risks of contracting.

The buyer must conduct the evaluation in an honest and fair manner. Vendors are very motivated to make a sale, and some will try to gain an advantage. Some buyers will give information to vendors so that they may improve their offer. This could cause confusion and frustration among other vendors. For the buyer to make clear to the vendors, ahead of time, just how the evaluation will be conducted is more professional and better for all parties. This includes the overall criteria of interest, the timetable, and whether there will be a best and final offer. If a vendor knows that there will be a best and final offer, there will probably be some margin left in the price so that a final cut can be made.

As we noted earlier, information should generally not be leaked to vendors during the evaluation process. As indicated in Figure 7.4, however, there would probably be additional cycles of requests and quotations to allow the buyer to establish the best set of offers from the vendors. A generally accepted approach for guiding the vendors when they are preparing a second round or best and final offer is to ask them questions about their proposals. These could be requests for clarifications of specifications or pricing. Vendors could also be asked for a quotation on an additional capability, one that had been included by one vendor but not by the others. This is the *equalizing* process often used by the government to cause all offers to become as similar as possible. Any responses to questions should be treated as if they were part of the original proposal.

The final report of the evaluation team would provide the comparisons of the offers in technical and financial terms. This would probably be a written document delivered to the management of the strategic unit. Because many people were probably involved in the evaluation, conducting a series of briefings would be useful so that all pertinent information could be transmitted to the decision-makers. Selection of the most advantageous offer and best vendor is probably of vital concern to the strategic unit.

7.4.6 Negotiating the Contract

The intention of the evaluation process is to select the most favorable offer from the best vendor. The next step is to negotiate the contract. The specification and statement of work are two key attachments or appendices to the contract, and they are usually negotiated by members of the technical evaluation team. The idea is to end up with documents which tightly define the hardware and services to be procured. Negotiation is justified because the original specification released with the RFP may not be completely consistent with the exact characteristics of the selected vendor's offering. Nonetheless, the vendor may be able to provide additional capability not contemplated in the RFP. The give and take of negotiation is very useful in arriving at a specification and SOW which allow the vendor to do the most fo the strategic unit.

The remaining part of the contract dealing with pricing and other T&Cs will usually require considerable give and take in negotiation. Financial terms are always adjustable in their amount and the timing of payments. The vendor will want as much money as possible up front to reduce financing costs. The buyer, however, may not be able to accommodate. This is particularly the case for a buyer who is in a start-up situation with limited capital.

There are a few basic forms that the contract can take, wherein each allocates financial risk between buyer and vendor. The *firm fixed price* (FFP) contract is the most popular in commercial contracting. The vendor agrees to deliver the system for a predetermined price and according to a specific schedule. The requirements must be spelled out and cannot change during the course of implementation. If there are any changes which the buyer wishes along the way, the added scope is estimated by the vendor and the price is increased accordingly. To allow changes after the contract is signed is usually not a good idea because the element of competition is almost always lost. The other broad type of contract is referred to as *cost plus contract* or *time and materials (T&M) contract,* wherein the vendor submits his actual expenses for reimbursement. The "plus" includes overhead charges and profit, which is usually established as a fixed percentage. Because the cost plus contract is like a blank check to the vendor, the buyer must keep watch on implementation progress. A detailed auditing process is necessary; otherwise, expenses which are unnecessary (or unrelated) may be included in the bill. The government uses another approach, called *cost plus incentive fee* (CPIF), where the vendor gets an additional share of savings relative to a target budget.

Much of the negotiation has to do with risks to either party from uncontrollable problems and the actions of others. Usually, the larger the organization is, the more lengthy the negotiation and the resulting contract. This book cannot go into detail on these matters, since they are generally of a legal nature. To involve contract professionals on both sides of the negotiation is usually best. Draft contract T&Cs are usually prepared during the negotiation, with updates made to reflect items which have already been negotiated. The final version may then be signed by officers of each organization, and thereafter the contract becomes binding upon the parties.

Management of the actual procurement process, including contract preparation and negotiation, could be in the hands of one of a number of different groups within the strategic unit. The purchasing or materiel department is normally involved with small to medium sized procurements, primarily of hardware and facilities. While purchasing personnel would not typically have technical backgrounds, they should be very experienced as buyers. Buying is a skill which itself can result in substantial savings when purchasing essentially standard products. The technical staff could provide the specifications and other details, while the buyer would obtain quotations from vendors. Acting as a team, the best vendor would be selected and the buyer could handle the negotiations dealing with delivery and price. Provided that the hardware or services are reasonably straightforward, this approach can be a good fit to the need.

On major procurements involving specialized or customized facilities, a senior individual within the strategic unit would normally handle the process. This might be the director or vice president of telecommunication, aided by members of the telecommunication staff. The leader could also be the head of the strategic unit if the procurement justifies. Again, a team would be formed to encompass the necessary skills. This would include one or more members of the legal staff to draft contracts and to assist, or even to lead, the detailed negotiations. Technical support would be very important because of the leverage that specifications and functionality have on the value of the facilities to the strategic unit. Another important function to be included is financial analysis, as discussed in Chapter 6. Some or all of the members of the proposal evaluation team should be included during the negotiation phase.

7.5 MITIGATING RISKS

Delineated in the following paragraphs are some of the risks borne by the strategic unit as a result of entering into a contract with a vendor to purchase facilities or services. Whenever a strategic unit consigns its future to another organization, it potentially loses control of vital aspects of its business. Also, there is the possibility that things will go wrong. The contract is expected to deal with many of the risks that are reasonably likely to occur and with which the vendor can deal. There are risks, however, that the vendor really cannot afford to assume. While the following discussion is not all-inclusive, the particular categories ought to stimulate some thought. We suggest ways of controlling or mitigating these risks.

7.5.1 Schedule Delay

There is almost always a period between the time that a contract or purchase order is issued and when the network capability is finally operational and ready for service. This time period encompasses perhaps the most critical of risks associated with implementing a new network capability: *schedule delay*. Whenever human beings are involved in building something new or substantially modifying an existing structure, there is the strong possibility (or probability) that the job will not be done on time. The reasons for this are many, and perhaps they are not even limited by human imagination. In one situation, schedule delay simply means that the new capability will not be available when planned, which would affect other aspects of the overall plans of the strategic unit. The effect could be minor were we talking about a situation where the new capability was simply an upgrade or expansion of an operating system. You would simply keep working in the old manner until the vendor or vendors finished their work. Alternatively, the strategic unit might be in a start-up mode, and implementation delays in the network could hurt the business prospects of the new venture. Such has happened in networks involved in information delivery services. A more serious problem resulting from schedule delay is where the existing network must be deactivated while the new network is turned on. During this time gap, the strategic unit is basically out of business because internal and even external communication can be blocked.

A basic rule in telecommunication projects is to assume immediately that there will be schedule delays of various kinds. All of the planning for implementation and cutover would then include the necessary contingency plans. The key is to have some type of network capability available at all times. This is like the situation when someone is remodeling a house. The family must still have a place to live, even though important rooms like the kitchen and bathroom are being taken apart and rebuilt in their entirety. This could involve moving out of the house and living in temporary quarters during part of construction. The analogous situation exists with telecommunication network implementation.

Some aspects of schedule delay risk can be considered in the contract. Financial penalties are common in the T&Cs so that the vendor stands to lose some of its profit if it is late. Vendors, of course, do not want to risk their profit, but at least the threat may cause them to be realistic about their schedules. Another approach is to provide an incentive for early delivery, which is similar to the CPIF type of government contract. Incentive and penalty can also be used together in the same contract. The one problem with the financial approach is that the financial threat or incentive may not be big enough if the problems causing the delay are serious. The vendor may wish to desert the project and accept a total loss. The strategic unit would then be left without a network.

As we indicated previously, the logical way to protect the strategic unit from the consequences of schedule delay is to provide an alternative. If the vendor can be encouraged to do this as part of the contract, that is all the better. In most instances, however, the strategic unit will need to make its own arrangements for backup and alternate facilities. A frequent approach is to obtain services from the common carriers on a month-to-month basis. Assume, for example, that the strategic unit is implementing a T1 backbone network using a combination of fiber optic facilities and satellite links, contracted on a long-term basis. To allow early cutover of services and to protect against schedule delay, the unit can order T1 services according to tariffs from the local telcos and AT&T. There would probably be a modest up-front construction fee to compensate the carriers for their expenses in attaching the locations to the digital network. While the monthly charges from tariffed T1 services are relatively high, there is the major advantage in that service can be terminated on short notice without penalty to the customer. Returning to the house remodeling example, the family could move to a temporary apartment and use it on a simple month-to-month basis with an open-ended departure date.

Any current operating network or telecommunication arrangement should not be abandoned until the new network capability has been operating for an extended period. Continuing to operate the old network may be expensive and laborious. Nevertheless, if all traffic is connected too quickly to the new network, problems can easily cause instability and breakdown. Users will be unfamiliar with the controls of the network. If services do not respond as expected, a great deal of dissatisfaction will result. Recall the example of SBS (cited in Chapter 2), where users of this all-digital private network were so dissatisfied that they refused to use it. (To make outside calls, they dialed 9 to reach the local telco instead of 8 to be connected to the private network.) A phased approach is almost always best, with the old network running alongside the new one. Among the more common causes of problems with new networks is the software which resides in switches, computers, and nodal processors of various types. Things work well during the trials and the hardware is very reliable. Software, however, will always have flaws, called "bugs," which are not discovered until user traffic has been flowing in significant amounts. This lobbies for extended trials and phased transfers, as previously mentioned. Waiting to get the needed software releases is also a significant cause of schedule delay.

7.5.2 New Technology

We have discussed at length the advantages of digital systems in information processing and communication. The telecommunication industry is moving very fast, with new devices and networking concepts being brought out of the laboratory and offered to the marketplace. Even the major equipment suppliers and public networks are moving extremely fast to gain a competitive advantage with technology. Likewise, strategic units are crying for new services and capabilities so that they, too, can compete more effectively and reduce their expenses at the same time. Therefore, all participants exhibit what can only be described as a "passion" for new technology in telecommunication networks. The fixation of buyer and vendor on advanced concepts must be tempered by the reality that, under worst-case assumptions, the network will not actually work as predicted. Alternatively, there will be a schedule delay as problems that were uncovered during development are corrected under real-world conditions with an operating network.

This should not discourage anyone from considering new technology in a telecommunication implementation project. We simply caution that technology hould be introduced for what it does, not as an end in itself. After the decision as been made to include technology development in the project, steps must be aken to provide flexibility during the program. The most direct result of technology mplementation problems is schedule delay because the new capability is not ready on time. This can probably be handled by taking the approach cited in the preceding ection.

Technology that has undergone sufficient development poses minimal risk in he commercial environment. IBM, for example, introduces products on a regular oasis and is in a position to assist the buyer with any start-up problems. Dealing vith an established equipment supplier is usually safe. Newly developed technology 'rom a start-up company, however, entails significant risk as the vendor may not aave the financial and human resources to resolve the situation to the buyer's satisfaction.

There are major industrial organizations in the business of developing advanced technology systems for major buyers such as the U.S. government and arge corporations. Examples include spacecraft manufacturers like Hughes Aircraft Company and GE Astrospace, computer systems houses like CSC and EDS, and electronics development firms like TRW Systems and Harris Corporation. Funding for development is mostly provided through progress payments. The period to complete development would last from perhaps as short as six months up to many years. The direct costs of developing new technology are very large because of the manpower and experience base of the developers and the capital investment needed in laboratory and specialized manufacturing facilities. The development and construction of a communication satellite typically will last three years from time of contract signing to the delivery of the satellite to the launch site. Such a lengthy development and manufacturing cycle means that the buyer must plan on this type of schedule. Fortunately, for many current users, there is a sufficient supply of standard satellite capacity already in orbit. For new technology like mobile satellite communication for vehicles and remote undeveloped areas, however, the necessary spacecraft hardware is still in the design phase. The trend is now for several potential users or network operators to form a consortium to fund the development of the new system. After development is completed, the same consortium continues with implementation and even operates the system as a commercial business.

A common practice in technology development is to pursue at least two possible approaches. This is because at any time a given approach may not succeed or it can fail to meet performance requirements in some important way. The Apollo Program was recognized for this type of conservatism. Conversely, parallel development is very expensive because there must be a separate team working on each approach. Today, budget constraints have forced technology developers to focus instead on the most practical approach with the greatest chance of success.

7.5.3 Acts of God

Acts of God, also called *force majeure,* is the term used to refer to major problems that cannot be anticipated and are totally outside of the control of the parties entering into the contract. A classic example is a major fire or earthquake, which stops the project or destroys all or a portion of what has already been constructed. Other examples include strikes, civil unrest, war, or government action such as confiscation or new laws which render illegal something that is critical to the project. Because of the unpredictability and effect of acts of God, the contract usually forgives the vendor of any delays which result. The buyer, however, would still be affected by the delay in the project as the vendor recovers from the losses of time and facilities.

Losses from acts of God can be covered by insurance policies purchased by the buyer or vendor. In any case, insurance will only provide monetary relief, and the buyer must still take care of any operational implications for the network. This means a schedule delay as the eventual consequence, and we can best deal with the risk by using the applicable techniques given previously.

7.5.4 Inadequate Maintenance or Network Management

The last category of risk has to do with the performance of the network after it has been turned over to the buyer by the vendor or vendors. Presumably, the network was in working condition at that time, meeting the applicable specifications. The strategic unit has the responsibility for operating and maintaining the network throughout its useful life. In Chapter 6, we discussed network management from the strategic unit's perspective. Our concept was that of treating the network as a business, in the same way as a communication service company or common carrier does. Importantly, a workable network management system will permit the operator to uncover problems when, or even before, they occur.

Of course, the strategic unit can turn responsibility for running the network over to the same vendors that implemented it. This is usually acceptable with local telcos and the major long-distance companies, where the facilities are merged with those of the carriers. In the case where the network is composed primarily of purchased components, the risk of serious problems arises. The risk can be compounded by inexperienced vendor personnel who are assigned to perform tasks under warranty or service contract.

Having a capable internal staff is fundamental to minimizing the risks of network maintenance and operational difficulties. Although vendors may be under contract to provide support functions, there will be times when immediate action is needed to respond to a catastrophic outage or one that is impending. The size of the internal staff need not be large. The important factor, however, is that they be trained and have the proper network management tools at their disposal. The tools include a network operations center with a monitor and control network, which extends to critical nodes and user locations. A field maintenance staff is also potentially valuable, but its cost can be prohibitive for many organizations. Fortunately, the overnight package delivery services and national airlines make it possible to dispatch the resources needed to effect repairs relatively expeditiously. The network can be maintained operational with redundant or standby facilities which are activated from the NOC.

Chapter 8
Strategies for Networks

In Chapter 1, we introduced the concept of the strategic unit, which is either a complete business entity or a governmental agency. Whether private or governmental, the strategic unit was established with the mission to serve a client or customer base. A strategic unit is comprised of a collection of functions that creates value for customers and supports the internal operation of the unit. The utility of the unit's offerings to the customer base is embodied in the value chain, a concept developed in this chapter. Among the important functions within the chain is telecommunication.

The concept of strategy relates to telecommunication networks just as it does to other aspects of business. Building a telecommunication network is a strategic decision because of the long time-frame involved first in developing the network and then for its operational service lifetime. The network is conceived in one era, and then lives on through the evolution or development of the business environment. The strategic unit thus takes on considerable financial risk because the network may not fit the actual requirements during a significant period of time. The telecommunication network should therefore be designed with a long-term strategy. We have discussed how owned facilities such as PBXs and high-level multiplexers can be combined with long-haul T1 services and a VSAT network to provide a cost-efficient and dependable network. This is an example of a strategy, but applicable only for a particular strategic unit with widely dispersed operations. A more consolidated business unit could base its strategy on leased CENTREX service at a few locations along with bulk calling packages and DDS for a limited number of data communication links.

The other point of view is that of *competitive strategy,* as developed in the general business context by Michael E. Porter [Porter, 1980] and for telecommunication by Peter Keen [Keen, 1988]. We introduced in Chapter 2 the concept of using a private telecommunication network to achieve competitive goals of a strategic unit. Competitive strategies for networks are as diverse as the organizations that implement them. The main purpose of the private network is to implement a portion of the business strategy of the organization. Typically, some forward-thinking unit acquires a telecommunication network capability to give it an edge in the marketplace. If the technology gives the unit a sustainable competitive advantage, then the other firms in the same industry (i.e., the rivals) usually imitate the strategy by building comparable networks. The competitive advantage could then be lost, unless the initiating firm developed another good idea. This leap-frogging continues to raise the stakes, and the cost of entry into the industry rises.

We will place greater emphasis on telecommunication in the context of competitive strategy. Our discussion begins with a review of competitive strategy in business, following Porter's structure, but primarily from the telecommunication perspective. We include this material to give the reader an appreciation for the role of telecommunication in the overall positioning of a strategic unit in its respective industry (which is probably *not* telecommunication). An individual who understands where an organization is going is probably better able to focus telecommunication resources on business needs, both present and future. Sometimes, an organization changes its top level strategy, which will inevitably change the requirements for internal and external telecommunication services. We move from this base into a discussion of some basic strategies for the network itself. These are approaches often taken by telecommunication management to provide a future direction for the network, making it more relevant to the organization's needs. The goal is to have a network strategy that truly supports the business in an effective way.

8.1 COMPETITIVE STRATEGY IN BUSINESS

Every organization which competes with others for the business of some customer base employs strategy. As stated by Porter [1985], "This strategy may have been developed explicitly through a planning process or it may have evolved implicitly through the activities of the various functional departments of the firm." Note that one of these functional departments is telecommunication, wherein the administrative needs for voice and data communication are satisfied. Again quoting Porter, "Left to its own devices, each functional department will inevitably pursue approaches dictated by its professional orientation and the incentives of those in charge. However, the sum of these departmental approaches rarely equals the best strategy." [Porter, 1985] In other words, individual departments may optimize their separate activities, but the performance of the overall strategic unit could still be suboptimal.

What, then, is competitive strategy? Simply stated, it is the particular set of business results which the strategic unit is trying to accomplish along with the approaches that it is using to achieve those results. Some organizations select their objectives, and the marketing and operational systems that they perceive will reach them. Others do not plan at all, allowing events to overtake the situation. Many businesses simply happen in such a manner and achieve a certain degree of success in spite of themselves. Our intent, however, is to help the organizations employ the telecommunication dimension of strategy to achieve those results which telecommunication can appropriately address. One general theme is that telecommunication reduces time and distance.

8.1.1 Value Chain

Private telecommunication appears to fit best in the context of a strategic unit's value chain. Illustrated in Figure 8.1 [Porter, 1985], the value chain is that collection of activities which effectively constitutes the organization's competitive strategy. These activities make possible production, marketing, delivery, and support of a product line, ideally, in a more satisfactory way than do those of competitors. Figure 8.1 shows that certain activities relate directly to customers and are performed basically in series (logistics, operations, manufacturing, marketing, and customer service), while there are others which are applied in general to support the entire enterprise. The leading edge of the value chain is the difference between price and the cost which the firm maintains as margin or profit. Basically, a strategic unit obtains adequate margin (so as to satisfy owners or investors) if the combination of the real activities produces something of perceived value to customers at a cost to the firm which is less than the pricing of the market.

Across the top of the value chain is a common layer called the *firm infrastructure*. Telecommunication facilities, both internal and external, add value to activity, including (as appropriate) customer prospecting and order placing, facility management, shop floor automation, materiel purchasing and processing, and financial management. The role of telecommunication as a general facilitator of value activities has led to it being called the "life's blood" of an enterprise. Every value activity in Figure 8.1 employs human operations, purchased inputs, and technology. Every activity also uses and creates information such as buyer data, performance parameters, and failure statistics of products, as well as inventory and financial information relative to receipts and disbursements. These functions can be performed electronically by using computers and software. More importantly, telecommunication technologies add the ability to compress distance and time. Anything that can be converted to data is a candidate for information processing and distribution. Even the lowly telephone has demonstrated new value for strategic units, primarily due to the switching capabilities of the latest generation of telecommunication devices.

In the generation of value for customers there are linkages between elements of the chain. As functions of the chain are better integrated and interrelated, the efficiency of meeting customer needs will increase. A properly engineered, implemented, and operated telecommunication network will improve integration of these

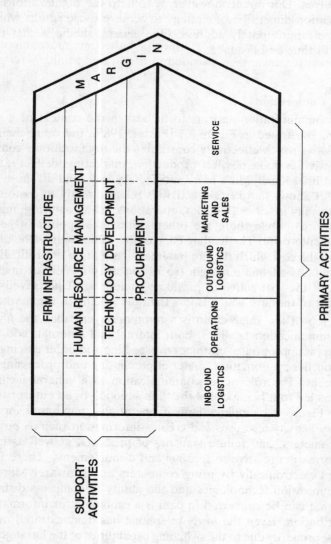

Figure 8.1 The value chain of activities of a strategic unit [Porter, 1985].

functions. For example, each activity generates or uses information which can be maintained in a common database to be accessed by appropriate internal personnel, as well as buyers and suppliers who gain access through a private network. The database can be centralized in a host or distributed to minicomputers at nodal locations, depending on what makes sense for the particular organization. The reliability and control of the network will inevitably become more important because of the competitive advantage provided to the strategic unit.

8.1.2 Versions of Competition

In trying to understand the composition and dynamics of competition in business, this author is particularly fond of Porter's graphic representation of the five competitive forces that determine industry profitability (see Figure 8.2) [Porter, 1985]. These are the definable aspects of industry structure. Underlying this picture is the basic tenet that a firm's profitability (and reason for being in business) is determined by all of these powerful forces, none of which are under the unit's control. The telecommunication manager needs to understand that the dynamics of this external business environment determine his or her company's ability to earn satisfactory rates of return on invested capital. Included in this capital is the investment in telecommunication facilities (which was presented in Chapter 6). We review each of the five competitive forces in the following paragraphs.

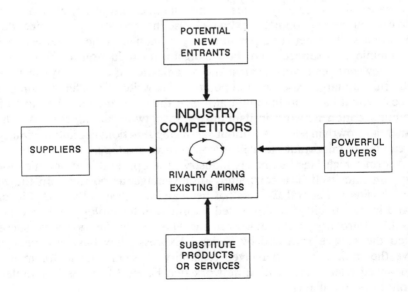

Figure 8.2 The five forces of competition which include firms in the same industry as well as other outside forces which can influence the performance of a given firm [Porter, 1985].

8.1.2.1 Rival Firms

We understand that competition inherently includes other rival firms that offer basically the same goods or services to the same market. The largest of the rival firms watch each others' actions, and respond to them when they think appropriate and in a manner that is within their ability. Looking at the telecommunication industry, which is characterized by strong rivalry, the competition among AT&T, MCI, and US Sprint for the long-distance customer is visible. Lately, these companies have introduced information and software-based products to gain advantage on their competitors. Examples include bulk calling packages, software-defined networks, and now ISDN. Consequently, the long-distance companies have become bigger consumers of information processing and data communication technologies. Each enhancement raises the stakes for others who consider entering the long-distance business.

Rival firms compete for a common customer base by using comparable product or service offerings. The firms are not primarily engaged in the telecommunication business. To the extent that they require telecommunication facilities, these firms must purchase them from outside suppliers. Some industries rely heavily on telecommunication, such as the airlines, banks, and nationwide brokerage firms. For these companies, the price paid for telecommunication is a significant expense of doing business and therefore greatly influences the value chain. The economic studies identified at the end of Chapter 6 are particularly germane. For example, a large national retailer may decide to implement an inventory control and point-of-sale telecommunication network, which also incorporates other voice, data, and video services. The network possibly may eliminate some expenses for public telecommunication services and show a significant reduction in these costs. Economics, however, can work against the organization if it overbuys facilities, and thereby runs up large expenses and potential tax write-offs. The burden of a large fixed cost would harm profitability, although the company had the most effective telecommunication network in its industry. A network architecture which allows incremental growth in services without sacrificing functionality and flexibility, however, can be an important element in competitive strategy.

A firm which lags behind its industry in adopting telecommunication technology can find itself in a completely disadvantageous position. In the package delivery business, Federal Express has been an innovator in the use of its airplane fleet and in its pursuit of sophisticated information technology. While other companies like Purolator, UPS, and even the U.S. Postal Service have successfully imitated the service provided by Federal Express, they have not been able to achieve the same level of perceived reliability. A significant factor may well be the advanced telecommunication network that Federal Express has implemented and continues to enhance.

8.1.2.2 Substitute Products

A strategic unit's market is continuously subject to competitive threat of substitute products or services of an entirely different nature. Since the customer is mainly interested in functionality and cost, the potential substitute is a threat if its value chain provides a better fit. As an example, information delivery by telecommunication facilities represents competition to print media like books and newspapers. The threat may not become real for some time, as in the limited success so far of videotex services. Telecommunication as an ingredient of another business could conceivably facilitate a substitute product overtaking its more entrenched counterpart. For example, television programming from broadcast stations, cable systems, and movie rentals can be a replacement for movie theatres. Movie studios have been successful at producing films of such quality that movie theatres have enjoyed a rebirth.

8.1.2.3 Suppliers

The fact may not be apparent, but suppliers represent a competitive threat. If a strategic unit is very dependent on a particular supplier, it is possible for that supplier to extract more than its share of the unit's cash flow. The corresponding reduction in margin could make the firm unprofitable against rivals in its competitive market. Again, in cases where telecommunication is an important ingredient, the strategic unit can now take steps to reduce the power of suppliers which may have monopoly control. One of the important underpinnings of private telecommunication is that the facilities can be installed and operated by the unit, thus bypassing the monopoly telecommunication supplier or suppliers. The telecommunication environment in the U.S. is becoming ever more competitive, as even the monopolies respond to their new rivals and substitute products (i.e., private bypass facilities).

Telecommunication and associated information technologies can play a role in bypassing intermediaries which do not add concrete value to a product or service. The question is asked [Keen, 1988], "Why pass important aspects of business over to intermediaries when they can be better handled directly?" A customer order entry system can also be used to perform credit verification and financing for a purchase. Also, a strategic unit can broker its own products or services, and thus avoid dealing with third parties who collect part of the overall margin but only perform a coordinating role. These kinds of tasks can be done with database systems and telecommunication network extensions, accessible either directly or indirectly by customers.

8.1.2.4 Buyers

One of the biggest competitive headaches for strategic units is caused by powerful buyers. As markets become more consolidated and corporations get larger, the buying power is becoming more centralized. This author was particularly impressed by a report on public television about a large North American retail chain called The Limited℠. You can find from one to four of their retail outlets in virtually every shopping mall. The stores themselves are rather small and trendy, with merchandise attractively displayed. On a routine and frequent basis, the merchandise is changed and rearranged, so that the store never appears exactly the same from the mall corridor. The company, however, has massive warehouse and distribution centers which are highly automated and very efficient. You can only imagine the buying power that a company of this size has in the apparel industry. This buying power translates into lower unit costs and a resulting competitive advantage in the retail market. Another such retailer is Benetton of Italy, a company which happens to use telecommunication in its worldwide business strategy [Istvan, 1988].

The largest buyers influence competition by forcing price and other concessions from their suppliers. Likewise, the largest buyers of telecommunication facilities and services usually can obtain them at a discount, which, along with the economy of scale in using telecom, further enhances the competitive edge that the powerful buyer has over its smaller competitors.

The American Business Networks (ABN) of Dallas, Texas, is a limited partnership created to increase the buying power of telecommunication users. By aggregating their needs for long-distance services and telecommunication equipment, ABN was able to extract significant cost concessions for its members. This activity, not unlike a cooperative, could represent a trend which may not be welcomed by suppliers in the telecommunication industry. The benefits for ABN members, however, are apparent.

8.1.2.5 Potential New Entrants

The category of competition relating to potential new entrants has a particular bearing on telecommunication. We stated earlier that the telecommunication ingredient in such industries as airlines and package delivery has "raised the stakes" for any company which intends to become a participant. This is the concept of a *barrier to entry,* which is erected by established firms as they evolve their own business strategies [Porter, 1980]. In the absence of barriers, a new entrant can quickly gain a foothold and seriously disrupt the profitability of the existing rivals. This situation is exacerbated if the new entrant has a strategy or technology that is decidedly different from the others, possibly involving the use of telecommunication facilities which the entrant had already in place from a former activity.

The concept of the *pre-emptive strike* using telecommunication and data processing systems is offered by Keen [1988]. When an advanced private telecommunication network is implemented, the strategic unit may consider how it can be employed to enter a related industry on an incremental basis. The inroads by Sears, Roebuck and Company into the national credit card business with their Discover Card℠ are being recognized. Likewise, American Express has become a significant factor in the travel business by exploiting their upscale credit card business and customer base. In both instances, the companies could wield considerable clout because of their solid telecommunication and data processing infrastructures. Any major successful business had better consider how a modern information-based competitor can have a measurable effect on its most prized markets. In fact, *competing in time* and *time-based competition* are the new names for such forward use of information technologies for competitive advantage.

8.1.3 Business Strategy

A strategic unit conducts its business in a particular way within a specific marketplace. The set of activities and approaches constitutes the business strategy. As we mentioned previously, the business strategy may have been created by careful intentional design, or just happened as the business grew [Porter, 1980]. From the telecommunication perspective, business strategy also includes the internal administrative procedures which facilitate the formation, implementation, and utilization of the external aspects of strategy. The previously given example of the airline reservation system is particularly germane, considering that this class of network is applied strategically as an important aspect of the firm's marketing through travel agents and large corporate customers.

Corporate strategy deals with the big picture for several business units which are viewed as investments in a portfolio, a concept pioneered by the Boston Consulting Group [Thompson, 1983]. A telecommunication network may play less of a strategic role in this context, except for the potential for cost containment through bulk buying. We provide more background information on strategy as it relates to single business units, which could be part of a larger enterprise.

Business strategies fall into three broad categories, which Porter defines as the generic strategies of cost leadership, differentiation, and focus [Porter, 1985]. In *cost leadership,* a major supplier builds its value chain so as to have the absolutely lowest cost of production. The firm is able to charge the lowest prices for its products or services, which are aimed at the broadest range of customers in the market. The low-cost leader usually exploits every avenue to obtain the lowest production cost. As we discussed previously, this strategy can include the telecommunication ingredient in the area of just-in-time manufacturing. In general, there is usually only room for one low-cost leader in a given industry.

Like cost leadership, a unit which employs the *differentiation strategy* tries to appeal to the broadest market. The unit does so, however, by offering something unique for which it can claim a premium over the lower price of the cost leader's product or service. The key is that the firm finds attributes that are different from those available on the general market so that it is perceived as unique and of particular value. Applying information technologies in a unique way can achieve differentiation. According to Keen, a differentiation strategy based on telecommunication can give a company an edge over its competitors in a mature market [Keen, 1988]. An example is McKesson, a wholesaler that distributes products to over 17,000 pharmacies. This firm's experience is that customers are not particularly price sensitive if you give them excellent service. McKesson provides the customer with a computer terminal, which greatly speeds up the process of locating items and then placing orders. This has had the added benefit of allowing the customer to do a significant part of the work in placing orders. Consequently, McKesson has been able to reduce staff while improving service reliability and speed.

The *focus strategy* is the process whereby a unit selects a certain segment or region of the larger market for exploitation. The unit would then attempt to serve that segment better than firms using the other strategies. One approach is to achieve a cost advantage by locating in close proximity to customers through reduced transportation and other costs. Telecommunication aids in the focus strategy because it can reach directly over long distances and reduce the time needed to respond to customer inquiries.

There is a *differentiation focus* ("niche") strategy where the firm develops products which precisely fit the needs of certain customers. The customer then is so well served as to accept a somewhat higher price, but the overall value is still greater than for the more generic product from the competition. Spectradyne is a company which uses its satellite network to deliver an entertainment product to its hotel client base. In addition, the company enhances its differentiation focus (it is not the only one selling movies to hotels) by including automated check-out services for hotel guests. The guest can use the control unit on top of the TV set in the room to review his or her bill and to complete the check-out process without going to the front desk.

One of the more fundamental questions executives ask is "What business are we in?" While this question has become rather overworked, it nevertheless provides a focal point for understanding the current strategy of a business. The important task of refining and redefining the strategy cannot begin until the present situation is understood. Business strategy breaks down into the areas of (1) the industry environment, (2) the competitive environment, and (3) the overall business situation of the strategic unit [Thompson, 1983].

For the telecommunication manager, each of the three areas should conjure up images of telecommunication's role in the three areas. The job of the general manager is to use the analysis for the purpose of setting the unit's goals. His or her responsibility is very broad, relating to marketing, technology development, operations, finance, and administration. Each of these areas is affected whenever a change of strategy is considered.

The telecommunication manager is in a unique position to inform the general manager of how telecommunication technology can be applied to business purposes. This author recalls his experience as a communication officer in the U.S. Army. Using this analogy, the general manager is effectively the commanding general of the army division with the telecommunication manger being his chief signal officer. All Signal Corps officers receive the same basic military training as their counterparts in the combat arms such as the infantry, artillery, and armor. Signal operations are conducted within the same battlefield as the fighting. Any military operation considers communication as a vital element because coordination of fighting units depends on the ability to communicate. This communication employs radio and wireline transmission, with links established between fighting units and from higher commands down to the fighting units. Under battlefield conditions, the communication service function is as essential to military operation as artillery or transportation.

Moving back to the corporate context, the telecommunication manager could be at the right hand of the general manager, providing guidance on effectively employing telecommunication in the pursuit of the strategic unit's business objectives. A thrust into a new market niche or a response to entry of a new competitor could include telecommunication as an important constituent. The current telecommunication network should be applied with little modification. Otherwise, a new network capability may be required.

The analysis of the business opportunity should therefore include the effects of telecommunication. In the military, every battle plan includes a section on the provision of communication to the operating units. So, too, should a business plan include the needed telecommunication resources and the manner in which they are to be employed. The cost of procuring those facilities would be a significant result of the proposed change of strategy. In addition, personnel will most likely be added to operate and to maintain the network. An organization which foresees the importance of telecommunication in the overall picture can have the competitive advantage that delivers business success.

8.2 BASIC NETWORK STRATEGIES

We move now into the area of strategy as it relates specifically to the private network. A well defined strategy is as important to the application of private telecommunication networks as it is to the strategic unit's business objectives. Figure 8.3 identifies four basic telecommunication network strategies. Each represents a particular focus for defining, implementing, and operating the required facilities. The four strategies of control of destiny, network reselling, use of public networks, and the fostering of customer dependence are common themes encountered in the context of private networks for strategic units.

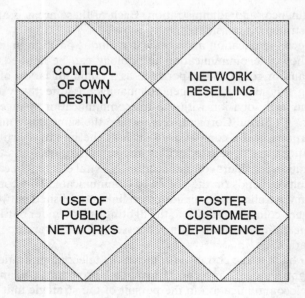

Figure 8.3 Basic network strategies which can be employed to enhance a unit's business strategy.

8.2.1 Controlling Your Own Destiny

The strategy that seems to recur the most in conversations with telecommunication users and in the literature is that of building a private network to gain control of one's own destiny. This phrase might be a rather pretentious way of referring to the simple idea that telecommunication should directly respond to the organization's various needs. The most fundamental way to gain effective control appears to be through ownership of facilities. The trade-off between ownership *versus* taking telecommunication as a service has already been examined from an economic point of view (i.e., see Chapter 6). For some strategic units, however, the telecommunication network is so important to business strategy that the ownership is demanded, even in the absence of cost savings.

Control of destiny can be exercised along two dimensions, as illustrated in Figure 8.4. The degree of actual ownership of facilities is displayed along the x-axis, where the origin represents use of public networks for all services and the far right quadrant corresponds to total ownership of the private network. Most real cases fall between these extremes, where the location to the right or left of center depends on the emphasis (with increasing investment in telecommunication facilities toward the right-hand end of the scale). The y-axis presents the other aspects of control which are technical and operational in nature. As discussed in

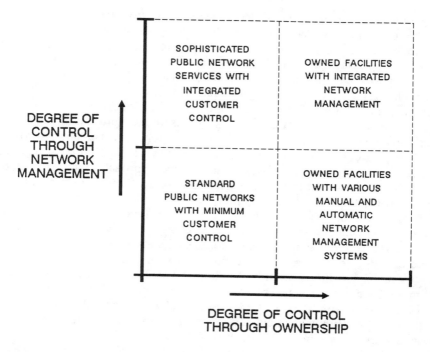

Figure 8.4 Control of network resources can be exercised along the two dimensions of ownership and network management.

Chapter 5, network management is an important aspect of private telecommunication. Manual approaches without remote monitoring and control systems are indicated at the origin. The high end of the scale corresponds to a well integrated network management environment, encompassing both the technical and administrative aspects of the problem. Again, most situations exist between these two extremes. A very important point to be made about this graphic representation is that the two dimensions are really independent of each other; that is, any combination on the two-dimensional grid can be obtained in practice.

A strategic unit would seek a large degree of control through ownership if it did not want to rely on other organizations to perform critical operational tasks for the network. In Chapter 1, we discussed how difficult it is to control resources provided over the public network in this age of divestiture and deregulation. A minimum of two, and more likely three, different, independent common carriers are required to complete a circuit from one city to another. The problem is aggravated when the services of a specialized carrier such as a PDN or satellite operator are added into the equation. To locate the source of a problem is difficult and more so to get the responsible service provider to perform corrective maintenance in a manner which is both timely, in the eyes of the user, and complete.

In instances where customer service and revenues are directly tied to the network, the strategic unit should strongly consider ownership or long-term leasing of the requisite facilities. A case in point is a cable TV programmer like Home Box Office (HBO), where subscribers pay for service only if they get it. HBO owns transponders on domestic satellites as well as a teleport on Long Island where the channels are uplinked (Figure 2.13). The satellite delivery of programming directly to cable systems is the optimum solution for current planning prospects.

Some strategic units wish to implement network capabilities which are not available from common carriers. Any innovative approach or novel technology, after being established through a special procurement and development project, may be proprietary. Turning the "family jewels," so to speak, over to an external company or companies is a concern for an organization which uses telecommunication as an element of its competitive strategy.

If ownership is not an objective nor even a consideration for a strategic unit, public network services can still be provided with a degree of control through network management capabilities. As discussed in Chapter 5, services of the telcos and major long-distance carriers now include customer controlled reconfiguration for T1 channels with DACS. AT&T and MCI have announced their plans for integrated network management services running on their mainframe computers. These carriers are in a unique position to facilitate network management for private networks using common carrier facilities, since both have reliable data communication networks running in parallel with circuit switched telephone networks. As long as a strategic unit is willing to turn its destiny over to the carrier, the centralized capability of network management could still deliver a measure of control.

8.2.2 Reselling

The selling of telecommunication services by the operator of a private network to other users is a strategy for reducing the cost base of owned or leased facilities. The belief is that any revenue from reselling excess capacity, over and above the cost of finding and connecting customers, is usually worth the effort. As will become apparent from our discussion, however, successful execution of the resale strategy is difficult to achieve in practice. Some private network operators even set up common carrier subsidiaries to conduct the business. For example, Sears and HBO offer generalized and specialized communication services using excess capacity on their private networks. In general, the FCC does not particularly care if a strategic unit resells capacity at a profit. Before making a deliberate move in this direction, we advise that you review the business plan with competent legal counsel who deals with FCC matters.

Any organization planning to pursue a telecommunication service marketplace should carefully consider the implications and the resources necessary to achieve the desired result. Competition from the local telcos and the major long-distance carriers is strong because of their economies of scale and widespread coverage of their networks. An exception to this rule is if the strategic unit controls C-band and Ku-band transponders on U.S. domestic satellites and wishes to sub-lease excess capacity. Third party customers may employ low-cost earth stations such as TV receive-only terminals and VSATs, which can be placed anywhere in the satellite footprint.

Aside from satellite services, reselling of capacity on a private network should probably be viewed as a sidelight and customer revenues gained from the sale of excess capacity considered found money. There would have to be a series of fortunate coincidences, not the least of which being that the customer geograph-ically located in proximity to the network connection points.

8.2.3 Foster Dependence among Customers

The strategy of fostering dependence among customers on the goods and services of the strategic unit is at the center of the value chain concept. Basically, the network allows the strategic unit to enter the value chain of its customers and to provide measurable benefit.

The information technologies allow many strategic units to enhance service to their customers with extended business hours through points of electronic access throughout a dispersed network. Customers can use their added convenience to enhance their own businesses, or, in the case of direct consumers, to improve their quality of life. Automation is extended from seller to buyer so that all may reap the benefit of reduced labor cost and potentially reduced financing cost if supply lines can be reduced as well (e.g., "just in time" manufacturing).

Telecommunication can be used to influence the customer's buying behavior [Runge, 1988]. Runge's thesis is that customers go through a buying cycle with the four stages indicated in Table 8.1. Customers need information about the product or service considered for purchase. Some of the questions are made prior to purchase (What do I need and what products are available?), some concurrent with it (How do I order the product and how do I pay for it?), and others after the transaction has been consummated (What is the delivery status and how can I get repair services?). A telecommunication network can be used very effectively to allow these types of information to flow between customer and seller. A strategic unit which builds its network to meet these needs can potentially have a competitive advantage over other sellers who do not.

An example is Buick's Electronic Product Information Computer (EPIC), used by dealers to assist prospective car buyers and thereby help them sell cars. First, EPIC helps the buyer to make an appropriate choice of car model. The terminal can display graphic images of various vehicles along with descriptions of optional features. After choices are made, the computer can locate the desired model and color, whether it is on the dealer's lot, en route to the dealer, or possibly

Table 8.1 Customer Questions in the Buying Cycle [Runge, 1988]

Stage	Puzzle	Need
Requirements	How much do I need?	Establish requirements
	What are my options?	Acquire information
Acquisition	Is what I want available?	Specify
	Where can I get it?	Select a source
	I'll take it!	Order
	How should I pay?	Authorize and pay
	How can I obtain it?	Acquire
Ownership	What's the status?	Monitor
	How can I upgrade it?	Manage
	What if it breaks?	Support
Disposal	How do I return it?	Terminate use
	How much did I spend?	Account for it

at another dealer. An order can be placed directly with the factory if the particular combination is not otherwise available. The clarity and rapidity of this type of telecommunication application helps the customer move toward the decision to buy the product, raising the dealer's sales rate.

Direct order terminals are provided to hospitals and drug stores, and airline reservation systems support travel agents and bulk-buying corporate customers. Access of customers to the seller's database of pricing and product availability could pose some competitive risks for the seller. For example, a potential buyer could use the network to determine how much excess capacity or inventory the seller has and then attempt to extract a substantial discount for a cash purchase. After careful consideration, a strategic unit could decide that this risk would be more than offset by the benefit of customer satisfaction with having reliable information and the ability to place orders quickly and conveniently. Also, the system could allow the seller to reduce staff and the costs of business.

The automated teller machine (ATM) used by many banks and savings and loan institutions in North America is a good example of private network extension, which ties customers to the strategic unit. This author prefers to bank by using the ATM of a particular savings and loan because of the speed and convenience. The network is almost always running, allowing the customer to make deposits or withdrawals of money day or night from many branch locations, including the one within walking distance.

As videotex services become more established in the U.S., to see which customer services become the most successful will be interesting. On the French Minitel-Teletel network, subscribers can place orders directly with merchandisers who arrange for delivery of products to the subscriber's home. The technology and variety of services of the Minitel system will eventually become a facet of

American consumerism. Those businesses which find a gainful way to use these networks to reach customers and to satisfy needs will certainly win in a strategic sense.

8.2.4 Strategic Use of Public Network Services

There are very few cases where a large private telecommunication network can be built and operated without relying on public telephone and data networks to some degree. As discussed under the control of one's own destiny strategy, public networks can be controlled through the greater degree of sophistication that digital technology provides. In Chapter 6, we presented some concepts for determining the quantity and mix of dedicated circuits and switched services from public networks. The proper balance between these two will be difficult to achieve at start of service, and will change over time due to the particular business dynamics.

Indeed, the competition among the long-distance carriers and PDNs is very fierce. Each carrier invests for expanding facilities to achieve the greatest economy of scale. Cost leadership is an effective competitive strategy, which cannot be achieved by multiple suppliers in the same markets at the same time [Porter, 1980]. Therefore, the carriers are sitting with excess transmission and switching capacity, waiting for the clever buyer to approach them. In addition, carriers are enhancing their networks with special features for bulk calling, telemarketing, and network management.

Low-cost switched services, deeply discounted wideband DDS, and private line channels at T1 rates should receive careful consideration in any network development project. Other carrier services like CENTREX and CO-LANs also could be economically justified for operations within a metropolitan area. In using carrier services, however, we must beware the possibility that the current buyer's market will probably yield to a sellers market. This could happen if the carriers cease construction and demand subsequently meets supply. Perhaps this may only happen in a certain segment or perhaps it will not happen at all in the foreseeable future. Binding bulk purchases with a long-term contract may be advisable, provided that some flexibility can still be obtained. The reader should be cautioned that the situation could change rather suddenly. If it costs a lot of money to make any changes, then the "good deal" from a motivated seller today easily evaporates with the passing of time.

A competitive advantage can be gained if the strategic unit is able to reduce its telecommunication cost. Reduced cost tends to increase the margin in a financial sense, or allows the unit to offer its product at a more competitive price. Cost containment is also important because telecommunication service providers may raise prices substantially as overall usage begins to meet capacity. Having some excess capacity on the private network should facilitate future growth in demand for services, obviating the need to go back to the market for equipment or services.

8.3 CASE STUDIES

The following case studies are examples of the ways in which telecommunication facilities and services are used to pursue business or network strategies. The concepts of the five competitive forces and the value chain should influence a strategic unit in its plans for implementing telecommunication networks. Gaining a competitive advantage, however, may involve going a step further by using one of the basic network strategies such as that of fostering customer dependence. Telecom periodicals such as *Network World, Communications Week, Telecommunications,* and *Business Communications Review* publish numerous articles on network implementations and strategies.

8.3.1 Corporate Digital Backbone

Corporate digital backbone networks are becoming very popular among dispersed organizations with heavy trunking requirements. The T1 transmission capacity available within and between North American cities is attractively priced, and the carriers are very interested in signing deals with users. Equipment suppliers have introduced a wide range of products employing digital technologies. These devices along with the service offerings of common carriers, as discussed in previous chapters, lend themselves to the implementation of a versatile private backbone network. Only the largest of industrial or government organizations, however, can justify the investment and annual expenses associated with such an endeavor. The rationale most often used is that of controlling or reducing costs over the long term. This implies that the organization has already established its need for private telecommunication, and is now moving toward optimizing its network design. An organization may also believe that control of destiny is equally important to the business strategy of the unit. Having ownership as well as technical control perhaps will protect the unit from the competitive forces exerted by suppliers of telecommunication services, and even rival firms which do not have equal control of their telecommunication resources.

Recall that the corporate digital backbone typically uses large switching or multiplexing nodes with either T1 or DS3 trunks in between. The transmission capacity would most likely be leased from an ensemble of common carriers. This, of course, brings in the potential problem of a loss of technical control of the facilities due to the interfaces between different entities (local telcos and long-distance carrier). In major cities, T1 channels may be employed bypassing the local telco. These could be owned by the strategic unit or obtained directly from the long-distance carrier. Recently, organizations have begun to offer T1 services within downtown sections of major cities like New York and Chicago, with Merrill Lynch being one example. There is evidence that intracity T1 services are being driven lower in price by this competition. The key to successfully making these deals and connections is to be sure that interfaces are properly defined and contracts correctly put into place for the desired term.

Switching equipment such as PBXs would normally be purchased and operated by the strategic unit. Today, these devices are implemented in such a way that new features can be added in the future. Characteristics of PBXs and the alternative, CENTREX service, are reviewed in Chapter 3. Since this case study assumes that the transmission between nodes is primarily on a private T1 backbone, owned PBXs will more likely be used. These should be equipped with a direct T1 interface to minimize per-line investment costs. Returning to the features of the PBXs, to consider the functions that the strategic unit would expect the telephone system to accomplish is best. Heavy use of telemarketing will effect the selection of the PBXs because the computational and software support are directly related. We would advise you to employ a common switch architecture so that additional systems like ACDs could be eliminated.

The other key nodal element is the high-level multiplexer. These devices (discussed in detail in Chapter 4), improve the loading of T1 and DS3 transmission "pipes" by such techniques as dynamic bandwidth allocation and voice compression. User interfaces can be configured for the transmission bandwidth corresponding to the data communication application and PBX interface. This is also very effective in achieving efficient loading of the network. Beyond the purely economic issue of efficient loading for lower cost per unit of bandwidth, the multiplexers provide a high degree of flexibility. Traffic flows can be rearranged to match the needs of the organization. The user services can be reconfigured locally or by remote control. Also, equipment can be physically relocated. Thus, we very strongly suggest a common equipment architecture and source of manufacture.

Network management is an important aspect of the digital backbone. With the wide variety of equipment and the diversity of T1 transmission facilities, the strategic unit must gain and maintain technical control. Integrated network management systems, such as NMCC and NetView, represent the most advanced approach in dealing with diverse facilities and telecommunication network architectures. To be able to equal the public switched networks in terms of reliability, the network management system must extend to every node and trunk in the private network. Referring to Figure 8.4 (Section 8.2.1), the corporate digital backbone would need to lie in the upper half of the matrix of control. The choice between facilities owned by the strategic unit or leased from common carriers is perhaps an economic decision with implications for the value chain.

The specifications for a private backbone project are developed for substantial trunking requirements between a relatively fixed set of locations. The network, once implemented, can carry both internal traffic as well as traffic destined for external locations. A backbone can be built within a metropolitan area, within a larger region such as a state, or nationwide. The nodes and links are basically the same. The network would not be application-specific because of its ability to integrate many services together, providing a common highway system for the administrative traffic of the organization. Customers would not generally be connected directly to this network due to the cost of extending T1 service to hundreds or thousands of locations. Instead, connection would more appropriately be the role of the local telco because the backbone interfaces with public networks through either multiplexer nodes or PBXs.

The backbone network provides basically point-to-point connections which can be installed and removed in response to user requirements. Packet switching systems can therefore use the network to interconnect PSNs with each other and to provide dedicated access lines to the user's data communication equipment. There is currently a great deal of interest in providing packet switching capabilities within the PBX itself. This is one of the approaches taken in ISDN (see Figure 8.5), which should be kept in mind when developing the architecture of the backbone.

Figure 8.5 Digital switching system: NEAX 61 which integrates digital voice with packet switched data. (Photograph courtesy of NEC Corp.)

The key to meeting evolving needs is the integration of all services on the common network. Users within the strategic unit would develop new requirements for particular services such as a multidrop data line or a wide bandwidth link for compressed digital video teleconferencing. The telecommunication management of the unit would convert this requirement into a circuit or equipment configuration to be implemented as a change to the corporate backbone. The engineering would be simplified by using the database of the network management system, and the service could be started perhaps with only software commands over the monitor and control system. In the ideal case, technicians would not even need to visit the sites in question. Being able to respond to user needs in this fashion could be the way of achieving competitive advantage in a particular market.

8.3.2 VSAT Transaction Network

Satellite communication is perhaps the only telecommunication medium which can reach locations without the use of public networks. VSAT technology (discussed in Chapter 4) reduces the cost and complexity of satellite communication to the point where direct service to a variety of remote locations can be economically justified.

In this case study, we consider the example of a VSAT network used in the hotel-motel industry. A national hotel-motel chain has implemented a transaction-based data communication network, exclusively by satellite, to provide reservation services and other assistance to franchised hotel-motel owners. The chain provides the VSAT to each owner and operates the main hub earth station, which is collocated with the headquarters' data processing center. Economic studies were performed to demonstrate that this private network would be lower in cost than using PDNs or leased lines. In terms of the value chain, the reduced expense of this network implementation gives the hotel-motel chain a competitive advantage with regard to the prices that are charged. Additionally, having every member tied into an efficient reservation network would assist the guests with making room arrangements during an extended trip.

Strategically, the VSAT is the lifeline of communication between the chain's management and the franchisee. Many of the members are in remote locations, where quality leased lines are not available. Satellite communication is inherently reliable, reaching every location with stable links which are in operation more than 99.9% of the time. With the network cost justified for the basic reservation application, other uses are obtained at very low incremental cost. The same ground antennas will be upgraded with the addition of an inexpensive receiver for specialized video programming. For example, training on the use of data communication links and services is easily provided. Announcements of new products or services would also flow over the video pathway. The same video system could be used to generate additional revenue for the members of the chain when other organizations pay to attend public or private teleconferencing sessions at the hotel's meeting facilities.

From the standpoint of the hotel-motel chain, the franchisee is the customer. Therefore, the VSAT network implements the strategy of fostering customer dependence, tying the customer and chain more tightly. The network is totally under the control of the chain management because it operates the critical central element: the hub earth station. Satellite networks are well integrated, providing nearly total control to the network operator. Any single VSAT location experiencing difficulties would employ the public telephone network for backup on a dial-up basis.

8.3.3 Reseller of Excess Capacity

In applying the resale strategy, listing examples of success stories is difficult and finding failures is easier. A private network is usually not conceived and implemented for the provision of services to outside customers. If it were, the network probably would not be optimum for the strategic unit. Contrast this with the situation of the common carrier which arranges its facilities to reach the largest possible customer base.

We mentioned previously that transponders on a domestic communications satellite could often be resold to other users. This was the situation for one particular data communication company that could only use a fraction of its purchased satellite capacity for internal needs. Through a fortunate set of circumstances, the excess transponders were found to be very desirable for the distribution of cable TV programming (which was not the owner's business, but a viable satellite application, as discussed in [Elbert, 1987]). The owner was able to sublet all of its excess transponders for reasonably high rates.

Conversely, a large aerospace company implemented a terrestrial microwave network to meet internal transmission needs in a major metropolitan area. Resale of excess capacity was planned in advance, and a common carrier subsidiary was set up to operate the system and to market T1 capacity on the links. Revenue from the captive customer (the parent company) was not sufficient to show a profit from the operation of the network, so other paying users were needed. The base revenues were low because capacity was also readily available from the BOC in the area, establishing a ceiling on pricing.

Marketing efforts over a number of years produced only a few orders, and even these were canceled after a relatively short operation term. The biggest problem was that many of the points of access to the network were on the premises of the parent company, and facilities from the local BOC would be needed to reach other customer locations. The alternative was to face heavy construction expenses to extend the network to satisfy the particular requirement. Under these circumstances, the only solution was to sell the network to another common carrier. This buyer (not the BOC) had other facilities in the same area and nationwide, and therefore could enhance the offering. One of the attractions to the buyer was the one large existing customer.

Another example of network reselling would be in the area of packet switched data communication. Several packet networks were implemented as part of an overall data communication application environment. Tymshare established its Tymnet PDN to connect time-sharing customers to its computers. The Tymnet PDN has proved to be a successful telecommunication business, having been bought and expanded by McDonnell Douglas. Computer Sciences Corporation is similarly expanding its business base by moving from data processing to data communication with its Infonet PDN. In 1988, CSC began selling interests in Infonet to foreign PTTs, thus assuring its extension throughout Europe. Both of these networks are actually much larger businesses than the time-sharing systems that they were designed to support (not unlike the tail wagging the dog). You should keep in mind that the origins of these network businesses were service businesses. The strategic unit was set up as a reseller of capacity, but in this case the capacity existed in computer systems rather than telecommunication facilities.

The success stories indicate two possible ways in which network reselling can prove of value to a strategic unit. With regard to transponders, the owner had acquired valuable assets. Transponders in orbit are like real estate condominiums. One of the basic rules of real estate also applies to satellites in orbit: the three most important things to remember are location, location, and location. In this instance, the location is that of the satellite, where a particular segment of the orbital arc over the U.S. is more desirable for video services. For the case of Tymnet and CSC, those organizations recognized the value of the business and made the commitment necessary to succeed. Both were already experienced and successful as facilities-based service organizations.

8.4 DIRECTIONS FOR TELECOMMUNICATION STRATEGY

Private telecommunication networks have important roles to play in helping organizations develop and implement business strategy. We see that information technologies can provide valuable business assistance if the strategic unit can take proper advantage. These media provide what could be described as electronic channels of distribution, paralleling the physical channels on which all businesses depend.

Strategic units are assisted by consulting organizations experienced in developing business strategies based on the application of information technologies. The principal providers of business consulting services are the major accounting firms, such as Arthur Andersen, Coopers and Lybrand, Peat Marwick Main, and management consulting organizations, such as Booz Allen and Hamilton, Boston Consulting Group, McKinsey and Company. Each of these firms maintains an information technology practice of considerable size. Relating specifically to telecommunication strategy, the DMW Group, the Gartner Group, and Telecommunications International, part of ESD, are some of the larger specialists. There are even so-called "boutique" consultant companies which serve specific segments or regions of the country. Examples include Communications Center (satellite communications), Telecom Resource Group (assistance to telecommunication marketers and users of telephone systems), and the Egan Group (space applications). One primary reason for using a qualified consulting organization is to obtain information on workable strategies being applied elsewhere.

Using these electronic channels can provide important benefits. The structure of the organization can change more rapidly in response to a changing business environment. As suggested by Keen, the availability of a modern telecommunication network can condense the time needed to win customers or deliver a product, both of which are important in today's economic environment. The fact that telecommunication networks extend worldwide means that services can be provided almost independently of distance. An efficient telecommunication infrastructure such as a digital backbone or VSAT network will allow the organization to extend its business incrementally at relatively modest cost. Such extensions could provide a competitive advantage for entry into a new market. Alternatively, information technology can add intangibles which customers appreciate, providing differentiation and binding them more tightly to the strategic unit.

Chapter 9
Conclusion

In this summary chapter, we review many of the key factors for implementing a successful private telecommunication network. The constituents of telecommunication networks were covered in sufficient detail in previous chapters to give the reader a reasonable feel for the tasks at hand. Our conclusions here are not a restatement of every principle covered in the book. Rather, we present elements of philosophy for effective telecommunication planning. The environment is the strategic unit, either in commercial business or government service, where use and service demands are growing and changing. We present the areas for primary focus of management's attention.

We then delve into the future possibilities for the technology and application of these networks. This author does not purport to be a telecommunication guru or visionary. There are others who spend their careers trying to predict the future of technology. My aim is to highlight some of the more salient trends in technology and its application. This view is based on current developments, so we cannot say anything about technologies that have not yet been invented or put into practice. An important part of this future is the development of the profession of telecommunication management. This field barely existed ten years ago, but today it is recognized as a prominent specialty within organizations that rely on telecommunication as a part of corporate strategy. Training of people for this field is still somewhat disjointed, with considerable confusion between the technical aspects (engineering) and the business concerns (management and finance). Recommendations are given for educational programs which can better bridge the gap.

9.1 WHAT ARE THE KEY FACTORS?

A good way to gain a foothold in a new area of study is to find a guiding light of some kind. Managers like to ask for the "bottom line" when reviewing new information. Rather than being financial, the bottom line is that concept or grouping of concepts which really matters in the particular situation. During this author's military experience, the bottom line or mission was to get the *commo* (communications) in, with point A and point B talking to each other. Sometimes, field or combat conditions made the simple matter of setting up a line-of-sight microwave link a nearly impossible task, but we kept the bottom line fixed in our minds so that the job was not done until A and B could talk.

The bottom line in private telecommunication networks today is not as simple as connecting A to B. We must understand the wide range of possibilities, including those of the applications as well as the means to implement them. Technology changes, new service offerings from common carriers often appear, and the regulatory environment bewilders. If only one guiding light were enough, it would be rather simple. We hope that the following paragraphs will help guide your way through the complexity.

9.1.1 Understanding Technology Trends

Traditionally, managers have an operational background, having moved up as buyers of telecommunication services from common carriers. Managers understand how to keep things going. The new environment demands technical comprehension on the part of telecommunication management. The fortunate individual has had exposure to the technical side of the field. Some may enjoy science or be tinkerers, but the most effective people are those who have an appreciation for what the technology is able to do. You need not be an electronic circuit designer to employ a functional device such as a multiplexer or packet switch. You should, however, have a feel for what one of these devices does within a network. Also, to compare devices which are supposed to perform the same function, you need to understand the performance parameters. To quote Goethe, "He who does not know the mechanical side of a craft cannot judge it."

A practical way to view the construction of a network is through a block diagram built from functional "black boxes." Most electronic systems are build up from functional elements such as modems, multiplexers with port cards, or radio transceivers, which are usually packaged individually. The block diagram shows which black boxes are used and how they are arranged. What, however, is the box supposed to do and what are the connections to it? The inputs and outputs must be correctly defined in the form of an interface specification. The performance of the box also needs to be specified. What goes on inside the box in detail is determined by the designer and manufacturer. Of course, electronic technicians, not managers, must be familiar with the internal workings of the electronic units so that alignments and repairs can be performed when necessary.

Due to the complexity of software-controlled equipment, block diagrams do not tell enough of the story for certain arrangements. Unfortunately, there is no easy solution for this. A given nodal processing element could have thousands of lines of computer code stored in its memory. The programmers developed this code so that the equipment could satisfy operational requirements. An important aim is that the system be "fail-safe," meaning that any disruption of operation will not cause the problem to spread like a virus throughout the network.

Most telecommunication black boxes employ microprocessors and software control, so it seems logical that a telecommunication professional should have a good understanding of computer systems. Therefore, some experience with computers is highly desirable. This can be obtained by working with small systems, such as personal computers, or with much larger systems used in scientific computing or business data processing.

Communication systems use transmission media to carry the information between nodes in the network. These media include electrical conductors (wires and cables), radio (broadcasting, terrestrial microwave, and satellite), or lightwave (optical fiber and lasers). The problems of installing a physical cable are decidedly different from establishing a radio communication path. Exposure to each of these fundamentally different approaches is useful when building a telecommunication network. The simple cordless phone or its more complex brother, the cellular telephone, provides a good example of radio communication. From the cable viewpoint, the home intercom involves basically the same concept. The building block approach is suitable for understanding transmission systems.

When an individual realizes that he or she needs to acquire a basic understanding of telecommunication technology, obtaining the necessary training is a relatively simple matter. This author has found it helpful to have an expert explain how an existing system of some type operates and works. One of the nice things about telecommunication is that if you can understand one particular system, it is relatively easy to gain a comparable understanding of another one. As you study more systems, your overall comprehension grows even more rapidly. The first exposure also provides the starting point for additional study, using reference books and seminars. A person with the interest and motivation can become educated through a combination of formal training and individual study. More detailed recommendations for telecommunication management education are provided at the end of this chapter.

9.1.2 Understanding the Requirements

Telecommunication management cannot develop a network strategy and implement a network capability without knowing the requirements that need to be satisfied. The absolutely wrong way to do things is to adopt a new technology like fast packet switching or VSATs without first seeing what people in the organization need. Quite obviously, some technologies are more effective than others when moving from the requirements domain to that of the network itself. Many of the chapters in this book, like others on telecommunication, provide considerable detail about technology. That is what people expect, but the real task is to make the proper match between needs and facilities.

Technologists understand how things work, but business people understand the function of things. This leads to a kind of schizophrenia, where a person who understands both views is torn between a curiosity about the bells and whistles of a technology and the practicality of what it can actually do. Obviously, you need a "bag of tricks" from which to draw ideas. The managers, however, must be in touch with the organization and its business strategy. This perspective was outlined in Chapter 8.

How do you find out what the users require? We discussed in Chapter 6 a few approaches for collecting requirements from users. An existing network, even one that is fragmented (and most are), can be a valuable source of information about requirements. If the network is not equipped to provide accurate readings of traffic flow and utilization, you can infer these characteristics by working backward using traffic engineering principles. If we are talking about a start-up situation, we can only use the survey technique, whereby the prospective users are asked questions either in writing or in person. The results are then tallied and the answers considered in the network design process. Of course, surveys suffer from the familiar problem of "garbage in, garbage out."

Many large organizations conduct user conferences where all who have a vested interested in telecommunication gather for a series of discussions and workshops. Everyone then has a chance to speak his or her mind and subsequently commit to the final architecture. There are two problems with this approach (although every approach has problems). First, the users will typically inflate their needs or have no idea of what they want. Second, someone must still rationalize the information after it has been collected. Solomon probably lacked the wisdom now required to make such decisions.

9.1.3 Performing the Trade-offs

A trade-off or trade study is an analysis of the measurable outcomes for each of a variety of approaches to performing the same tasks. We are accustomed to comparing things on the basis of their value to us. For any possible approach to a problem or need, there are a number of ways in which to proceed. Each approach has a unique set of ramifications or costs to bear. Using trade-offs, we attempt to quantify or measure the consequences of each alternative so that we can compare them side-by-side. In the context of private telecommunication networks, the most straightforward criterion to use is cost, treated in Chapter 6. There are many strategic business issues that may be just as important as cost, which forces us to introduce qualitative factors into the comparison. For example, when we look at network architecture, we should introduce aspects of flexibility of use and adaptability for changing requirements.

There will be many times when significant changes are being considered for an existing network. Performing the trade-offs in this case involves the detailed configuration of the network in terms of its nodes, transmission systems, and user equipment. Ideally, the current network is specified in a database of the type discussed in Chapter 5, and can be analyzed in an on-line computer system. The "what ifs" can be checked rather conveniently by making the changes in the stored configuration without touching the actual network.

Major changes in the network or the creation of an entirely new network are approached differently. The lack of precise information should not deter us from trying to perform the trade-offs. A *scenario* is a particular set of assumptions with all of the predicted significant results laid out. There is obviously the potential for significant inaccuracy when we make projections. Computer spreadsheets would be used to examine alternative scenarios. If the alternatives treat dissimilar approaches, such as assuming an ISDN-based terrestrial architecture with services provided by telcos *versus* an integrated network of private satellite earth stations and leased lines, then things ought to be kept as simple as possible. If we make the model too complex, the people using it will not be able to comprehend the model, much less the results. The comparison should be made over a range of requirements to determine the sensitivity to critical factors in the network architecture.

9.1.4 Management Must Manage

The phrase, "management must manage," was coined by Harold Genneen, former chairman of the board of ITT [Genneen, 1980]. Mr. Genneen is well recognized for the deliberate and highly organized style of management that he advocated and practiced during his years at the helm of that multinational corporation. He demanded of himself and his senior managers full understanding of activities within their organizations, not just in general terms, but in specific details, particularly financial. In the telecommunication context, this phrase says that the managers of the network must become involved with the details of the technical operations and finances. We are not suggesting that the telecommunication manager understand the design of each element at the circuit level, nor that he or she must know the cost of every piece part. We suggest instead that the leadership be familiar and comfortable with the functional aspects of each significant element, how the elements work together to implement the network capabilities, and the nature of the cost drivers. A *cost driver* is a technical or operational factor under the organization's control that has a direct result on the cost of doing business or running the network [Porter, 1985].

Senior managers must delegate the detailed responsibilities for doing the work to those in a better position to do so. There are skill levels in electronics, telecommunication, and finance which the managers cannot be expected to have, but the leaders must be involved in the initial stages, during network design and development, and through the period when the network establishes itself as an important part of the organization's value chain. Hence, the network really cannot be left under the control of operations staff. New issues will continue to arise as the business environment changes.

We suggested earlier in this chapter that an understanding at the block diagram level was sufficient for those in the ranks of management. This understanding must come with a thorough comprehension of the functions performed in those boxes. A critical aspect of the network is how it will react to problems and failures. Major disruptions will definitely occur, and the manager will find himself or herself getting a lot of attention from others. If a major switching center experiences a fire and connectivity through that location is lost, something must be done; otherwise, a large part of the network will go down. Contingency planning, also called *business recovery planning* (BRP), is an important function to be performed at a senior corporate level. The banking industry, which relies heavily on data processing and telecommunication, has begun a careful study of how to handle serious disruptions of their facilities. The consequences for one institution, or the banking system as a whole, would be very dire if alternative facilities could not be found to carry the network. Fortunately, BRP can exploit alternatives like cellular telephones, VSATs, and private microwave systems, which can be used to bypass terrestrial facilities that have been lost in a disaster.

Returning to "management must manage," the job of the telecommunication manager is very diverse, involving technology on one hand and corporate strategy on the other. This individual has important management functions to perform just to keep the network in good working order and responsive to user needs. On the other hand, a close association with corporate management is essential if telecommunication is an important ingredient of the overall business strategy. As discussed in Chapter 8, the business strategy delineates how the organization positions itself in its industry. The proper approach is to have a deliberate strategy for doing business, which also requires that telecommunication be approached in the same way. As indicated elsewhere, some corporations and government agencies recognize the importance of telecommunication by designating the chief information officer as a top management position.

9.1.5 Working with Vendors and Consultants

Working with support groups outside of the organization is appropriate for a strategic unit which itself is not in the telecommunication business. Vendors provide the technology, facilities, and services, which are actually used to implement the network. Therefore, we cannot proceed without vendors. In Chapter 7, we provided a framework for dealing with vendors by way of the process of buying their wares. With the wide variety of products and services on the market (and the regulatory freedom that we now have with which to employ them), almost everything that we may want to do apparently can be done with off-the-shelf items.

Much has been written in the trade press about developing a "strategic relationship" with a vendor. The concept is to work essentially as business partners with a particular vendor who has an appropriate technology base. From another perspective, this relationship is like contracting with the vendor to develop a new technology. We reviewed in Chapter 8 that new technology development has its risks, both financial and functional. There may be cases, however, where the business needs can only be met with a new technology or repackaging of an existing technology. A special relationship with a vendor may be the best way to accomplish this aim.

Organizations that do not have a cadre of experienced specialists can use consultants to fill the gap on either a temporary or more permanent basis. Consulting help in engineering or business fields can be obtained from individuals acting on their own or from larger consulting companies. The individual consultants can be found by referrals from former clients or through professional organizations. Consulting companies are often associated with the large accounting firms, as discussed in Chapter 8. Others operate as independent consulting organizations in telecommunication or information technologies. A good consulting team is worth its weight in gold! Assuming that ongoing support is not required, the expert services of a consultant can be obtained at significant savings as compared with hiring and maintaining the staff necessary to provide the same help. Also, exposure of the internal staff to enough diverse project activity to allow them to expand their understanding of new networking architectures and strategies may not be feasible.

There is still the problem of selecting a particular consultant and verifying his or her technical competence. Use of a major consulting organization will supposedly eliminate the risk of selecting someone who is unqualified. The quality of the consultant's output, however, is very much a function of his or her personal ability. Probably the best way to evaluate a consultant is to try the person on a small project before making a commitment to a larger and important one. Then, you simply ensure that the same qualified person is maintained on the job.

9.2 WHERE ARE WE GOING?

Before embarking on a trip to the future, let us look backward in time. James Martin, a respected author who has written many books on telecommunication and data processing, assembled a thorough picture of the future in the late 1970s [Martin, 1977]. Interestingly enough, Martin identifies essentially all of the telecommunication technologies which form the basis of today's private network picture. At that time, the preponderance of networking techniques and architectures were mainly in the developmental stage. Martin, however, saw the possibilities for applying the concepts that were technically feasible within the corporate context.

James Martin's crystal ball, however, was not completely clear. Some items appeared to have high utility, but have since proved to be suboptimal as compared to other approaches. The one that stands out is long-distance transmission by waveguide. The concept was actually to bury a continuous line of circular waveguide between repeater stations. Because the transmissions within the waveguide were at millimeter wavelengths, the bandwidth obtained was many times greater than that possible on a line-of-sight microwave link. The waveguide system would have been very expensive to install because sharp bends and mechanical distortions must be carefully controlled. In the age of fiber optics, this does seem impractical, but, back in the 1970s before fiber technology had been fully proven, the waveguide approach was at least as practical. There had been many years of research and trials by Bell Telephone Laboratories and numerous technical papers were published on the subject. Plans were being assembled for the implementation of the first operational systems, but events caught up with this expensive and complicated approach, and it was eventually abandoned for commercial use.

In general, we have two problems when predicting the future. First, we tend to be too conservative because we view the future with the eyes of today. Back in the mid-1950s, a prominent scientist is reported to have stated with confidence that a radio receiver would never be built smaller than the size of cigar box. The reasoning was that if you attempted to make it smaller, it would not be possible to fix it were it to fail. (Curiously, a layman cartoonist, Chester Gould, "invented" Dick Tracy's wrist radio many years before.) What the expert did not foresee was the *microcircuit,* which allows us to build a receiver on a chip that can subsequently be replaced in its entirety. Our second problem is that there is a massive telecommunication plant investment worth billions of dollars. For service providers and private network operators to discard this installed base while still functional would be uneconomical.

The technologies Martin foresaw to have commercial value were offshoots of other activities. Packet switching was developed for use over radio links and international phone lines of low quality and poor reliability. VSATs are compact digital earth stations similar in design to large INTELSAT facilities used in the first trials of TDMA and for demand-assigned telephony. For telephone switching, the sophisticated PBX is a marriage of the minicomputer and the digital telephone switch, first introduced in long-distance networks. The integration of voice and data in a single network was conceived in the late 1960s and early 1970s by researchers working for the U.S. military trying to find ways to improve the utility of battlefield communication. During many years of research and development, then initial trials, precursors to ISDN were evaluated by the government.

Exploitation of these technologies has taken more than a decade to occur. Necessity is the mother of invention, as the saying goes, and many approaches were certainly developed to meet specific needs. We need only look at the space programs, such as Apollo and the Space Shuttle, to see where literally hundreds of innovative ideas were deployed to implement a system which had not existed before. Nevertheless, how many of these space age innovations have diffused to other areas of the economy? Significant commercial application always appears to take a while longer, probably because of the sheer inertia of the embedded investment in current technology [Martin, 1978]. Nonetheless, if the approach is viable and has significant advantage over the current technology, it will most likely be adopted.

To illustrate the concept of telecommunication evolution, Figure 9.1 presents a path in time starting around 1978 when Martin's futuristic book was published in its second edition. In the center are the major developments in the telecommunication environment, while on either side are other aspects of the evolution of today's picture. The environment at the time of the writing of this book is somewhere around the center of the flow chart. The regulatory environment, by virtue of deregulation and divestiture, creates the pull for the application of the technologies. Equipment is more capable and less expensive today than predicted in Martin's time frame. Competition among long-distance carriers has lowered the ceiling on pricing for the critical transmission element of most networks.

Many of us think that the focus will now be on improving the interfaces between the various networking architectures and customer devices through the process of developing national and international standards. The big push by the telecommunication carriers toward a uniform architecture through ISDN will do much to rationalize today's confusion. This, however, is but one facet — a mere steppingstone toward a truly universal architecture. Another important emerging trend is competition at the local loop, where the telcos hold a monopoly.

All of these capabilities and trends have created what can be called the new corporate telecommunication utility, where services are added to a network in response to needs. No longer must a strategic unit build a new network every time it needs to implement a capability. The resources are available today to accomplish this aim, particularly through integration of services on a common digital backbone. More work is needed in the area of standards and local loop development, however, before the truly flexible utility can exist.

This brings us to the final point in the evolutionary picture of the future, shown as the last element of Figure 9.1. We can expect that a new organizational structure of the strategic unit will evolve as a consequence of the corporate tele-

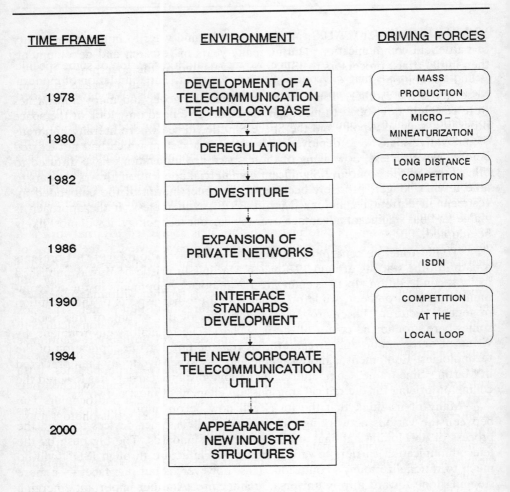

Figure 9.1 An evolutionary picture of the future of private telecommunication.

communication utility. Already, some organizations have experienced dramatic change as a result of tying the business strategy to telecommunication. The airline business is one in which telecommunication networks play a key role, almost defining the service end of their business. In banking, telecommunication is the life's blood of the system of multiple branches and nationwide banking. The companies with the greatest extension of private telecommunication are those which will be influenced most by this trend, but no organization will be immune. The belief is that the new telecommunication technologies must be adopted, or the strategic unit will lose out to ever increasing competition from others who understand and apply those technologies in an appropriate way.

9.2.1 The Digital Environment

From this overview of the future, we move into a discussion of some specific technological trends that will continue to be important. The foundation of modern telecommunication rests upon the digitization of all signals. We have provided information in Chapters 2, 3, and 4 on digital communication approaches and applications. The telecommunication industry is expending great effort to offer more capabilities, which will eventually tend to increase the quantity and quality of options. The digital approach, however, is permanent, and we may expect that any service which is analog today will be converted to digital. Digital technology is already synonymous with high quality and low noise. Witness compact audio disc and digital long-distance services, which have captured the consumer's attention and confidence.

Digital processing of information before it is connected to a network is a domain which has only begun to show results. We have reviewed the process of digital bandwidth compression, which reduces the number of bits per second needed to send a clear signal. There is another aspect of digital processing which will improve its quality and utility. Perhaps the reader is familiar with image enhancement used to restore the quality of television pictures of distant planets sent back from deep space probes. The same can be done for ordinary television pictures to render an image of very high quality. This approach will probably revolutionize television broadcasting. Similar improvements have already appeared in facsimile transmission by using machines that operate over ordinary telephone lines. With widespread deployment of direct digital connections to the circuit switched network, documents sent by facsimile will match the quality of a photocopier. The technology used today in the newspaper business (Figure 4.2) sends photographs, graphic images, and entire pages ready for printing, but the cost of the equipment is impractical for normal commercial use. Laser facsimile equipment targeted for the business customer will appear on the market, and so introduce these capabilities in offices.

The rapid trend toward the integration of voice, data, and video services will certainly continue until it becomes the norm. Organizations are already using bulk T1 transmission for site-to-site communication within metropolitan areas and between cities. The commercial market is flooded with high level multiplexers that use proprietary signaling and framing structures. As this technology becomes more defused, the importance of standardization emerges. Users currently must compromise some of the functionality (bells and whistles) in exchange for the ability to interconnect T1 networks without regard to the particular manufacturer. This limitation should disappear with the acceptance of networking standards. The local and long-distance networks already have T1 networking standards, but these are still too inflexible to meet the needs of many strategic units.

Some of the more visible ingredients of the new telecommunication service "pie" are indicated in Figure 9.2. For transmission, T1 bulk capacity is clearly attractive because of its cost-effectiveness. Couple this with user interfaces along the lines of the ISDN *basic rate interface* (BRI). Packet switched networks, either from public carriers or in the form of private packet networks, represent an efficient means of integrating dissimilar data requirements. At the trunk level, packet nodes can employ the T1 superhighways. VSAT networks also fit into the pie, where many remote locations must be tied to a high quality network utility. We can anticipate improvements of VSAT technology in the bandwidths that can be carried and in the cost per installation.

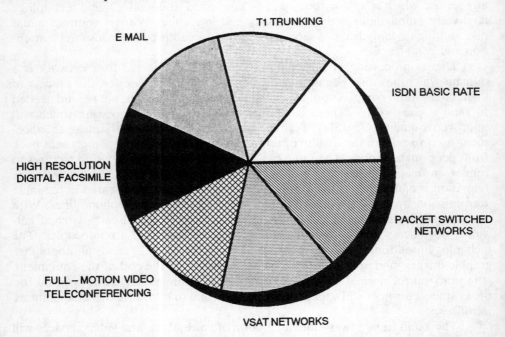

Figure 9.2 Some of the key ingredients of the evolving private telecommunication future.

With respect to applications, E-mail has proved itself as a common mode of achieving office automation for multiple sites. E-mail has the advantage that, with suitable interface and routing standards (i.e., X.400), messages can be created by anyone and sent to both internal and external destinations. Generalized development of applications has been hampered by incompatibilities in computer architectures and data communication network structures. The U.S. government is attempting to resolve this problem through standardization (e.g., GOSIP). Since the TCP/IP protocol suite is so pervasive in the Department of Defense, the government has chartered the development and use of gateways from TCP/IP to

OSI. Software available from MITRE Corporation can perform this gateway function "on the fly," meaning that protocol conversion and reformatting is done in real time without use of disk space. What this suggests for the future is that the gateways will be performed automatically without the user recognizing that it is happening. The conversion will be transparent, activated selectively depending on the systems on both sides of the interface. This is analogous to drivers found in modems and computer access ports which automatically adjust to the user's data rate and character format. In general, sophisticated conversion gateways may ultimately eliminate the need for a universal standard architecture for data networks.

Teleconferencing still represents a deluxe application due to the high cost of signal processing equipment and meeting room facilities. Perhaps we can expect to see a low-cost version whereby individual videoconferences can be conducted. The integrated voice-data-video terminal or work station may well prove useful in business applications of the future. The facsimile approach for sending reproductions of printed pages and graphic images has, however, nearly reached critical mass. Now, the direction for facsimile will most likely be toward better reproduction quality and faster speed of transmission.

Operators of private telecommunication networks accept the concept of integration at the trunk (T1 or DS-3) level, allowing the purchase of bandwidth at a bulk discount. The only problem with this is that the strategic unit does not always need this amount of bandwidth between every two points in the network. Introduction of the ISDN BRI has the benefit of giving users a smaller chunk of bandwidth. With two 64 kb/s circuit switched lines and a 16 kb/s packet switched line, many of the needs of remote and branch office locations can be served. Currently, there is a search for viable applications for the BRI, but this is a search ahead of its time. Among pioneering users of BRI, in 1988 the University of Arizona committed to installing almost 15,000 ISDN lines throughout 117 buildings. One motivating factor was the desire for campuswide connectivity with a nationwide 56 kb/s scientific research network. Installation is being provided by U.S. West, using AT&T's No. 5 ESS. More applications will evolve as the importance of telecommunication grows in the business strategy. Until such time that the BRI is universally available, VSATs probably offer the only uniform means of delivering ISDN-like services throughout a nation or hemisphere.

An area of technological research and development is that of broadband ISDN. At present, we are unclear as to the capabilities that will be introduced into public networks using very high speeds of digital transmission, both at the user interface and on trunk lines. Fast packet switching can be performed at hundreds of megabits per second, or even at the gigabit level. At these speeds, true integration of all digital services — no matter what their individual bandwidths or message statistics — becomes possible. Advances in the fast packet approach provide the means of performing information switching and multiplexing within a single facility.

Broadband ISDN on the subscriber loop is a way to bring virtually any communication service to end users via a common fiber optic circuit. Today, these bandwidths are just not needed as the present copper wire approach is adequate for the analog and digital applications currently foreseen. Cable television employs conventional coaxial, and this is a potential target for local telcos, which may wish to expand their domains by using fiber loops. Many businesses, however, would relish the thought of having fiber installed at every location, obviating the need for special construction. An example of a broadband ISDN terminal is shown in Figure 9.3.

9.2.2 Advanced Network Management Concepts

Network management systems are becoming a facet of private telecommunication, like the PBX and statistical multiplexer of the past decade. In Chapter 6, we reviewed the architecture and functionality of network management, focusing on the use of an integrated environment with a centralized *network operations center* (NOC). This field is relatively new, but the common carriers have been pursuing it for several years already for better management of their businesses. All of the current approaches rely upon human operators. When problems occur, personnel at the NOC must be familiar with the network and the means to correct problems.

The area of promise in network management is that of *artificial intelligence* (AI). The particular segment of AI that applies is called *expert systems* wherein the computer contains the knowledge and analytical ability of one or more human experts. (Expert systems were introduced in Chapter 6.) When programmed with the needed information for the particular application, the expert system can be used to diagnose problems and to recommend the most appropriate solution. The effort needed for gathering and organizing the basic "expert" knowledge is immense. This task is compounded by the fact that the trouble-shooting requirements of a typical system are continually changing, thus making programming of the expert system an ever incomplete job. The potential for simplifying the human problems of maintaining a complex, multivendor network is still probably worth the effort. The all-digital network facilities of the future, with both public and private elements, should permit users to monitor and control their respective network resources. The ideal situation would be for an expert system to recognize the existence of a problem or failure, to take remedial corrective action, and then to inform the appropriate party of what was done. Users would not need to call as sensors throughout the network would respond to trouble much more quickly than anyone could notice.

For this we will need network management standards so that messages containing vital monitor and control information can be routed between the points

Figure 9.3 Broadband ISDN demonstration. (Photograph courtesy of NEC Corp.)

where an event occurs and where action can be taken. This is potentially one of the more important applications being developed for the OSI model, as mentioned at the end of Chapter 5. Agreement will be needed among domestic and international common carriers, equipment vendors around the world, and large users of telecommunication services with an investment in their own facilities. The fact that a usable standard for E-mail, X.400, has now appeared through this same process implies that a network management application protocol is likely to become practical. Interested parties will then specify these message formats for all network management functions so that equipment suppliers will be obliged to comply.

9.2.3 Mobile Communication

The cellular radiotelephone is not just a status symbol — it is an important tool of business. Although constrained by limited bandwidth, the mobile radio systems have unequaled flexibility and independence from the physical terrestrial infrastructure. Here, the trade-off between functionality and quality falls on the side of the former. The more important aspects of mobile communication have to do with access to points where information is provided or needed, and integrating the entire operation of a strategic unit. Public safety services like police and fire have come a long way in their use and integration of mobile communication. The "dial-911" emergency systems are now at the leading edge of network development.

In the corporate sector, Federal Express maintains its leadership in the overnight package delivery business with their mobile and hand-held data communication units, which extend their network all the way into the hands of field personnel.

Many organizations already allow their employees and customers to access information services over the telephone network. Current cellular service is hampered by difficulty with data throughput caused by variations in the radio path. This problem can be overcome by tailoring modem and terminal characteristics to the performance of the cellular radio channel. At the present time, however, the cellular radio system employs analog transmission. Conversion to an all-digital network using ISDN-like principles is probably well in the future. A new generation of digital radio telephones could connect to digital cell site equipment operating in parallel with their analog counterparts (which are needed to support the existing base of analog mobile units).

Paging systems, on metropolitan and national bases, are an established part of our telecommunication structure. An added dimension would be for the network to provide *position location* in connection with paging and broadcast data transmission. Position location (or *radiodetermination*) is potentially valuable to the trucking and transportation industries. A satellite-based approach, like that being developed by Geostar, appears to have an advantage over an approach using radio towers, since total coverage of a nation is highly desired. An interesting application using Inmarsat is being developed in Europe to aid in police work. A burglar alarm installed in trucks and railroad cars can signal intrusion through the satellite link. This kind of innovation will provide more of a base for newer technologies such as position location which is currently in its infancy.

The combining of cellular radiotelephone with paging and radio location is the objective of the mobile satellite program. A single entity in the U.S., the American Mobile Satellite Consortium, has been established with authority from the FCC to pursue this opportunity as a commercial enterprise. A pair of large, high powered satellites operating in the 1.2 GHz L-band are planned for the early 1990s. Studies are still ongoing to identify the most attractive applications and the technology to support them. The type of terminal will be similar to the current cellular radiotelephone, although the requirement to cover the sky will result in a vehicular antenna that is somewhat different in appearance from the familiar helical whip. Because conventional cellular radiotelephone service is not available across most of the geographic area of the U.S., the satellite system will raise the overall standard of mobile communication for the nation.

9.2.4 Regulatory Change

The significant regulatory changes of the 1970s and 1980s have had a pronounced effect on the telecommunication industry and the way in which all organizations conduct business. For some these changes have opened incredible opportunities, for others they have created incredible problems. Every strategic unit must go forward on its own to create and implement a network strategy. We can anticipate more changes in the regulatory environment as governments continue to relieve their grasp on telecommunication resources and markets. Organizations are now exposed more than ever to the intricacies of telecommunication systems by virtue of all of the options that now present themselves.

Even with deregulation, there is a vital role for the FCC to play. The Commission's regulation of radio frequencies is essential to maintaining an orderly usage of the spectrum. Furthermore, they are the U.S. representative in the International Telecommunication Union (ITU), where frequency allocations and rules for their use are made and enforced. In this interdependent world, bodies like the ITU also support efficient electrical communication within and between countries. Particularly important are technical standards for telecommunication networks which must interface with each other. In the past, strategic units could only deal with the international community through their respective governments, but the new environment forces the largest users to become active on the international scene.

More effort will be required to overcome some of the lingering problems which still restrict private telecommunication network development. Oddly enough, local codes and ordinances make it difficult to implement VSAT networks. Many communities prohibit customer-owned antennas on rooftops. A reason for this may relate to esthetics, but steps can be taken to conceal antennas. Another aspect is that local regulatory bodies may be operating from incorrect assumptions, such as fear that microwave radiation from an antenna can pose a health hazard. As companies increase reliance on satellite networks to reach remote locations, the local zoning boards and regulators will likely develop more enlightened attitudes.

We have mentioned the trend toward deregulation of the local telcos. When this becomes more concrete, the competition for the local customer ought to produce interesting results. Imagine the possibility of having a choice as to who provides your local telephone service! The RBHCs and major independents obviously have the critical mass and economy of scale to remain the dominant forces in their respective areas. There will be instances where competition will be meaningful. For example, when a new commercial or residential real estate development is completed, the telecommunication aspect is also brand new. Several telcos could compete for the right to serve the area and the most attractive offer of services and price would be in the favorable position.

Regulatory change in other Western (and Eastern) countries has become apparent, and the results can be even more dramatic than in the U.S. For the time being, the regulators and PTTs in Europe define deregulation as allowing users to connect their own terminals and subscriber equipment to the public network. A possible exception is the United Kingdom where use of leased lines for private networking is increasing. The trend, however, has been established, and users will demand and get more freedom to install switching and transmission systems. Private networks as we know them in the U.S. will eventually be more common in the other developed countries. As an example, satellite communication is a technology which facilitates private network development. The Cable News Network (CNN) is now delivered by European satellite directly into cable television systems and hotels, requiring only a three-meter antenna to receive the signal. Only a few years ago, this would have been thought to be extremely difficult to arrange.

9.2.5 Better Educated Telecommunication Management

Over the years, the profession of telecommunication management has been defined as a multidisciplinary field involving technical and business abilities. This is consistent with the emphasis throughout this book on developing network strategy in concert with business strategy. In the remaining paragraphs, we review how the requisite training can be provided to those individuals who wish to develop their careers as the industry continues to evolve. Quite obviously, we see a need for a solid grounding in telecommunication technology and business management.

9.2.5.1 University Programs

Some say that nothing can beat a well rounded university education, and this author certainly will not argue that point. A technical education provides a scientific underpinning for a career in technology, which telecommunication most certainly is. Total concentration in engineering, mathematics, and the sciences, however, will overlook the management and business perspectives of the field that we are trying to develop. Some universities offer a degree program in communication, including University of Southern California, University of Colorado, and Golden Gate University. An important consideration is that the technical aspects which are essential to effective use (or marketing) of telecommunication facilities and services need to be adequately covered. The ideal would be to provide a solid technical foundation through courses which teach the theory behind the systems that we use. Laboratory experiments, where students work with actual hardware and software, provide a valuable teaching experience.

A good example of a comprehensive degree program in telecommunication is found at the State University of New York (SUNY) College of Technology, at Utica-Rome. This is a new multidisciplinary program which provides a working knowledge of the history and methodology of the field, as well as an awareness of current issues and advances in systems, policies, and applications. Every area in telecommunication systems is examined, including network design, products and systems, vendor selection, implementation, diagnostics, and management. The classroom studies are reinforced with laboratory exercises; in addition, many students gain valuable practical knowledge through off-campus work experience. The success of the program is measured by the fact that 100% of the first graduating class were placed in telecommunication jobs.

Contrast this with the concentration of business administration programs on topics such as accounting and finance, organizational behavior, marketing, and business strategy. This is, by definition, a broadly defined program designed to graduate generalists in business. Business education does not prepare the student for work in a particular industry such as automobile manufacturing or retailing. The graduate headed for industry will want to work for a viable company and then learn on the job what he or she needs to know. This would be particularly applicable if the job were in telecommunications.

The ideal university program in telecommunication management would include a thorough grounding in appropriate technical disciplines. These currently come under the headings electronic engineering and computer science. A number of universities offer programs in communication engineering, a specialty which tends to be theoretical and heavily mathematical. Some of this material could be shaped for telecommunication management courses since the student ideally should comprehend what makes the critical elements tick. It would be wonderful if a supervisor could expect that a new employee in telecommunication would have learned the correct principles of telecommunication engineering during his or her university education. The telecommunication manager cannot operate in the new telecommunication environment with a "monkey see, monkey do" approach to problem solving. Therefore, a solid technical curriculum in telecommunication technology is valuable.

Of potentially equal importance to technology is management, which is normally the central area of study in a school of business. Telecommunication managers must be comfortable with the operational, financial, and strategic aspects of management. Consequently, the telecommunication management program ought to adopt the approach taken in the field of medicine, where the practitioner gets as heavy a dose of theory in conjunction with practice.

The U.S. military trains new officers as communication managers, whether they have technical backgrounds or not. In fact, the majority of Signal Corps officers graduated from universities with liberal arts degrees. As discussed previously, every officer is trained for combat duty and to lead a military unit. Therefore, the basic aspects of strategy and management are covered during basic officer training. Indoctrination in communication is done through a series of classes in theory and operation of wireline and radio networks. Classes are supplemented with laboratory periods where students learn how to operate the same facilities (switching, cable transmission, radio links, microwave, *et cetera*) that they will ultimately be responsible for as leaders. The program culminates in a field exercise by the students operating as a unit to install a working network. The real strength of the military training approach is that the student gains confidence in his or her ability to employ telecommunication facilities. The students therefore become better leaders as they understand that with which their subordinates must deal.

The comparable approach in civilian education is the old idea of the work-study program. The student is exposed to principles and theory in the college classroom. Then, during an extended period of several months, he or she has an actual work assignment with a company or government agency. This type of program can provide exposure to the telecommunication field from the inside. Until such time as the education requirements for the profession become better defined, the work-study approach perhaps will be the most effective way to deliver to industry an individual who can contribute in a relatively short period.

9.2.5.2 Industrial Training Courses

The organizations that supply facilities and services often provide specialized training courses for customers. IBM has long been a strong proponent of quality customer training on the computers, data communication systems, and software which the company manufactures and markets. On an elaborate scale, IBM uses its own Interactive Satellite Education Network (ISEN) to connect dozens of classrooms around the U.S. to as many as four different instructors. IBM's use of education enhances the marketing of its products, since customers benefit directly from the high quality and professionalism of the training. IBM is not alone in this regard. The larger suppliers of telecommunication facilities provide appropriate training programs for what they offer. The motivation is that an educated customer will know how to employ the capability properly, and is less likely to call on the contractor to correct routine problems.

In a major new telecommunication project, such as a satellite network, a special training program is usually organized and delivered by a team of instructors. Interestingly, the first students to go through such training are often the best and brightest that the buyer can locate. The reason for this is that, with new technology being used, the buyer cannot afford to fail during the initial operation of the network. Subsequently, this first "graduating class" grows with the system to become the leaders during future expansion.

The *video cassette recorder* (VCR) is playing an important part in providing good quality and consistent training on complex subjects. While the cost of production of video training tapes is significant, the benefits from capturing an excellent presentation for multiple viewing are well worth the expense. The next generation of recorded training products will employ video and data stored on compact disc. In research at the MIT Media Laboratory, an integrated computer work station permits the student to move through programmed instruction, involving text and graphics, supported by full-motion video. The material is arranged in any sequence which the student selects, and the work station is capable of asking questions to maximize comprehension. This technology will eventually affect corporate technical training.

There is a perception that vendor training programs are focused on specific aims and do not provide enough background in the overall architecture or technology of the particular application. Of course, the time and effort to expand a training program in this way is somewhat out of proportion with the need to allow the customer to use the product or service with reasonable facility. The more basic understanding can be developed through seminars and home study courses, as discussed in the next section.

9.2.5.3 Seminars and Home Study Courses

A reasonably thorough education can also be obtained by self-study. There are many publications on most aspects of telecommunication. For many of us, self-study is difficult because of the need to find the time for reading. For others, it may be unpleasant to absorb the material in written form. This problem is the basis for seminars offered by national training organizations like Control Data Corporation, Datapro Research, and Telestrategies or by nonprofit professional organizations like the International Communications Association (ICA), the Tele-Communications Association (TCA), and the Institute of Electrical and Electronic Engineers (IEEE). Some major universities, including University of California at Los Angeles (UCLA) and the George Washington University, offer short courses intended to provide a quick education in specific topics of telecommunication applications and engineering. The mails are currently stuffed with announcements of seminars on such subjects as T1 networks, pay phones, alternative operator services, VSATs, network management, testing of telecommunication networks, and ISDN. The organizers of these seminars usually include speakers from industry who are working in a particular area. In some cases, telecommunication managers from organizations that have implemented private networks are brought in to present their experience with a particular technology or service. This information can be of great value to a strategic unit considering such a development.

These seminars are between one and three days in duration. The information is organized in a topical format, usually beginning with some kind of introduction for someone unfamiliar with the subject matter. Then the information is presented in a relatively concentrated fashion. The seminar organizer may provide a pre-printed document, called the proceedings, so that the student can concentrate on learning rather than taking detailed notes. This approach is often essential because of the rapid pace with which the information is covered.

For experienced telecommunication people, seminars offer the opportunity to obtain current information on a new technology or service capability not yet described adequately in the journals or books. For example, before the phrase VSAT was coined, the first VSAT exposition was conducted by Telestrategies in 1984. The seminar was a commercial success and, from this author's perspective, the event delivered what was promised. After an introduction by the seminar's organizer, Dr. Jerry Lucas, many experienced speakers discussed the technology, the marketplace, and the applications to which VSATs had already been applied.

In addition to seminars, there are annual conferences run by recognized publications and professional groups. The Communications Networks show, the ICA convention, the TCA convention, the Interface show, and the IEEE International Conference on Communications (primarily for practicing communication engineers) are among the most respected in the U.S. *Satellite Communications* magazine runs an annual conference, called the Satellite Communications User's Conference, which has a business and applications focus, and the American Institute for Aeronautics and Astronautics (AIAA) has its biannual Communications Satellite Systems Conference, mainly for technical professionals. The ITU conducts an international telecommunication exposition, called TELECOM, in Geneva, Switzerland, every four years. This exposition receives much attention from major vendors of telecommunication and computer equipment, and is attended by the telecommunication authorities of most of the world's governments. The booths and exhibits at TELECOM are considered second to none.

9.2.5.4 *Future Education Trends*

The information technologies will affect the manner in which education is conducted. University systems in Virginia, West Virginia, Oklahoma, California, and other states around the country already extend their graduate programs by using satellite transmission to remote, off-campus locations. This approach eases the working student's problem of going to class in the evenings and also tends to increase class size since more seating space is obtained from the remote sites. We have already mentioned IBM's ISEN and MIT's education work station, which are prototypes of advanced education media. Classroom study will probably never disappear, but it is too inflexible in the corporate context to be maintained as the primary delivery vehicle. There is, however, no alternative to hands-on training in a real network environment. Education needs to supplement job experience in the application of information technologies.

9.3 TOWARD THE NEW TELECOMMUNICATION

Imagine that you are a Rip Van Winkle type of character who falls asleep (we hope not while reading this book) and wakes up ten years in the future. What would the telecommunication world look like in the year 2000? James Martin did an excellent job of judging today's perspective more than ten years ago. The new telecommunication will have at its base a digital infrastructure with broadband common carrier services like ISDN and fast packet switching. Such public networks will perform better and at lower cost than the combination of public and private networks described in previous chapters of this book. Strategic units and equipment vendors, however, will invent new applications which will again raise the stakes. There will be new requirements for innovation through private telecommunication networks, even in the next century, simply because the public network will be designed to satisfy the majority of users who are accustomed to standard (albeit advanced) services.

Many of us continually try to grasp for that guiding star which would allow us to comprehend where we are and where we are going. In telecommunication, the quest is no different. Private telecommunication networks are important today because of the rapid change experienced by the supporting industries. We suggested that public networks may again deliver the widest range of capabilities through concepts like ISDN. At present, we are not sure which structure will do all things for all people. As telecommunication professionals, we can only strive to stay on top of developments and be willing to try something new which promises to give us a strategic edge. We continually take calculated risks — on technology, on vendors, and most of all on people. Telecommunication is, ultimately, a means of allowing people separated by distance to transact their business as if they were not separated at all. This book hopes to make a contribution to the field and the reader, as you pursue your goals in this rapidly changing field.

Glossary

ABN	American Business Networks
ACD	Automatic Call Direction
ACD	Automatic Call Distribution
A/D	Analog-to-Digital Conversion
ADPCM	Adaptive Differential Pulse Code Modulation
AI	Artificial Intelligence
AIAA	American Institute for Aeronautics and Astronautics
AM	Amplitude Modulation
Americom	GE American Communications
AMSC	American Mobile Satellite Consortium
AO	Administrative Operations
AP	Associated Press
APPC	Advanced Peer-to-Peer Communications
ARPAnet	Advanced Research Projects Agency Network
ASCII	American Standard Code for Information Interchange
ASK	Amplitude Shift Keying
AT&T	American Telephone and Telegraph Company
ATM	Automated Teller Machine
ATT-COM	AT&T Communications
BCR	*Business Communications Review*
BELLCORE	Bell Communications Research
BER	Bit Error Rate
B-ISDN	Broadband Integrated Services Digital Network
BOC	Bell Operating Company
BRI	Basic Rate Interface
BRP	Business Recovery Planning
BSA	Basic Service Arrangement
BSC	Bisynchronous Communications

BSE	Basic Service Element
BT	British Telecom
CCR	Customer Controlled Reconfiguration
CCS	Hundreds of Call Seconds
CCSS-6	Common Channel Signaling System No. 6
CCSS-7	Common Channel Signaling System No. 7
CDMA	Code Division Multiple Access
CDR	Call Detail Recording
CIM	Computer Integrated Manufacturing
CIO	Chief Information Officer
CMIP	Common Management Information Protocol
CNN	Cable News Network
CO	Central Office
CO-LAN	Central Office Local Area Network
Codec	Coder-Decoder
COMSAT	Communications Satellite Corporation
CONUS	Continental United States
CPIF	Cost Plus Incentive Fee
CRC	Cyclic Redundancy Check
CSC	Common Signaling Channel
CSC	Computer Sciences Corporation
CSCU	Circuit Switched Control Unit
CSMA-CD	Carrier Sense Multiple Access with Collision Detection
CSU	Channel Service Unit
DACS	Digital Access and Cross-Connect System
DAMA	Demand Assignment Multiple Access
DARPA	Defense Advanced Research Projects Agency
DBS	Direct Broadcast Service
DCU	Delay Compensation Unit
DDCMP	Digital Data Communications Message Protocol
DDD	Direct Distance Dialing
DDS	Dataphone Digital Service
DEC	Digital Equipment Corporation
DES	Digital Encryption Standard
DID	Direct Inward Dialing
DNA	Digital Network Architecture
DNS	Digital Network Service
DOD	Direct Outward Dialing
DP	Data Processing
DSI	Digital Speech Interpolation

DSU	Digital Service Unit
DTMF	Dual-Tone Multiple Frequency
DTS	Digital Termination Service
EBS	Emergency Broadcast System
EDI	Electronic Data Interchange
EDS	Electronic Data Systems Corporation
EPIC	Electronic Product Information Computer
ESF	Extended Super Frame
ETN	Electronic Tandem Network
FCC	Federal Communications Commission
FDM	Frequency Division Multiplex
FDMA	Frequency Division Multiple Access
FEP	Front-End Processor
FFP	Firm Fixed Price
FM	Frequency Modulation
FSK	Frequency Shift Keying
FTP	File Transfer Protocol
FX	Foreign Exchange
GDI	General DataCom, Inc.
GE	General Electric Company
GEISCO	GE Information Systems Company
GM	General Motors Corporation
GOSIP	Government Open Systems Interconnection Profile
GPSS	Global Positioning Satellite System
GSA	General Services Administration
GTE	GTE Corporation
HAC	Hughes Aircraft Company
HBO	Home Box Office
HCI	Hughes Communications, Inc.
HDLC	High-Level Data Link Control
HDTV	High Definition Television
HNS	Hughes Network Systems
HP	Hewlett-Packard
IBM	International Business Machines Corporation
IBS	INTELSAT Business Service
ICA	International Communications Association
IDNX	Integrated Digital Network Exchange

IEEE	Institute of Electrical and Electronics Engineers
IMT	Inter-Machine Trunk
IMTS	Improved Mobile Telephone Service
INM	Integrated Network Management
INTELSAT	International Telecommunications Satellite Organization
IPX	Integrated Packet Exchange
IRS	Internal Revenue Service
ISDN	Integrated Services Digital Network
ISEN	Interactive Satellite Education Network
ISO	International Standards Organization
ITFS	Instructional Television Fixed Service
ITU	International Telecommunication Union
IWS	Integrated Work Station
IXC	Interexchange Carrier
JCSat	Japan Communications Satellite Company
KDD	Kokusai Denshin Denwah
KSU	Key Service Unit
LAN	Local Area Network
LATA	Local Access and Transport Area
LCR	Least Cost Routing
LED	Light-Emitting Diode
LOS	Line of Sight
LU	Logical Unit
M&C	Monitor and Control
MAN	Metropolitan Area Network
MAP/TOP	Manufacturing Automation Protocol–Technical Office Protocol
MCI	Microwave Communications, Inc.
MELCO	Mitsubishi Electric Company
MFJ	Modified Final Judgement
MIND	Modular Interactive Network Designer
MIT	Massachusetts Institute of Technology
MLSS	Multipoint Line Simulator
Modem	Modulator-Demodulator
MSS	Mobile Satellite Service
MTS	Message Telephone Service
MTSO	Mobile Telephone Switching Office

NBS	National Bureau of Standards
NCP	Network Control Program
NET	Network Equipment Technologies, Inc.
NM	Network Management
NM&C	Network Monitoring and Control
NMCC	Network Management Control Center
NMP	Network Management Protocol
NOC	Network Operations Center
NTI	Northern Telecom, Inc.
NTT	Nippon Telegraph and Telephone
O&M	Operations and Maintenance
OLTP	On-Line Transaction Processing
ONA	Open Network Architecture
OSI	Open System Interconnection
PABX	Private Automatic Branch Exchange
PacBell	Pacific Bell
PAD	Packet Assembler-Disassembler
PBS	Public Broadcasting Service
PBX	Private Branch Exchange
PC	Personal Computer
PCM	Pulse Code Modulation
PDN	Public Data Network
PM	Phase Modulation
PO	Purchase Order
POTS	Plain Old Telephone Service
PRI	Primary Rate Interface
PS/2	Personal System/2
PSK	Phase Shift Keying
PSN	Packet Switched Node
PTT	Post, Telephone, and Telegraph Agency
PU	Physical Unit
PUC	Public Utilities Commission
RAM	Random Access Memory
RBHC	Regional Bell Holding Company
RBOC	Regional Bell Operating Company
RCA	Radio Corporation of America
RDSS	Radiodetermination Satellite Service
RELP	Residual Excited Linear Prediction

RF	Radio Frequency
RFI	Radio Frequency Interference
RFP	Request for Proposal
RFT	Radio Frequency Terminal
RFT	Request for Tender
RJE	Remote Job Entry
ROI	Return on Investment
RPS	Repetitive Pattern Suppression
S/N	Signal-to-Noise Ratio
SAA	Systems Applications Architecture
SBS	Satellite Business Systems
SCA	Subsidiary Carrier Authorization
SCPC	Single Channel per Carrier
SCUC	Satellite Communications User's Conference
SDLC	Serial Data Link Control
SDN	Software Defined Network
SMDR	Station Message Detail Recording
SNA	System Network Architecture
SNG	Satellite News Gathering
SOW	Statement of Work
SPCC	Southern Pacific Communications Company
SPI	Spectrum Planning, Inc.
SS-7	Signaling System No. 7
SSCP	System Service Control Point
SSOG	Satellite System Operating Guide
SSP	Service Switching Point
SSPA	Solid-State Power Amplifier
STAT MUX	Statistical Multiplexer
T&Cs	Terms and Conditions
T&M	Time and Materials
TASI	Time Assigned Speech Interpolation
TCA	Tele-Communications Association
TCP/IP	Transmission Control Protocol–Internet Protocol
TDM	Time Division Multiplex
TDMA	Time Division Multiple Access
TIMS	Transmission Impairments Measuring Set
TOP	Technical Office Protocol
TOPO	Topological Design
TSI	Time-Slot Interchange
TSO	Time Sharing Option

TV	Television
TVRO	Television Receive-Only
TWTA	Traveling Wave Tube Amplifier
UCD	Uniform Call Director
ULP	Upper Layer Protocol
UNMA	Unified Network Management Architecture
VAN	Value Added Network
VCR	Video Cassette Recorder
VPI	Vertical Blanking Interval
VPN	Virtual Private Network
VSAT	Very Small Aperture Terminal
VTAM	Virtual Terminal Access Method
WAN	Wide Area Network
WATS	Wide Area Telecommunications Service

Bibliography

[ABRAMSON, 1973] N. Abramson, "The Aloha System," *Computer Networks,* N. Abramson and F. Kuo, editors, Prentice Hall, Englewood Cliffs, NJ, 1973.

[ATT, 1988] AT&T Network Management Protocol Specification — Transport Through Application Layers, Technical Reference TR 54004, AT&T, Basking Ridge, NJ, 1988.

[BERTSEKAS, 1987] D. Bertsekas and R. Gallager, *Data Networks,* Prentice Hall, Englewood Cliffs, New Jersey, 1987.

[BROWNE, 1986] T. E. Browne, "Network of the Future," *Proceedings of the IEEE,* Vol. 74, No. 9, September 1986.

[CASWELL, 1988] S. A. Caswell, *E-Mail,* Artech House, Norwood, MA, 1988.

[CCIR, 1985] CCIR, *Handbook on Satellite Communications,* ITU, Geneva, 1985.

[CCITT, 1985] CCITT Red Book, Volume VIII — Fascicle VIII.3, *Data Communications Networks, Interfaces,* Recommendations X.20-X.32, VIIIth Plenary Assembly (Málaga-Torremolinos), ITU, Geneva, 1985.

[CYPSER, 1978] R. J. Cypser, *Communications Architecture for Distributed Systems,* Addison-Wesley, Reading, MA, 1978.

[DATAPRO, 1988] Datapro Research Corporation, Reference Series, Delran, NJ, 1988.

[ELBERT, 1987] B. R. Elbert, *Introduction to Satellite Communication,* Artech House, Norwood, MA, 1987.

[FAULHABER, 1987] G. R. Faulhaber, *Telecommunications in Turmoil — Technology and Public Policy,* Ballinger, Cambridge, MA, 1987.

[GLASGAL, 1983] R. Glasgal, *Techniques in Data Communications,* Artech House, Norwood, MA, 1983.

[GREEN, 1986] J. H. Green, *Handbook of Telecommunications,* Dow Jones–Irwin, Homewood, IL, 1986.

[ISTVAN, 1988] Rudyard L. Istvan, "Competing in the Fourth Dimension — Time," *Chief Information Officer Journal,* Vol. 1., No. 1, Summer 1988

[KEEN, 1988] P. G. W. Keen, *Competing in Time — Using Telecommunications for Competitive Advantage* (Updated and Expanded), Ballinger, Cambridge, MA, 1988.

[MARTIN, 1977] J. Martin, *Future Developments in Telecommunications* (Second Edition), Prentice Hall, Englewood Cliffs, NJ, 1977.

[MORGAN, 1988] W. L. Morgan and D. Rouffet, *Business Earth Stations for Telecommunications,* Wiley Interscience, New York, 1988.

[NOLL, 1986] A. M. Noll, *Introduction to Telephones and Telephone Systems,* Artech House, Norwood, MA, 1986.

[NOLL, 1988] A. M. Noll, *Introduction to Telecommunication Electronics, Artech House, Norwood, MA, 1988.*

[PETERS, 1985] T. Peters and N. Austin *A Passion for Excellence,* Random House, New York, 1985.

[PORTER, 1980] M. E. Porter, *Competitive Strategy — Techniques for Analyzing Industries and Competitors,* The Free Press, New York, 1980.

[PORTER, 1985] M. E. Porter, *Competitive Advantage — Creating and Sustaining Superior Performance,* The Free Press, New York, 1985.

[ROTHBLATT, 1987] M. A. Rothblatt, *Radiodetermination Satellite Services and Standards,* Artech House, Norwood, MA, 1987.

[RUNGE, 1987] D. A. Runge, "Capturing Customers with Ties that Bind," *Business Communications Review,* November-December 1987.

[SCHWARTZ, 1980] M. Schwartz, *Information Transmission, Modulation and Noise,* McGraw-Hill, New York, 1980.

[SCHWARTZ, 1987] M. Schwartz, *Telecommunication Networks,* Addison-Wesley, Reading, MA, 1987.

[SCHWEBER, 1988] W. L. Schweber, *Data Communications,* McGraw-Hill, New York, 1988.

[SHIMASAKI, 1987] N. Shimasaki (Editor in Chief), NEC Research and Development, Special Issue on ISDN, NEC Corporation, 1987.

[SLUMAN, 1988] C. Sluman, "Network and Systems Management in OSI," *Telecommunications,* January 1988.

[SMIDT, 1970] O. Smidt, *Engineering Economics,* Telephony Publishing Corps., Overland Park, KS, 1970.

[SMITH, 1985] D. R. Smith, *Digital Transmission Systems,* Van Nostrand Reinhold, New York, 1985.

[THOMPSON, 1983] A. A. Thompson and A. J. Strickland, III, *Strategy Formulation and Implementation — Tasks of the General Manager* (Revised Edition), Business Publications, Inc., Plano, TX, 1983.

[VAN NORMAN, 1988] H. J. Van Norman, "A User's Guide to Network Design Tools," *Data Communications,* April 1988.

[WRIGHT, 1988] R. Wright, "The Open Systems Future," *Telecommunications,* January 1988.

Index